Neuroscience FOR DUMMIES®

by Frank Amthor, PhD

John Wiley & Sons Canada, Ltd.

Neuroscience For Dummies®

Published by
John Wiley & Sons Canada, Ltd.
6045 Freemont Boulevard
Mississauga, Ontario, L5R 4J3
www.wiley.com

For general information on John Wiley & Sons Canada, Ltd., including all books published by John Wiley & Sons, Inc., please call our warehouse, Tel 1-800-567-4797. For reseller information, including discounts and premium sales, please call our sales department, Tel 416-646-7992. For press review copies, author interviews, or other publicity information, please contact our marketing department, Tel 416-646-4584, Fax 416-236-4448.

For technical support, please visit www.wiley.com/techsupport.

Wiley publishes in a variety of print and electronic formats and by print-on-demand. Some material included with standard print versions of this book may not be included in e-books or in print-on-demand. If this book refers to media such as a CD or DVD that is not included in the version you purchased, you may download this material at http://booksupport.wiley.com. For more information about Wiley products, visit www.wiley.com.

Library and Archives Canada Cataloguing in Publication

Amthor, Frank

Neuroscience for dummies / Frank Amthor.

Includes index.

Issued also in electronic formats.

ISBN 978-1-118-08686-5

1. Neurosciences–Popular works.

2. Brain–Popular works. I. Title.

QP376.A57 2011 612.8'2 C2011-902699-6

ISBN 978-1-118-08686-5 (paper); 978-1-118-08967-5 (ePDF); 978-1-118-08968-2 (ePub); 978-1-118-08966-8 (Mobi)

Printed in the United States of America

2 3 4 5 RRD 17 16 15 14 13

About the Author

Frank Amthor is a professor of psychology at the University of Alabama at Birmingham, where he also holds secondary appointments in the UAB Medical School Department of Neurobiology, the School of Optometry, and the Department of Biomedical Engineering. He has been an NIH supported researcher for over 20 years and has also been supported by the U.S. Office of Naval Research, the Sloan Foundation, and the Eyesight Foundation. His research is focused on retinal and central visual processing and neural prostheses. He has published over 100 refereed journal articles, book chapters and conference abstracts.

Dr. Amthor's career has been devoted to understanding neural computation, both for its own sake, and for the sake of making neural prosthesis that restore and augment human function. His specific research has been to investigate complex neural computations in retinal ganglion cells, the first locus in the visual system of highly specific and nonlinear analyses such as motion and directional selectivity. The investigative techniques he has used include virtually the entire suite of single cell neurophysiological techniques, including single cell extracellular recording, sharp electrode intracellular recording and staining, patch clamp recording, optical imaging with both calcium and potentiometric dyes, dual electrode recording, and, most recently, microelectrode array recording. His current research interests involve further translating basic research on the retina to the development of neural prostheses both for the visual system and for other disabilities.

No professional neuroscientist is competently up to date on the entire brain or nervous system, and what we know and understand about it is constantly changing. Although a great deal of time and effort has gone into making sure the material in this book is accurate and up to date, any mistakes within are mine alone. If you find an error or would like to make any other comments about this book, feel free to contact me at `amthorfr@gmail.com`.

Dedication

To Becky, my partner in life, and the mother of my three wonderful children.

To Philip, Rachel and Sarah, for being the world's best kids and now the world's hope for the future.

To my parents, Agnes and Ryder, and my stepfather, Jim, who launched me into this world and gave me guidance for living in it.

To all my teachers who thought I was someone worth investing time in.

Author's Acknowledgments

This book owes its existence first to my agent, Grace Freedson (The Publishing Network). I thank Robert Hickey, acquisitions editor, for working with me to develop the framework for the book, and Wiley's Graphics Department for their work on the illustrations throughout this book. Really special thanks go to Tracy Barr, the project editor, who improved virtually every sentence in the book. I cannot thank her enough for her kind diligence.

Publisher's Acknowledgments

We're proud of this book; please send us your comments at http://dummies.custhelp.com. For other comments, please contact our Customer Care Department within the U.S. at 877-762-2974, outside the U.S. at 317-572-3993 or fax 317-572-4002.

Some of the people who helped bring this book to market include the following:

Acquisitions and Editorial

Editor: Robert Hickey

Project Editor: Tracy Barr

Production Editor: Pauline Ricablanca

Editorial Assistant: Katie Wolsley

Technical Editor: Bernice Grafstein, PhD

Cover Photo: © iStock Sebastian Kaulitzki

Cartoons: Rich Tennant (www.the5thwave.com)

Composition Services

Project Coordinator: Kristie Rees

Layout and Graphics: Carl Byers, Noah Hart, Lavonne Roberts, Corrie Socolovitch, Christin Swinford

Proofreaders: Laura Albert, BIM Indexing & Proofreading Services, Lauren Mandelbaum

Indexer: Potomac Indexing, LLC

John Wiley & Sons Canada, Ltd.

Deborah Barton, Vice President and Director of Operations

Jennifer Smith, Publisher, Professional and Trade Division

Alison Maclean, Managing Editor

Publishing and Editorial for Consumer Dummies

Kristin Ferguson-Wagstaffe, Product Development Director, Consumer Dummies

Ensley Eikenburg, Associate Publisher, Travel

Kelly Regan, Editorial Director, Travel

Publishing for Technology Dummies

Andy Cummings, Vice President and Publisher

Composition Services

Debbie Stailey, Director of Composition Services

Contents at a Glance

Introduction 1

Part I: Introducing Your Nervous System 7

Chapter 1: A Quick Trip through the Nervous System 9
Chapter 2: All about the Brain and Spinal Cord 23
Chapter 3: Understanding How Neurons Work 47

Part II: Translating the Internal and External World through Your Senses 65

Chapter 4: Feeling Your Way: The Skin Senses 67
Chapter 5: Looking at Vision 83
Chapter 6: Sounding Off: The Auditory System 103
Chapter 7: Odors and Taste 119

Part III: Moving Right Along: Motor Systems 135

Chapter 8: Movement Basics 137
Chapter 9: Coordinating Things More: The Spinal Cord and Pathways 149
Chapter 10: Planning and Executing Actions 161
Chapter 11: Unconscious Actions with Big Implications 179

Part IV: Intelligence: The Thinking Brain and Consciousness 197

Chapter 12: Understanding Intelligence, Consciousness, and Emotions 199
Chapter 13: How the Brain Processes Thoughts 223
Chapter 14: The Executive Brain 243
Chapter 15: Learning and Memory 259
Chapter 16: Developing and Modifying Brain Circuits: Plasticity 283
Chapter 17: Neural Dysfunctions, Mental Illness, and Drugs That Affect the Brain 305

Part V: The Part of Tens 317

Chapter 18: Ten Crucial Brain Structures 319
Chapter 19: Ten Tricks of Neurons That Make Them Do What They Do 327
Chapter 20: Ten Amazing Facts about the Brain 335
Chapter 21: Ten Promising Treatments for the Future 343

Index 351

Table of Contents

Introduction 1

About This Book 1
Conventions Used in This Book 2
What You're Not to Read 3
Foolish Assumptions 3
How This Book Is Organized 4
Part I: Introducing Your Nervous System 4
Part II: Translating the Internal and External World through Your Senses 4
Part III: Moving Right Along: Motor Systems 4
Part IV: Intelligence: The Thinking Brain and Consciousness 5
Part V: The Part of Tens 5
Icons Used in This Book 5
Where to Go from Here 6

Part I: Introducing Your Nervous System 7

Chapter 1: A Quick Trip through the Nervous System 9

Understanding the Evolution of the Nervous System 9
Specializing and communicating 10
Moving hither, thither, and yon — in a coordinated way 10
Evolving into complex animals 11
Enter the neocortex 11
Looking at How the Nervous System Works 12
The important role of neurons 12
Computing in circuits, segments, and modules 13
What a charge: The role of electricity 14
Understanding the nervous system's modular organization 15
Looking at the Basic Functions of the Nervous System 15
Sensing the world around you 16
Moving with motor neurons 16
Deciding and doing 17
Processing thoughts: Using intelligence and memory 18
When Things Go Wrong: Neurological and Mental Illness 19
Revolutionizing the Future: Advancements in Various Fields 20
Treating dysfunction 20
Augmenting function: Changing who we are 22

Chapter 2: All about the Brain and Spinal Cord 23

Looking Inside the Skull: The Brain and Its Parts 24
The neocortex: Controlling the controllers 25
Below the neocortex: The thalamus 31
The limbic system and other important subcortical areas 32
Transitioning between the brain and the spinal cord 35
Looking at differences: Size, structure, and other variations 37
The Spinal Cord: The Intermediary between Nervous Systems 40
Looking at the spinal reflex 41
Getting your muscles moving 42
Fighting or Fleeing: The Autonomic Nervous System 42
How We Know What We Know about Neural Activity 43
Examining problems caused by brain injuries 43
Using technology to image the brain: From early EEGs to today 44

Chapter 3: Understanding How Neurons Work 47

Neuron Basics: Not Just Another Cell in the Body 47
Sending and receiving info between neurons: Synaptic receptors 49
Receiving input from the environment: Specialized receptors 50
Ionotropic versus metabotropic receptors 51
The three major functional classes of neurotransmitters 52
How Shocking! Neurons as Electrical Signaling Devices 53
Yikes, spikes — The action potential 55
Closing the loop: From action potential to neurotransmitter release 57
Moving Around with Motor Neurons 59
Non-neuronal Cells: Glial cells 59
Astrocytes 60
Oligodendrocytes and Schwann cells 60
Microglial cells 61
Recording Techniques 61
Single extracellular microelectrodes 61
Microelectrode arrays 61
Sharp intracellular electrodes 62
Patch-clamp electrodes 62
Optical imaging devices 62

Part II: Translating the Internal and External World through Your Senses 65

Chapter 4: Feeling Your Way: The Skin Senses 67

How Do You Feel? The Lowdown on the Skin and Its Sensory Neurons 68
General properties of the skin 68
Sensing touch: The mechanoreceptors 69

How mechanoreceptors work 70
Sensing temperature and pain 72
Sensing position and movement: Proprioception and kinesthesis 73
Skin Receptors, Local Spinal Circuits, and Projections to the Brain 73
Somatosensory receptor outputs 74
Locating the sensation: Specialized cortical sensory areas 75
Understanding the Complex Aspects of Pain 77
Reducing — or overlooking — pain 78
Pain-free and hating it: Peripheral neuropathy 79
Chronic pain and individual differences in pain perception 80

Chapter 5: Looking at Vision........83

The Eyes Have It: A Quick Glance at Your Eyes 84
The retina: Converting photons to electrical signals 85
Catching photons: Light and phototranduction 85
Getting the message to the brain 86
Processing signals from the photoreceptors: Horizontal and bipolar cells 87
Sending out and shaping the message: Ganglion and amacrine cells 89
From the Eyes to the Vision Centers of the Brain 91
Destination: Thalamus 91
Other destinations 94
From the thalamus to the occipital lobe 95
Impaired Vision and Visual Illusions 98
Looks the same to me: Color blindness 99
Understanding blindness 99
Visual illusions 101

Chapter 6: Sounding Off: The Auditory System........103

The Ear: Capturing and Decoding Sound Waves 104
Gathering sound: The outer ear 105
The middle ear 106
Playing chords to the brain: The inner ear 107
Making Sense of Sounds: Central Auditory Projections 110
Stops before the thalamus 110
Off to the thalamus: The medial geniculate nucleus 112
Processing sound in the brain: The superior temporal lobe 112
Handling complex auditory patterns 113
Locating Sound 115
Computing azimuth (horizontal angle) 115
Detecting elevation 116
I Can't Hear You: Deafness and Tinnitus 117
Hearing loss 117
Oh those bells bells bells bells bells bells bells: Tinnitus 118

Chapter 7: Odors and Taste 119
What's That Smell? 120
Sorting things out through the olfactory bulb 121
Projecting along different paths 121
Getting more specific in the orbitofrontal cortex 125
Having Good Taste 126
The discriminating tongue: The four basic tastes 127
Sending the taste message to the brain: Taste coding 129
Identifying and remembering tastes 131
The Role of Learning and Memory in Taste and Smell 132
Lacking Taste and Smelling Badly 133
Smelling poorly or not at all 133
Satiety 134

Part III: Moving Right Along: Motor Systems 135

Chapter 8: Movement Basics 137
Identifying Types of Movement 138
Movements that regulate internal body functions 138
Reflexive movements 138
Planned and coordinated movements 139
Controlling Movement: Central Planning and Hierarchical Execution 140
Activating non-voluntary muscle movements 140
Activating the withdrawal reflex 141
Stepping up the hierarchy: Locomotion 142
Using your brain for complex motor behavior 144
Pulling the Load: Muscle Cells and Their Action Potentials 144
Muscle and Muscle Motor Neuron Disorders 147
Myasthenia gravis 147
Motor neuron viral diseases: Rabies and polio 147
Spinal cord injury 148

Chapter 9: Coordinating Things More: The Spinal Cord and Pathways 149
The Withdrawal Reflex: Open-Loop Reflexes 150
Hold Your Position! Closed-Loop Reflexes 150
Opposing forces: Extensor-flexor muscle pairs 151
Determining the correct firing rate with the comparator neural circuit 152
The Modulating Reflexes: Balance and Locomotion 154
Maintaining balance: The vestibulospinal reflex 154
Do the locomotion 155

Correcting Errors: The Cerebellum....157
Looking at cerebellar systems....157
Predicting limb location during movement....158

Chapter 10: Planning and Executing Actions....161

Making the Move from Reflexes to Conscious or Goal-Generated Action....162
How the frontal lobes function....162
Planning, correcting, learning: Prefrontal cortex and subcortical processors....164
Working memory....165
Initiating actions: Basal ganglia....166
In the middle of things: Supplementary and pre-motor areas....168
The cerebellum: Where you coordinate and learn movements....169
Putting it all together....169
Where Are the Free Will Neurons?....170
Which comes first: The thought or the action?....170
Contemplating the study results....171
You're still accountable!....172
Discovering New (and Strange) Neurons....173
Mirror neurons....173
Von Economo neurons....174
When the Wheels Come Off: Motor Disorders....176
Injuries to the spinal cord and brain....176
Degeneration of the basal ganglia....176

Chapter 11: Unconscious Actions with Big Implications....179

Working behind the Scenes: The Autonomic Nervous System....180
Understanding the functions of the autonomic nervous system....180
Dividing and conquering: Sympathetic and parasympathetic subsystems....181
Controlling the autonomic nervous system....183
Crossing signals: When the autonomic nervous system goes awry....186
Sweet Dreams: Sleep and Circadian Rhythms....187
Synchronizing the biological clock with light exposure....187
Looking at the different stages of sleep....189
Functional associations of brain rhythms....192
Controlling the sleep cycles....194
Not so sweet dreams: Fighting sleep disorders....195

Part IV: Intelligence: The Thinking Brain and Consciousness 197

Chapter 12: Understanding Intelligence, Consciousness, and Emotions 199

Defining Intelligence 200
Understanding the nature of intelligence: General or specialized? 201
Components of intelligence 204
Looking at the different levels of intelligence 207
Intelligence about Emotions 208
Tapping into memories of strong emotional reactions 209
Emoting about the limbic system 209
Understanding Consciousness 213
Looking at assumptions about consciousness 213
Types of consciousness 214
Studying consciousness 215
Two camps and a middle ground 218
Unconscious processing: Blindsight, neglect, and other phenomena 219

Chapter 13: How the Brain Processes Thoughts 223

The Brain: Taking Command at Multiple Levels 224
All about the Neocortex 225
The four major lobes of the brain and their functions 225
Gray matter versus white matter 226
Universal versus small-world connectivity 227
Minicolumns and the six degrees of separation 228
Defining the six-layered structure of the cortex 229
Hail to the neocortex! 232
Controlling the Content of Thought: Sensory Pathways and Hierarchies 232
Sensory relays from the thalamus to the cortex 233
The hippocampus: Specializing for memory 235
Dividing and Conquering: Language, Vision, and the Brain Hemispheres 236
Specialized brain systems for language 237
Seeing the whole and the parts: Visual processing asymmetries 239
Where Consciousness Resides 239
Language and left- or right-hemisphere damage 240
Understanding the "left side interpreter" 240

Chapter 14: The Executive Brain 243

Getting the Brain You Have Today: The Neocortex versus Your Reptilian Brain.....244
My neocortex is bigger than yours: Looking at relative sizes.....244
The relationship between prefrontal cortex size and the ability to pursue goals.....246
Working Memory, Problem-Solving, and the Lateral Prefrontal Cortex.....248
Brain processes managing working memory.....248
The limits of working memory.....250
Perseveration: Sticking with the old, even when it doesn't work anymore.....252
Making Up and Changing Your Mind: The Orbitofrontal Cortex.....253
Feeling it in your gut: Learned emotional reactions.....254
Gambling on getting it right: Risk taking, aversion, and pleasure.....254
Case-based reasoning: Thinking about social consequences.....255
Are We There Yet? The Anterior Cingulate Cortex.....256
Logging errors and changing tactics.....257
Acting without thinking.....257
Who's minding the store? Problems in the anterior cingulate cortex.....258

Chapter 15: Learning and Memory 259

Learning and Memory: One More Way to Adapt to the Environment.....260
Developmental adaptations.....260
Classical learning.....261
Sending More or Fewer Signals: Adaptation versus Facilitation.....262
Adaptation.....262
Facilitation.....263
Studying habituation and sensitization in sea slugs.....264
Exploring What Happens during Learning: Changing Synapses.....264
Neural computation: Neural AND and OR gates.....265
The McCulloch-Pitts neuron.....267
Rewiring your brain: The NMDA receptor.....268
The Role of the Hippocampus in Learning and Memory.....271
Going from short- to long-term memory.....272
A matrix of coincidence detectors.....272
Remembering as knowing: Cortical mechanisms.....275
Knowing versus knowing that you know: Context and episodic memory.....276
Losing Your Memory: Forgetting, Amnesia, and Other Disorders.....278
Getting Brainier: Improving Your Learning.....280
Distributing study time over many shorter sessions.....280
Getting enough sleep.....281
Practicing in your mind.....281
Rewarding and punishing.....281

Chapter 16: Developing and Modifying Brain Circuits: Plasticity. . . . 283

Developing from Conception 284
Arising from the ectoderm: The embryonic nervous system 284
Adding layers: The development of the cerebral cortex 287
Wiring it all together: How axons connect various areas of the brain to each other 290
Learning from Experience: Plasticity and the Development of Cortical Maps 291
Mapping it out: Placing yourself in a visual, auditory, and touching world 292
Firing and wiring together: Looking at Hebb's Law 293
Environmental effects: Nature versus nurture 295
Taking the Wrong Path: Nervous System Disorders of Development 296
Looking for genetic developmental errors in mutant mice 297
Environmental effects on development of the human brain 299
The Aging Brain 300
Living long and well: Lifespan changes in brain strategy 301
Accumulating insults: Aging-specific brain dysfunctions 302
Autoimmune diseases 303
Strokes 303
Tumors 304

Chapter 17: Neural Dysfunctions, Mental Illness, and Drugs That Affect the Brain 305

Looking at the Causes and Types of Mental Illness 305
Genetic malfunctions 307
Developmental and environmental mental illness 308
Mental illness with mixed genetic and developmental components 309
The Promise of Pharmaceuticals 314
Typical and atypical antipsychotic medications 314
Drugs affecting GABA receptors 315
Drugs affecting serotonin 315
Drugs affecting dopamine 316
Some natural psychoactive substances 316

Part V: The Part of Tens 317

Chapter 18: Ten Crucial Brain Structures 319

The Neocortex 319
The Thalamus, Gateway to the Neocortex 320
The Pulvinar 320
The Cerebellum 321
The Hippocampus 321
Wernicke's and Broca's Areas 322
The Fusiform Face Area 323

The Amygdala 323
The Lateral Prefrontal Cortex 324
The Substantia Nigra (Basal Ganglia) 325
The Anterior Cingulate Cortex 325

Chapter 19: Ten Tricks of Neurons That Make Them Do What They Do 327

Overcoming Neurons' Size Limit 327
Getting the Biggest Bang for the Buck with Dendritic Spines 328
Ligand-Gated Receptors: Enabling Neurons to Communicate Chemically 329
Getting Specialized for the Senses 329
Computing with Ion Channel Currents 330
Keeping the Signal Strong across Long Distances 331
The Axon: Sending Signals from Head to Toe 331
Speeding Things Up with Myelination 332
Neural Homeostasis 332
Changing Synaptic Weights to Adapt and Learn 333

Chapter 20: Ten Amazing Facts about the Brain 335

It Has 100 Billion Cells and a Quadrillion Synapses 335
Consciousness Doesn't Reside in Any Specific Area of the Brain 336
It Has No Pain Receptors 337
Cutting the Largest Fiber Tract in the Brain Produces Few Side Effects 337
Einstein's Brain Was Smaller than Average 338
Adults Lose a Hundred Thousand Neurons a Day with No Noticeable Effect 338
Pound for Pound, It Takes a Lot of Energy 339
It's a Myth That We Use Only 10 Percent of Our Brains 340
Brain Injuries Have Resulted in Savant Skills 341
Adult Brains Can Grow New Neurons 342

Chapter 21: Ten Promising Treatments for the Future 343

Correcting Developmental Disorders through Gene Therapy 343
Augmenting the Brain with Genetic Manipulation 344
Correcting Brain Injury with Stem Cells 345
Using Deep Brain Stimulation to Treat Neurological Disorders 345
Stimulating the Brain through TMS and tDCS 346
Using Neuroprostheses for Sensory Loss 347
Addressing Paralysis with Neuroprostheses 347
Building a Better Brain through Neuroprostheses 348
Engaging in Computer-Controlled Learning 349
Treating Disease with Nanobots 350

Index 351

Introduction

The central mystery about the brain is simply this: How can a bunch of interconnected cells make each of us what we are — not only our thoughts, memories, and feelings, but our *identity*. At present, no one can answer this question. Some philosophers think it is not answerable in principle.

I believe we *can* understand how the brain makes us what we are. This book, while surely not containing the complete answer, points the way to what the answer looks like: In short, the brain is made of neurons, each of which is a complex little computer. Parts of the nervous system make suggestions to the rest of it about what you should do next. Other parts process the sensory inputs you receive and tell the system how things are going so far. Still other parts, particularly those associated with language, make up a running dialog about all of this as it is going on; this is your consciousness.

Those concepts aren't too difficult to grasp, but people think of neuroscience as hard. And why? Because in order for your nervous system to perform these functions, it takes 100 billion neurons and a quadrillion connections structured over billions of years of evolution and all the human years of development and learning that resulted in who you are and where you are now.

You need to know three things to understand how the nervous system works. The first is how the neurons themselves work. The second is how neurons talk to each other in neural circuits. The third is how neural circuits form a particular set of functional modules in the brain. The particular set of modules that you have make you human. The content of your specific modules make you, you.

Our nearest animal relative, the chimpanzee, has pretty much the same neurons and neural circuits that you and I do. They even have most of the same modules. We humans have a few extra modules that permit consciousness. Understanding this is what this book is about.

About This Book

Let's face it. Neuroscience is a complex topic. How could it not be since it deals with the brain, the most complex structure in the known universe and the heart — if you don't mind mixing a few body parts — of the nervous system. But in this book, I explain some very complex ideas and connections in a way that both students enrolled in introductory neuroscience courses and those who are just interested in the topic for fun can understand.

To use and understand this book, you don't have to know anything about the brain except that you have one. In this book, I cover as much of the basics as possible with simple language and easy-to-understand diagrams, and when you encounter technical terms like *anterior cingulate cortex* or *vestibulospinal reflex,* I explain what they mean in plain English.

This book is designed to be modular for the simple reason that I want you to be able to find the information you need. Each chapter is divided into sections, and each section contains information about some topic relevant to neuroscience, such as

- The key components of the nervous system
- How neurons work and what the different kinds of neurons are
- What systems are involved in planning and executing complex actions
- The role of the neocortex in processing thoughts

The great thing about this book is that *you* decide where to start and what to read. It's a reference you can jump into and out of at will. Just head to the table of contents or the index to find the information you want.

Note: You can use this book as a supplemental text in many undergraduate courses because I discuss the neuron and brain function as a system. Typical undergraduate perception courses, for example, give short (and usually unsatisfactory) introductions to neurons and neural processing and little if any coverage of cognition. Cognitive psychology and neuroscience courses typically cover cognition well but often don't ground cognition at the level of neurons. Behavioral neuroscience courses sometimes ignore cognition and neurophysiology almost altogether while doing a decent job explaining heuristics and phenomenology of behavior and learning. You can also use this book as an adjunct to graduate or health profession courses where the nervous system or mental illnesses or disorders are mentioned but little explicit coverage is given of the nervous system and the brain.

Conventions Used in This Book

To help you navigate through this book, I've set up a few conventions:

- *Italic* is used for emphasis and to highlight new words or terms that are defined.

- **Boldfaced** text indicates the action part of numbered steps and the keywords or phrases in bulleted lists.
- `Monofont` is used for Web addresses.

What You're Not to Read

What you're not to read is whatever you don't want to read or don't need to read. Because this book is modular and headings clearly indicate what each section is about, you can go right to the appropriate section and bypass the others at whim. In addition, I don't assume you have any particular prior knowledge about biology or neuroscience, so each topic starts with the basics and works up to more complicated explanations. If you do know a lot about neurons or brain anatomy, you can skip some of the early chapters and head right to the later ones that deal with major brain systems. If you're just curious and really want to skip most of the details, you can go to the last four Part of Tens chapters and get a very brief overview of brain organization, function, and future developments that are likely to occur in neuroscience and the treatment of neural and brain disorders.

In addition, sidebars (text in the shaded box) and paragraphs marked with Technical Stuff icons are "skippable" — that is, they contain nonessential info that, while interesting, you can ignore without impairing your understanding of the topic at hand. Typical nonessential info includes discussions that explain some of the behind-the-scenes details that you may (or may not) be interested in, or interesting historical notes about early ideas in neuroscience.

Foolish Assumptions

In writing this book, I made some assumptions about you. To wit:

- You're not a professional neuroscientist or neurosurgeon but may be a beginning student in this field. (If you notice that your neurosurgeon thumbing through a copy of this book before removing parts of your brain, you might want to get a second opinion.)
- You're taking a course that relates to brain function, cognition, or behavior and feel that you would do better if you had a firm grasp of the how the nervous system and its components work.
- You want information in easy-to-access and easy-to-understand chunks, and if a little humor can be thrown in, all the better!

If you see yourself in the preceding points, then you have the right book in your hands.

How This Book Is Organized

To help you find information that you're looking for, this book is divided into five parts. Each part covers a particular topic related to neuroscience and contains chapters relating to that part.

Part I: Introducing Your Nervous System

In Part I, I explain the overall structure of the brain and nervous system and how its component neurons function. Chapter 1 gives an overview of neuroscience, including some of the methodology used to study the nervous system and how it is evolving. Chapter 2 gives an overview of the brain, spinal cord, and peripheral nervous systems, including major modules and some fiber tracts. Chapter 3 is an introduction to neurons, the all-important universal components of the nervous system.

Part II: Translating the Internal and External World through Your Senses

How the various senses work is the subject of Part II. Chapter 4 treats the skin senses, from receptors to cortical processing. Chapter 5 covers the visual system starting at photoreceptors and tracing the visual signal through numerous brain areas that process this sense. Chapter 6 does the same for the auditory system. Taste and smell are the subject of Chapter 7. These chapters give an overview of many of the topics that would be the subject of a college perception course.

Part III: Moving Right Along: Motor Systems

Movement and motor control are treated in Part III. Just as the sensory system can be covered from receptor to high-order cortical processing, the motor system can be looked at as a hierarchy from muscle and motor neuron, to spinal reflex and pattern generator circuits, to cortical motor control, to frontal and prefrontal planning and execution. I cover movement basics in Chapter 8. Chapter 9 deals with spinal cord level coordination and its communication with the brain stem and higher centers. Chapter 10 takes up the cortex and action planning and sequencing. Chapter 11 deals with unconscious aspects of neuronal processing associated with behavior.

Part IV: Intelligence: The Thinking Brain and Consciousness

Part IV covers the big enchiladas: intelligence, thought, and consciousness. Chapter 12 discusses what intelligence is and why we may have it. How the brain processes thoughts is covered in Chapter 13. The executive brain, as a hierarchy of controllers, is the subject of Chapter 14. Chapter 15 discusses neural aspects of learning and memory. Chapter 16 treats plasticity in the brain and development. Chapter 17 covers neural dysfunctions, mental illness, and drugs that affect the brain.

Part V: The Part of Tens

In this part, I offer you a condensed overview of important and interesting facts, discoveries, and ideas about the brain. Chapter 18 in surveys ten crucial brain areas and what they do. Knowing just these ten might allow you to strike up or follow a conversation if you find yourself suddenly surrounded by neuroscientists in your favorite tavern (which is generally where we neuroscientists like to hang out). Chapter 19 gives a concise explanation of how neurons can do what they do. Chapter 20 lists ten amazing facts about the brain, providing sure bet winning material from reading only a few pages. Chapter 21 surveys promising and some possibly not-so-promising treatments and brain modification technologies coming in the near future.

Icons Used in This Book

The icons in this book help you find particular kinds of information. They include the following:

Looking at things a little differently or thinking of them in a new way can make potentially confusing concepts easier to understand. Look for this icon to find these "think of it this way" types of discussions.

This icon appears next to key concepts and general principles that you'll want to remember.

In a subject as complicated as neuroscience, it's inevitable that some discussions will be very technical. Fortunately for you, you don't need to know the detailed where- and whyfors, but I include this info anyway for those who are voraciously curious or gluttons for punishment. Read or skip paragraphs beside this icon at will.

Over the ages, a lot of myths have sprung up about the brain and how its functioning impacts behavior, movement, psychological states, and more. This icon highlights those misunderstandings.

We wouldn't know what we know about how the brain works without the diligent work of scientists and researchers who asked important questions and sought the answers. Look for this icon to find information about key studies or observations that profoundly affected what people now know about the brain and its function.

Where to Go from Here

Finally, the purpose of this book is to get you up to speed fast in understanding neurons and the nervous system, particularly the brain, but there are many important neuroscience topics that fall well beyond the scope of this book. Here's just a sampling: intra-neuronal metabolism and second messenger cascades, association of neurological deficits with lesions in specific tracts and nuclei, traditional learning theory, and modern genetics. You can find detailed discussion of most of these subjects in Kandel, Schwartz, and Jessel's *Principles of Neural Science,* 4th Edition (McGraw-Hill, 2000), the bible of neuroscience books.

Part I

Introducing Your Nervous System

"Information is moved via neurotransmitters from neuron to neuron via the synapses into the brain where it is then retrieved by the memory via a slap on the back of the head."

In this part . . .

We use many things that we don't understand entirely; elevators, cars, and computers are just a few examples. The brain (and the rest of the nervous system) is surely at the top of this list because it is the seat of understanding itself.

In this part, I introduce you to the nervous system, covering its main divisions and how the neurons that compose it work. You won't be ready to start a neurosurgery residency after you finish Part I, but you'll have a good idea how your brain is put together and what the various parts do.

Chapter 1

A Quick Trip through the Nervous System

In This Chapter

- Following the evolution of the nervous system
- Understanding how the nervous system works
- Listing the basic functions of the nervous system
- Looking at types of neural dysfunction
- Peeking into neuroscience's future contributions

"My brain: it's my second favorite organ."

— Woody Allen (*Sleeper,* 1973)

The brain you are carrying around in your head is by far the most complicated structure known in the universe, and everything you are, have been, and will be arises from the activity of this three-pound collection of 100 billion neurons.

Although this book is about *neuroscience,* the study of the nervous system, it's mainly about the brain, where most of the nervous system action takes place, neurally speaking. If your brain functions well, you can live a long, happy, and productive life (barring some unfortunate circumstances, of course). If you have a brain disorder, you may struggle to overcome every detail of life, a battle that will take place within your brain. So read on for an introduction to the nervous system, how it works, what it does, and what can go wrong.

Understanding the Evolution of the Nervous System

The earth formed 4.5 billion years ago. Evolutionary biologists believe that single-celled *prokaryote* life (cells without a cell nucleus) appeared on earth less than one billion years after that. What's remarkable about this date is that geophysicists believe this was the earliest point at which the planet had cooled enough to sustain life. In other words, life appeared almost the instant (in geological time) that it was possible.

For unknown reasons, it took more than another billion years for *eukaryotic* life (cells with nuclei) to appear, another billion years for multicellular life to evolve from eukaryotic cells, and *another* billion years for humans to appear — which we did less than a million years ago. The processes that lead to multicellular life all took place in the earth's oceans.

Specializing and communicating

In multicellular organisms, the environment of cells on the inside of the cell group is different from the environment of the cells on the outside of the group. These different environments required the cells in these multicellular life forms to develop a way to specialize and communicate. Understanding this specialization is one of the keys to understanding how the nervous system works.

Imagine a ball of a few dozen cells in a primitive ocean billions of years ago. Because these cells aren't exposed to the seawater, those on the inside of the ball might be able to carry out some digestive or other function more efficiently, but they don't have any way to get the nutrients they need from the seawater, and they don't have a way of ridding themselves of waste. To perform these tasks, they need the cooperation of the cells around them.

For this reason, multicellular life allowed — in fact, mandated — that cells specialize and communicate. Eukaryotic cells specialized by regulating DNA expression differently for cells inside the ball of cells versus those on the outside. Meanwhile, some of the substances secreted by cells became signals that other cells responded to. Cells in multicellular species began specializing and communicating.

Moving hither, thither, and yon — in a coordinated way

Currents, tides, and waves in Earth's ancient oceans moved organisms around whether they wanted to be moved or not. Even organisms specialized for photosynthesis developed buoyancy mechanisms to keep themselves in the upper layer of the ocean where the sunlight is.

Some multicellular organisms found an advantage in moving more actively, using flagella. But moving flagella on different cells on different sides of a multicellular organism without coordination is not the best way to direct movement (picture a sculling team that has every member rowing in a different direction). Without some form of communication to synchronize their activity, the boat — the organism in this case — would go nowhere fast. The result? Networks of specialized cells with gap junctions between them evolved. These networks allowed rapid electrical signaling around ring-like neural nets that became specialized for synchronizing flagella on the outside of the organism.

Evolving into complex animals

Balls of cells with nervous systems that had become capable of moving in a coordinated fashion in the oceans evolved into complex animals with sensory and other specialized neurons.

About half a billion years ago, invertebrates such as insects crawled onto the land to feast on the plants that had been growing there for millions of years. Later, some vertebrate lung fish ventured onto the land for brief periods when tidal pools and other shallow bodies of water dried up, forcing them to wriggle over to a larger pool. Some liked it so much they ended up staying on the land almost all the time and became amphibians, some of which became reptiles. Some of the reptiles gave rise to mammals, whose descendants are us.

Enter the neocortex

When you look at a human brain from the top or sides, almost everything you see is neocortex. It's called "neo" because it is a relatively recent invention of mammals. Prior to mammals, animals like reptiles and birds had relatively small brains with very specialized areas for processing sensory information and controlling behavior.

What happened with the evolution of mammals is that a particular brain circuit expanded enormously as an additional processing layer laid over the top of all the older brain areas for both sensory processing and motor control.

Neuroscientists are not exactly sure how and why the neocortex evolved. Birds and reptiles (and dinosaurs, for that matter) did pretty well with their small, specialized brains before the arrival of the neocortex. However it happened, once the neocortex appeared in mammals, it enlarged tremendously, dwarfing the rest of the brain that had evolved earlier. This occurred despite the fact that large brains are expensive, metabolically. The human brain consumes about 20 percent of the body's metabolism despite being only about 5 percent of body weight.

Looking at How the Nervous System Works

Look at just about any picture of the brain, and you see immediately that it consists of a number of different regions. The brain does not appear to be an amorphous mass of neural tissue that simply fills up the inside of the skull.

Given the appearance of the brain, you can ask two very important and related questions:

- Do the different regions of the brain that look different really do different things?
- Do the regions that look the same do the same thing?

The answer to both questions? Sort of. The next sections explain.

The important role of neurons

The nervous system, explained in detail in Chapter 2, consists of the central nervous system (the brain and spinal cord), the peripheral nervous system (the sensory and motor neurons), and the autonomic nervous system (which regulates body processes such as digestion and heart rate).

Fields and bumps: Early theories about how the brain works

The early history of neuroscience saw a number of brain function theories. Two of the more interesting are *phrenology* and the *aggregate field* theories.

The aggregate field theories supposed that, for the most part, the brain is a single, large neuronal circuit whose capabilities are related mostly to its total size. These theories assumed that the brain's internal structure is of little consequence in understanding its function.

At the other extreme were the phrenologists who believed that almost every human characteristic, including attributes such as cautiousness, courage, and hope, are located in specific parts of the brain. These folks believed that the development of these attributes can be determined by measuring the height of the skull over those areas (bumps), the presumption being that the underlying brain grows and pushes the skull upward for traits that are highly developed. You can read more about phrenology in Chapter 12.

All the divisions of the nervous system are based universally on the functions of neurons. *Neurons* are specialized cells that process information. Like all cells, they are unbelievably complicated in their own right. All nervous systems in all animal species have four basic types of functional cells:

- **Sensory neurons:** These neurons tell the rest of the brain about the external and internal environment.
- **Motor (and other output) neurons:** Motor neurons contract muscles and mediate behavior, and other output neurons stimulate glands and organs.
- **Communication neurons:** Communication neurons transmit signals from one brain area to another.
- **Computation neurons:** The vast majority of neurons in vertebrates are computation neurons. Computational neurons extract and process information coming in from the senses, compare that information to what's in memory, and use the information to plan and execute behavior. Each of the several hundred brain regions contain very approximately several dozen distinct types of computational neurons that mediate the function of that brain area.

What really distinguishes the nervous system from any other functioning group is the complexity of the neuronal interconnections. The human brain has on the order of 100 billion neurons, each with a unique set of about 10,000 inputs, yielding about a quadrillion synapses — a number even larger than the U.S. national debt *in pennies!* The number of possible distinct states of this system is virtually uncountable.

You can read a detailed discussion on neurons and how they work in Chapter 3.

Computing in circuits, segments, and modules

The largest part of the brain, which is what you actually see when you look at a brain from above or the side, is the neocortex. The neocortex is really a 1.5 square foot sheet of cells wadded up a bit to fit inside the head. The neurons in the neocortex form a complex neural circuit that is repeated millions of times across the cortical surface. This repeated neural circuit is called a *minicolumn.*

The brain contains many specialized areas associated with particular senses (vision versus audition, for example) and other areas mediating particular motor outputs (like moving the leg versus the tongue). The thing determining the function or association of an area isn't the area of the brain itself or the particular structure of the minicolumns that are in it, but the kind of info that travels into those minicolumns and the place where that info originates.

So even though the cell types and circuits in the auditory cortex are similar to those in the visual and motor cortices, the auditory cortex is the auditory cortex because it receives inputs from the cochlea (a part of the ear) and because it sends output to areas associated with processing auditory information and using it to guide behavior.

Many other parts of the nervous system also are made up of repeated circuits or circuit modules, although these are different in different parts of the brain:

- **The spinal cord** consists of very similar segments (cervical, thoracic, lumbar, and so on), whose structure is repeated from the border of the medulla at the top of the spinal cord to the coccygeal segments at the bottom.
- **The cerebellum,** a prominent brain structure at the back of the brain below the neocortex, is involved in fine-tuning motor sequences and motor learning. Within the cerebellum are repeated neural circuits forming modules that deal with motor planning, motor execution, and balance.

All the modules that make up the central nervous system are extensively interconnected. If you were to take a section through about any part of the brain, you'd see that the brain has more *white matter,* or *axon tracts* (the neural "wires" that connect neurons to each other) than *gray matter* (neural cell bodies and dendrites, which receive inputs from other neurons and do the neural computations). Here's why: The brain uses local interconnections between neurons to do *computations* in neural circuits. However, any single neuron contacts only a fraction of the other neurons in the brain. To get to other brain modules for other computations, the results of these computations must be sent over long distance projections via axon tracts.

What a charge: The role of electricity

Most neurons are cells specialized for computation and communication. They have two kinds of branches: *dendrites* (which normally receive inputs from other neurons) and *axons* (which are the neuron's output to other neurons or other targets, like the muscles) emanating from their cell bodies.

Neuronal dendrites may be hundreds of micrometers in length, and neural axons may extend a meter (for example, axons run from single cells in the primary motor cortex in your brain down to the base of your spinal cord). Because the neuron is lengthened by the dendrites and axons, if the neuron is going to process signals rapidly, it needs mechanisms to help that intracellular communication along. That mechanism? Electricity.

Neurons use electricity to communicate what is happening in different parts of the neuron. The basic idea is that inputs spread out all over the dendrites and cause current flow from the dendrites into the cell body. The cell body converts this changing electrical current into a set of pulses sent down its axon to other neurons. To find out more about how neurons communication in general, head to Chapter 3. The chapters in Part II explain the specific details for each of the sensory systems.

Understanding the nervous system's modular organization

The nervous system has an overall modular organization. Neurons participate in local circuits consisting of several hundred neurons composed of a dozen or two (or three, or sometimes four!) different types of neurons. These local circuits perform neural computations on inputs to the circuit and send the results to other circuits as outputs.

Local circuits form modules that perform certain functions, like seeing vertical lines, hearing 10,000 Hz tones, causing a particular finger muscle to contract, or causing the heart to beat faster. Groups of similar modules form major brain regions, of which there are several hundred, give or take. Modules in the brain, spinal cord, peripheral nervous system, and autonomic nervous system all work together to maintain your survival by regulating your internal environment and managing your interaction with the external environment. Of course, humans do more than just survive. We have feelings and memories and curiosity and spiritual yearnings. We are capable of language, self-reflection, technology, and curiosity about their place in the universe.

Looking at the Basic Functions of the Nervous System

Animals have nervous systems, but plants don't. The question is why not? Both plants and animals are multicellular, and many plants, such as trees, are far larger than the largest animals.

The key difference, of course, is movement. All animals move, but almost no plants do. (Venus Flytraps have a bi-petal leaf that snaps shut on insects, but we won't count that.) Nervous systems enable movement, and movement is what separates plants and animals.

Eat your brain out, you (sea) squirt!

Sea squirts are filter feeders (they filter nutrients out of ocean water) that live on the ocean floor. What's interesting about these organisms, despite the many shapes and colors they come in, is that they have a mobile, larval form with a cerebral ganglion that controls swimming, but the adult form is *sessile* (anchored, like a plant).

During metamorphosis to its adult form, the sea squirt digests this central ganglion and thus "eats its own brain."

Sensing the world around you

Sensory neurons detect energy and substances from inside and outside our bodies. Energy detectors include photoreceptors in the eye that detect light (Chapter 5), auditory hair cells in the cochlea that detect sound (Chapter 6), and mechanoreceptors in the skin that detect pressure and vibration (Chapter 4). Sensory cells that detect molecules include olfactory neurons in the nose and taste buds in the tongue (Chapter 7).

We also have detectors inside our bodies that detect body temperature, CO_2 levels, blood pressure, and other indications of body function. The central and autonomic nervous system (discussed in Chapter 11), use the outputs of these internal sensors to regulate body function and keep it in an acceptable range *(homeostasis)*. This typically occurs without our conscious awareness.

Sensory neurons are the most specialized of all neurons because they have unique mechanisms for responding to a particular type of energy or detecting a particular substance (as in smell and taste receptors). For example, some animals can directly sense the earth's magnetic field. They do this because they have cells that have deposited little crystals of magnetite in their cytoplasm that react to the magnetic field force of the earth to generate an electrical signal in the cell. This electrical signal is then communicated to other cells in the animal's nervous system for navigation.

Moving with motor neurons

Most neurons are computation neurons that receive inputs from other neurons and have outputs to other neurons. However, some neurons, like those listed in the preceding section, are different:

- **Some neurons are specialized for sensation.** The input for these neurons comes from the world, not other neurons.
- **Some neurons send their output to muscles, glands, or organs instead of other neurons.** In this way, they spur action, which can be anything from secreting a particular hormone to regulate a bodily process to darting out the front door and across the lawn when you hear the ice cream truck.

Our bodies execute two very different types of movement. *Voluntary movement,* which is what most people normally think of as movement, is controlled by the central nervous system whose motor neurons innervate *striated muscles* (these same muscles and neurons are involved in reflex, too). We also have *smooth muscles* controlled by neurons in the autonomic nervous system, such as in the digestive system or those that control the pupil of the eye. Movement is such an important topic in neuroscience that I devote all of Part III to it.

Deciding and doing

Central nervous systems are complex in mammals because large areas of the neocortex conduct motor control, sensory processing, and, for lack of a better term, what goes on in between. Devoting a large amount of brain tissue to motor control allows sophisticated and complex movement patterns. Large brain areas processing sensory inputs can allow you to recognize complex patterns in those inputs.

Large amounts of brain not devoted directly to controlling movement or processing sensory input have traditionally been called *association cortex.* Although lumping all non-sensory, non-motor cortex together under this term is not very accurate, it is clear that association cortex allows very complex contingencies to exist between what is currently being received by the senses and what behavior occurs as a result. In other words, a large neocortex allows a lot of deciding to go on about what it is you will be doing.

Among mammalian species, those that we tend to think of as the most intelligent, such as primates, cetaceans, and perhaps elephants, have the largest neocortices. Well, it's not just the neocortex that impacts intelligence; it's the size of the frontal lobe, too. The most intelligent among the animals just listed (primates) have the largest frontal lobes relative to the rest of the neocortex.

The most anterior part of the frontal lobe is called the *prefrontal cortex.* This area is highly expanded in primates and particularly in humans. The prefrontal cortex is responsible for the most abstract level of goal planning.

If you don't have large frontal lobes, your behavior tends to be dominated by your current needs and what is currently going on in the world around you. If you're a lizard, you're either hungry or cold or hot or seeking a mate or in danger of being caught by a predator. You have a number of behavioral repertoires, and your brain selects among them. For example, you may be seeking a mate, in which case you're following the looking-for-love motor program, when you spot a hawk circling overhead, at which point you switch to the avoiding-hawks motor program and seek a rock to crawl under.

Mammals, with their frontal lobes, have the capacity to plan complex, multistep action sequences. They can avoid hawks and still remember where the potential mate was and return to mate pursuit after the hawk leaves. Mammals can interact in large social groups in which their relationship to every other member is individualized, not just based on who's bigger or smaller or receptive to sexual advances at the moment.

Processing thoughts: Using intelligence and memory

When thinking about intelligence, we typically think about the differences between humans and animals, although some animal behavior is certainly acknowledged as being intelligent. Two attributes — our capability for language and our episodic memory — are associated with human intelligence. The following sections give a very brief outline of key points related to memory, language, and intelligence. For a complete discussion of the hierarchy of intelligence — and the key discoveries and remaining conundrums — head to Chapters 12 through 15.

Language

One attribute associated with human intelligence is language, which, when defined as the use of sign sequences within a complex grammar, appears to be uniquely human. What's interesting about language — at least from a neuroscientist's perspective — is that it resides on only one side of the brain (the left side in most right-handers).

What makes it mind-boggling is that the two sides of a human brain appear nearly identical in both large- and small-scale organization. In other words, there appears to be no physical difference between the two halves. Neuroscientists know of no circuit or structure or cell unique to the left side of the brain that would explain its language capacity compared to the lack of it on the right side. Yet, as seen in patients whose left and right brain halves have been disconnected for medical reasons, the left side is capable of carrying on a conversation about recent experience, but the right side is not.

Episodic memory

Another, less appreciated distinction between human and animal intelligence is human's capacity for episodic memory. *Episodic memory* is the memory of a particular event and its context in time. It can be contrasted with *semantic memory,* a kind of associative memory involving the general knowledge of facts or associations. It's the difference between knowing *when* you learned the capital of Alabama was Montgomery (episodic) versus knowing the fact that the capital of Alabama is Montgomery (semantic).

Even primitive animals can form associative memories, such as in classical or operant conditioning (does the name Pavlov ring a bell?), but there is virtually no accepted evidence that animals other than humans have episodic memories, which depend on the operation of working memory in prefrontal cortex.

The prefrontal cortex is larger in humans than other primates, but even non-primate mammals have prefrontal cortices, so the question becomes, does episodic memory depend on language? What neuroscientists do know is that the complex planning that humans are capable of depends on executive functions in the prefrontal cortex.

When Things Go Wrong: Neurological and Mental Illness

Given the enormous complexity of the brain, it should not be surprising that sometimes it gets broken. Mental disorders range from those with a clear genetic basis, such as Down and Fragile X syndromes, to disorders with high but not complete heritability, such as schizophrenia and autism, to conditions that may be almost completely attributed to life events, like some types of depression.

Some mental disorders are also associated with aging, such as Alzheimer's and Parkinson's diseases. These diseases have no clear genetic basis, although increasing evidence points to associations between some genetic constituencies and risk for these diseases. Huntington's disease is genetic, but its symptoms typically don't appear until adulthood.

What can go wrong with the brain can occur at multiple levels. The following is just a sampling of mental and neurological illnesses that can occur:

- **Developmental errors in gross structure:** Genetic mutations or environmental toxins can lead to defects in gross brain structure. Defects can include missing or abnormally small brain areas, such as the cerebellum, or missing axon tracts connecting brain areas.
- **Developmental errors in specific local circuits:** Some recent theories for autism suggest that, in people with autism, the balance between short and long range neural connections is skewed towards an excess in the short range. This is hypothesized to lead to over-attention to details and inability to respond well to the big picture.

- **Dysfunctional neural pathways:** Mutations in genes that specify neurotransmitter receptors may lead to brain-wide processing deficits. While some brain areas may compensate with other neuronal receptors, other areas may not. Excitatory/inhibitory receptor balance may be implicated in epilepsy and some forms of depression.
- **Environmentally caused organic dysfunctions:** The brain can be damaged by overt injury, such as by a blow to the head. It can also be damaged by toxins such as lead and mercury that produce retardation and other mental problems without overt signs of brain damage.
- **Environmentally caused psychological dysfunctions:** Sometimes mental illnesses, such as some types of depression, occur after environmental triggers in people who have had no previous indications of mental problems. A crucial question in mental illnesses such as depression is whether non-organic causes, such as loss of a loved one, produce depression primarily by changing brain neurochemistry.

For more information on these types of diseases and disorders, head to Chapter 17.

Revolutionizing the Future: Advancements in Various Fields

Revolutions in neuroscience that will have significant ramifications on humanity will occur within 20 years in these two areas:

- Treatments and cures for dysfunctions
- Augmentation of the brain beyond its heretofore "normal" capabilities

I discuss both in the following sections.

Treating dysfunction

Until the last quarter of the 20th century, attempts to treat brain problems were a lot like trying to fix a computer with a hammer and a hacksaw. We simply lacked the appropriate tools and the knowledge about how to use them. Research on the brain has started to change this, and the change is now happening very rapidly.

Pharmacological therapies

Most major mental disorders, including depression, schizophrenia, anxiety, and obsessive-compulsive disorder, are currently treated primarily with drugs. Most of these drugs target neurotransmitter systems.

Pharmacological therapies vary in their effectiveness and side effects. Lessons learned from first- and second-generation drugs are being used to design and screen third- and higher-generation agents. Although the cost of bringing a major new drug to market is currently on the order of one billion dollars, there are extensive international, private, and publicly funded efforts to develop new drugs. Drugs that are effective in eliminating most mental illness, substance abuse, or sociopathy would transform humankind.

Transplants

Neural transplants offer great hope for treating neurological disorders such as Parkinson's disease, which are caused by the death of relatively small numbers of cells in specific brain areas (the *substantia nigra*, in the case of Parkinson's). Transplants may consist of either donor tissue or stem cells that can differentiate into the needed cell types when transplanted into the affected region.

Many laboratories are working on transplanting tissue containing foreign secretory cells shielded from the recipient's immune system by membranes that allow the secretory products out but not the host's immune cells in. If the encapsulated cells respond to circulating levels of neurotransmitters in the host in an appropriate way, they may be able to regulate the levels of what they secrete more accurately and effectively than can be done by taking pills.

Electrical stimulation

Deep brain stimulation (DBS) is a technique in which the balance of activity in a neural circuit involving several brain areas is altered by continuous stimulation of neurons in one part of the circuit. This technique evolved partly from attempts to achieve the same ends by surgically removing brain areas that were thought to be over-activated in the basal ganglia circuit in Parkinson's disease.

DBS has seen extensive success in treating Parkinson's disease and certain kinds of tremors. DSB has also shown promise in treating certain kinds of depression.

Neural prostheses

Paralysis from spinal cord and brain injuries has been almost impossible to treat because the motor neurons that would activate the muscles were either all killed in the original injury or degenerate from lack of use afterward. A long-time rehabilitation dream has been to intercept brain signals commanding movements, relay them past the interruption, and drive muscles directly with electrical stimulation.

Another type of neural prosthesis is for sensory replacement. By far the most successful of these is the cochlear implant for deafness. Over 80,000 of these have been implanted worldwide at the time of this writing. In most cases these prostheses allow the recipient to carry on normal conversations, even on the telephone.

Prostheses for vision have been less successful. This is partly due to the fact that the information channel is so much larger (1 million ganglion cell axons versus 10,000 auditory nerve fibers), and partly because the cochlea presents a unique environment suitable for the introduction of a stimulating prosthesis. Demonstration projects for visual prostheses have implanted them in both retina and visual cortex, but neither approach has achieved clinically relevant effectiveness. Work continues, however.

Augmenting function: Changing who we are

We humans are now beginning to augment ourselves. This augmentation will go far beyond the vaccines, surgical procedures, and prosthetics that alter our bodies, because it will involve our brains being directly connected to electronic circuits and, through those circuits, to the universe.

Using likely extensions of current technology, imagine using an implanted neural prosthesis to access the Internet just by thinking about it. Similar prostheses could translate languages in our heads or allow us to do complex mathematical calculations. They could allow us to communicate with anyone on earth simply by thinking about that person.

Sound far-fetched? Consider that neuro-prostheses consisting of hundreds of electrodes have already been experimentally implanted in a few people who were either paralyzed or blind. The principles involved in recording from or stimulating individual neurons in the brain are well within current technology. What remains to be done is achieve better resolution and signal processing and longer lasting implants, which no doubt will happen in 20 years.

Chapter 2

All about the Brain and Spinal Cord

In This Chapter

- Looking at the hemispheres and lobes in the brain
- Getting the scoop on gender-based processing differences
- Understanding the spinal cord's function in the central nervous system
- The role of the autonomic nervous system
- Imaging technology

Right now, as you're reading, your brain is busy processing visual images from this page into letters, words, and sentences. You understand what these images mean because of your memories of English words and grammar and because your general knowledge of the world is activated by those words. But your brain is always doing much more than this, no matter how hard you may be concentrating on the reading. The brain constantly controls and monitors *all* bodily functions, allowing you to sneeze, shift in your chair, lift your coffee cup for a drink, swallow, and breathe.

The brain is part of your nervous system, which is a network of subsystems and interrelated parts. The brain and spinal cord together (with the retina in the eye) constitute the *central nervous system*. The spinal cord connects to muscles, organs and receptors via the *peripheral nervous system,* which includes the autonomic nervous system that regulates involuntary movements (think heart rate, digestive processes, and internal temperature regulation, to name just a few). The nervous system consists of *neurons*, which do the work of processing and communicating signals, and *glial cells* that form the structures in which neurons reside and that do various housekeeping chores that keep the neurons functional and happy.

This chapter presents the gross anatomy of the brain and spinal cord (no, not gross as in disgusting, but gross as in big), examines the controversial hypothesis that male and female brains process information differently, and outlines the technology used to understand how the brain works.

Looking Inside the Skull: The Brain and Its Parts

The most complex structure in the universe (that we know of) is the three pound mass of cells within your skull called the brain. The brain consists of about 100 billion neurons, which is about the same number as all the stars in our Milky Way galaxy and the number of galaxies in the known universe.

Like any complex machine, the brain contains a lot of parts, each of which have subparts, which have themselves subparts, all the way down to the "nuts and bolts" — the neurons.

Figure 2-1 shows the brain from the left rear. To get a view like this, you have to remove the brain from the skull by cutting through the *ventral* aspect of the brain (where the brain merges with the spinal cord) and the 12 cranial nerves that connect the brain to the eyes, ears, nose, facial muscles, skin, and some glands.

The following sections take a look at the different parts of the brain, from top to bottom.

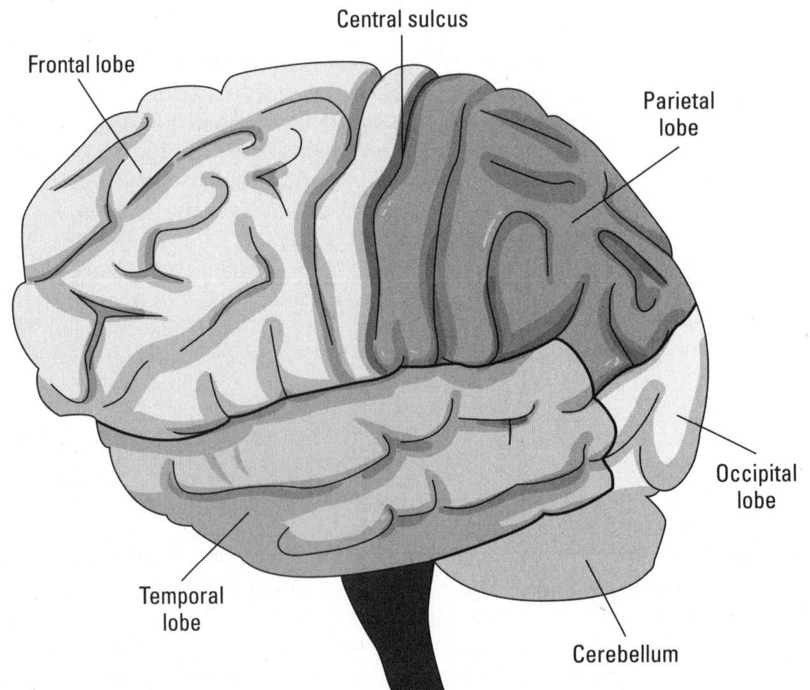

Figure 2-1: The neocortex.

The neocortex: Controlling the controllers

Almost all the exposed brain you see in Figure 2-1 is the *neocortex. Neo* means "new," and *cortex* means something like "husk" or "outer shell."

Although the neocortex looks like a three dimensional blob with some grooves in it, it's more like a large sheet of tissue that's been wadded up to fit into the skull. The actual area, if you were to smooth it out, is about 1.5 square feet, roughly equivalent to one and a half standard floor tiles. Obviously a mass this large wouldn't fit inside your skull without some folding!

A large neocortex distinguishes mammals from all other animals. The human neocortex is so large that it completely covers all the rest of the brain except for a bit of *cerebellum* that sticks out from the back. The neocortex enables the most complex mental activity that we associate with being human.

Think of the neocortex with respect to the rest of the brain as the controller of controllers. Species that existed before mammals could clearly move, sense their environment, and exhibit many complex behaviors such as those that are now seen in birds and lizards. These abilities were all enabled by brain structures older than, and hierarchically below, the neocortex. What the neocortex allowed was a new level of advanced behavior — particularly social behavior — culminating in humans with tool making and, ultimately, language and high-level consciousness.

The following sections introduce and explain the key parts of the neocortex you need to be familiar with: the four main lobes — frontal, parietal, temporal, and occipital — into which the neocortex is divided, and some of the grooves and ridges in the neocortex's tissue. It also explains the terminology used to refer to positions: dorsal, ventral, anterior, and so on.

Finding your way around the neocortex: Dorsal, ventral, anterior, and posterior

Imagine a shark. The *dorsal* fin sticks out of the water from the shark's back, so dorsal refers to the shark's back, which is on top. The opposite of dorsal, *ventral*, refers to bottom. It's the same with a four-legged animal, where *dorsal* refers to top and *ventral* refers to bottom. Things get a little complicated with humans and other primates, though, because we stand up on two legs. When referring to the spinal cord, *dorsal* refers to the back and *ventral* to the front, which, when we're standing, isn't up versus down.

So how do we refer to directions in the human brain? There, *dorsal* refers to the top and *ventral* to the bottom. The term *anterior* refers to the front (forehead) and *posterior* refers to the back (back of the head). ***Note:*** *Anterior* and *posterior* have more or less replaced (in the brain) the older equivalent terms *rostral* and *caudal,* respectively, but you may still hear those terms.

Grooves and ridges: The sulci and gyri

The grooves in the neocortex are called *sulci* (singular: *sulcus),* and the ridges between the grooves are called *gyri* (singular: *gyrus).* No real difference exists between the neocortical tissue visible on the surface gyri versus that down in the sulci (see Figure 2-2).

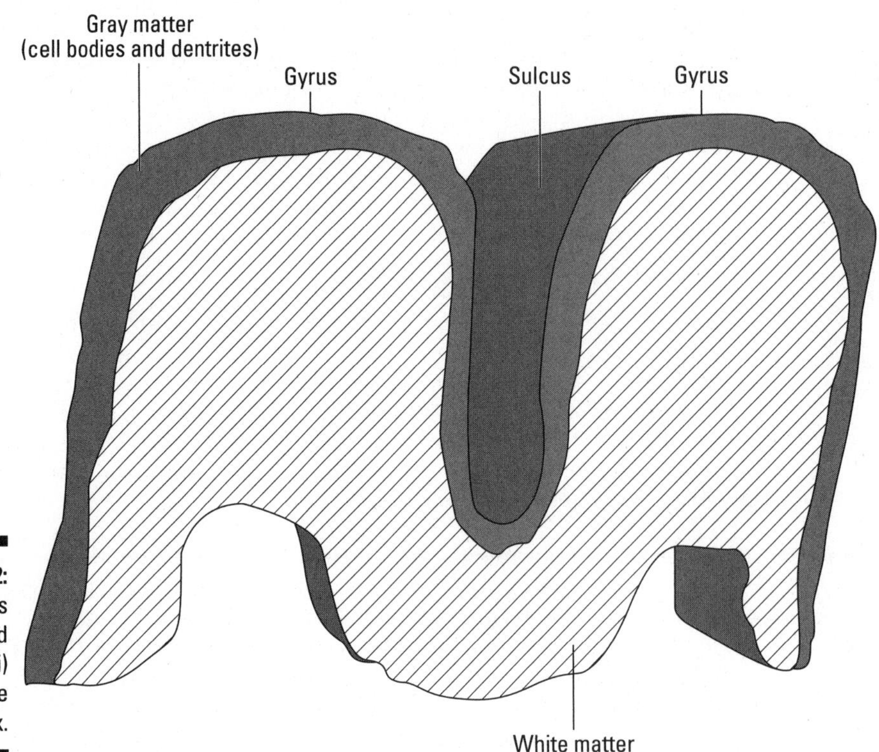

Figure 2-2: The grooves (sulci) and ridges (gyri) in the neocortex.

In some cases, sulci separate cortical lobes or functions, but in many cases, they don't. In the brain, areas that interact extensively with each other are often close together, but ultimately, any brain area can interact with any other area, producing a very complex theoretical packing problem beyond the scope of this book. So don't worry, you won't be asked to deduce the system of gyri and sulci from first principles on Exam 1!

Here's what you do need to know: Some of the sulci separate parts of the brain that do different things (such as the central sulcus that separates the frontal lobe from the parietal lobe), and some gyri seem to be associated with particular functions (such as processing auditory input in the superior temporal sulcus), but most sulci and gyri merely represent how the brain is folded to fit in the head and don't correspond to any specific, isolated function. Moreover, the pattern of gyri and sulci vary considerably not only between human brains, but even between the two sides of the same brain. The neocortex is the majority of what's called the *telencephalon*, which means something like "highest" brain area).

The left and right hemispheres

The most obvious aspect of the visible brain is that it is composed of two nearly mirror-image lobes called the *left* and *right hemispheres.* The left hemisphere receives most inputs from and controls mostly the right side of the body. This hemisphere in humans is also specialized for language, rule-based reasoning, and analytic skills. The right hemisphere deals with the left side of the body, and it is better at visual pattern recognition and more holistic kinds of perception.

In most tasks, the two hemispheres use a divide-and-conquer strategy, where the left hemisphere processes the details, and the right takes in the big picture. The two hemispheres are connected by the largest fiber tract in the brain, the *corpus callosum*, which contains 200 million fibers.

How do scientists know about the differences in the abilities of the left versus right hemispheres? One important source of evidence has come from people with intractable epilepsy who have had their corpus callosum cut to stop the spread of seizures from one side of the brain to the other. Stimuli can be presented to only one hemisphere of these patients. When stimuli are presented to the right (non-language) hemisphere of these patients, they can't verbally report about them, but they can point with their left hand (controlled by the right hemisphere) to a picture of what was presented.

A few years ago, a rash of pop psychology babble suggested that people should shun their in-the-box, analytical left hemispheric processing and learn to use their holistic, creative right hemisphere. The truth is that we use both hemispheres for virtually all tasks, and that even left hemisphere-oriented tasks, like writing poetry, certainly involve creativity.

The four major lobes

The neocortex is divided into four major lobes: frontal, parietal, temporal and occipital:

- **The frontal lobe,** as its name implies, includes all the neocortex from the front, most anterior part of the brain to a major sulcus, called the *central sulcus,* that runs from side to side at about the middle of the brain.
- **The parietal lobe** goes straight back from the central sulcus to almost the most posterior tip.

- **The occipital lobe** is the lobe at the most posterior tip. There is no clear continuous border between parietal and occipital lobes in most brains.
- **The temporal lobe** is the tongue-like extension from the border between the occipital and parietal lobes that extends in the anterior direction.

The next sections explain each lobe in more detail.

The frontal lobe

The frontal lobe (the brain has one on each side, of course) of the neocortex is, as its name implies, in the front. The frontal lobe extends from the front of the skull to a sulcus that divides this lobe from the parietal lobe about midway towards the back of the brain (the occipital lobe). This sulcus is named the *central fissure* (formerly known as the fissure of Rolando). The frontal lobe gyrus just anterior to (in front of) the central fissure is the *primary motor cortex.*

Cells in this region send their axons (output cables) down the spinal cord and connect directly to motor neurons that activate muscles. Researchers know this, because, for example, neurosurgeons such as Wilder Penfield (Wilder Graves Penfield 1891 – 1976 Montreal Neurological Institute) found that electrically stimulating this area during exploratory surgery for epilepsy caused muscle movements. Penfield and others showed that there is an orderly (although distorted) mapping between the point stimulated in the primary motor cortex and the location in the body of the muscle that would twitch.

This "motor map" (called a *homunculus*) has some interesting properties. One is that the area of primary motor cortex devoted to driving muscles in the face and hands, for example, is proportionally much larger than that devoted to driving much larger muscles such as in the leg. This is because the muscles in areas like the face are more finely differentiated to allow very complex and subtle patterns of activation (think of all the myriad facial expressions we use), whereas many large muscles mostly just vary in the amount of contraction of the entire muscle. Another interesting aspect of the homunculus map is that the motor innervation of some areas of the body, such as the face and hands, are close together in the homunculus, but they are not very close together in terms of skin distance. You can read more about this map in Chapters 4 and 10.

Figure 2-3 shows the location of the motor cortex, and some other prominent divisions of the frontal lobe.

The most anterior portion of the frontal lobe is called *prefrontal cortex.* In humans, the prefrontal cortex takes up the majority of the lobe. Moving from the primary cortex in the anterior direction (forward, toward the front of the skull), you have a series of regions that perform particular functions. Closest to the primary motor cortex are regions that are concerned with coordination and sequencing of movement of several muscles.

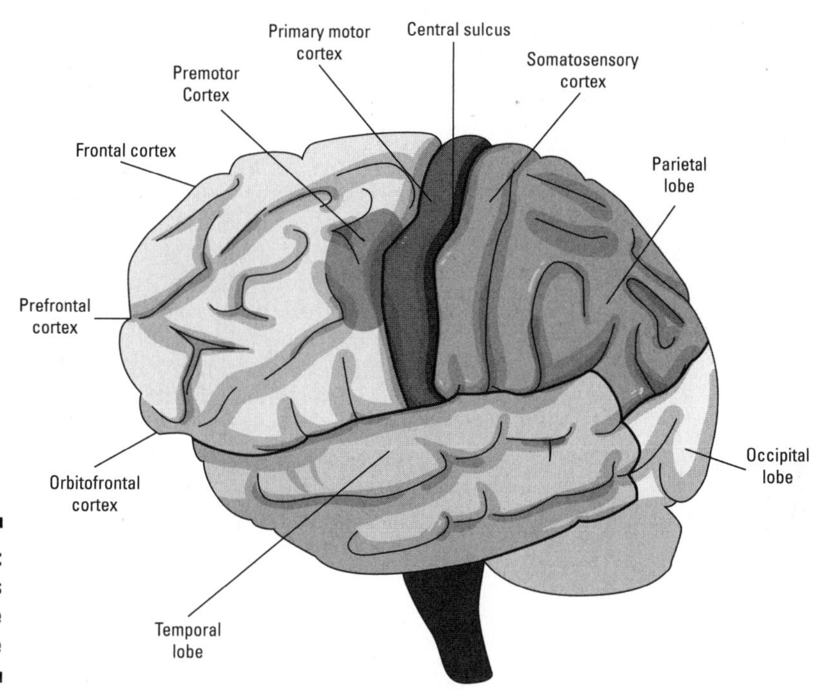

Figure 2-3: Divisions within the frontal lobe

Just in front of those regions are areas dealing with the most abstract areas of planning. Consider playing tennis: When the ball comes at you, you can choose (if you have time) to hit either a forehand or backhand, with topspin or slice. You can choose to stand there and not hit the ball at all if you think it might be out. These kinds of decisions are made in the most anterior parts of prefrontal cortex.

Think of the frontal lobe, which is concerned with executing behavior, as "polarized" from anterior (front) to posterior (back). Farthest back, at the central sulcus, are neural wires going almost directly to muscles. In front of that are areas that organize and sequence movements. In front of that are abstract planning levels. At these abstract levels, for example, you select from a variety of different strategies that may involve completely different muscles, muscles sequences, or, as in the tennis shot, the decision to not move at all.

The prefrontal cortex tends to be larger in primates than other mammals, and it's larger in humans than in other primates. This is correlated with the amount of high level planning done by members of different species. Most mammals operate mostly on instinct and don't live in complexly differentiated social groups. Primates, on the other hand, have complex male and female hierarchies and may hatch plots against each other that span years of planning. Humans build tools, modify their environments for their own purposes, and have specific relationships with up to hundreds of other individuals (and this was even before Facebook).

The parietal lobe

Just posterior to (behind) the central fissure is the *parietal lobe.* This lobe extends from the central fissure back to a not very well structurally-defined border with the occipital lobe (refer to Figure 2-1).

The parietal lobe contains neurons that receive sensory information from the skin and tongue, and processes sensory information from the ears and eyes that are received in other lobes. The major sensory inputs from the skin (touch, temperature, and pain receptors) relay through the thalamus to the gyrus just posterior to the central fissure, where a map of the skin exists. This map is distorted so that areas of the skin with high sensitivity, such as the face and fingertips, are disproportionately larger than areas with low sensitivity, such as the back. If this reminds you of something going on in the primary motor areas, just on the other side of the central sulcus, you're right. The skin homunculus map closely resembles the primary motor cortex map. In order to execute a complex motor task like tying a knot, you need high touch sensitivity to guide the finally differentiated motor control. Chapter 4 has more on this skin map.

The occipital lobe

The back of the brain is the occipital lobe. This area processes visual input that is sent to the brain from the retinas. The retinas project in an orderly, *spatiotopic* way onto the posterior pole of the occipital lobe, called *V1* (for visual area one), so that activity in different areas of V1 is related to whatever is in the image around your current point of gaze. In other words, V1 is another map.

The occipital lobe has many other visual maps derived from V1 (V2, V3, V4, and so on), all almost entirely devoted to processing vision. Other subareas beyond V1 specialize in visual tasks such as color detection, depth perception, and motion detection. The sense of vision is further processed by projections from these higher occipital lobe areas to other areas in the parietal and temporal lobes, but this processing is dependent on early processing by the occipital lobe. Researchers know this because damage to V1 causes blindness in that part of the visual field that projects there.

The fact that the visual system gets an entire lobe for processing emphasizes the importance of high visual acuity and processing among our senses.

The temporal lobe

The temporal lobe combines auditory and visual information. The superior (upper) and medial (central) aspect of the temporal lobe receives auditory input from the part of the thalamus that relays information from the ears. The inferior (lower) part of the temporal lobe does visual processing for object and pattern recognition. The medial and anterior parts of the temporal lobe are involved in very high-order visual recognition (being able to recognize faces, for example) as well as recognition depending on memory.

Below the neocortex: The thalamus

So what's underneath (and hierarchically below) the neocortex? The thalamus. The cortex interacts with the rest of the brain primarily via a structure called the *thalamus,* which you can see, along with other structures, in Figure 2-4.

The thalamus and another structure called the *hypothalamus* (which controls *homeostatic* body functions such as temperature and circadian rhythms) constitute the *diencephalon.* The diencephalon is the name given to the thalamus and hypothalamus because of their position just below the neocortex and because of their interrelation during embryonic development.

The root of the word thalamus comes from a Greek word (*tholos*) related to the entrance room to a building, so you can think of the thalamus as the gateway to the cortex. Virtually all signals from the senses are relayed through the thalamus, as are the signals from other subcortical areas. Many areas of the neocortex also communicate with each other through the thalamus.

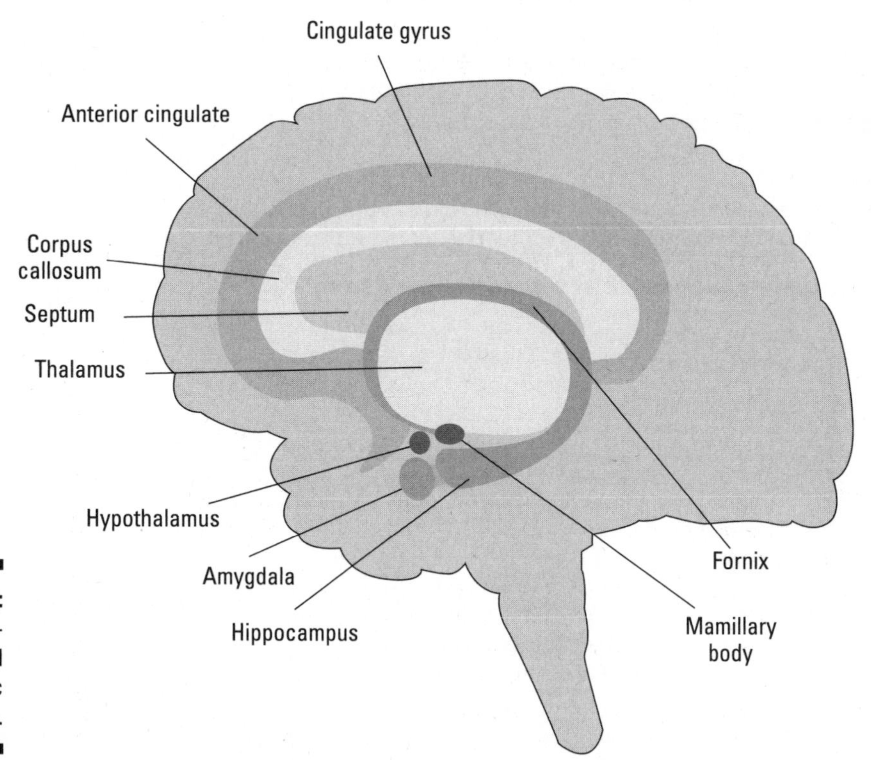

Figure 2-4: The thalamus and the limbic system.

So, what does the thalamus do, exactly? It functions like a command center that controls what information goes between different parts of the neocortex and the rest of the brain. While the neocortex can do very fine-grained analysis of the patterns you're looking at, the thalamus controls where you look. When your neocortex is damaged, you lose particular skills. If your thalamus is damaged sufficiently, you lose consciousness.

The limbic system and other important subcortical areas

Below the neocortex and the thalamus are several important subcortical brain areas. One of the most important is a network of distinct, phylogenetically old nuclei called the *limbic system.* (Saying that these limbic system nuclei are phylogenetically old means that they existed in species much older than mammals, such as lizards, birds, and probably dinosaurs). Several important structures are within the limbic system (refer to Figure 2-4).

The limbic system evolved to incorporate memory into overall behavior control via the *hippocampus* and *amygdala*. These memory structures interact not only with the neocortex, but also with an older type of cortex called the *cingulate gyrus*. The cingulate gyrus is what's called *mesocortex,* a type of cortex that evolved as a high-level controller, before the neocortex, to control behavior.

Here's how many brain scientists think this works. Animals like lizards (that resemble animals long before mammals) clearly have the capability for memory, despite not having a neocortex. Memory modifies behavior in such animals via what we think of as emotions. When a lizard loses a fight with another lizard, it becomes *fearful* of that lizard when it sees it again. When a lizard finds itself in an area where it has fed before, it becomes *hungry* and may begin stalking for tasty bugs.

General patterns of behavior in animals can be thought of as being organized around particular goals: You're either seeking food, avoiding predators, seeking a mate, or avoiding cold, for example. Many of these goal-states are produced by instincts and homeostatic mechanisms. However, brain areas within the limbic system add memory of past experiences and circumstances to modify pure instinct: If you haven't eaten for a while, you're likely to be hungry, but maybe not if a hawk is circling overhead. The top integrating area for all this limbic computation may once have been the cingulate gyrus.

Mammals, like humans, add the thalamus and neocortical system on top of all the subcortical areas possessed by our non-mammalian vertebrate ancestors. Rather than replace these older systems, we add additional layers that are capable of more computational subtlety. For example, many predatory mammals hunt in packs or family groups that take turns driving prey towards rested members of the pack, using very complex and flexible cooperative strategies. You never see lizards do this. Although lizards have limbic systems and memory areas such as the hippocampus, they lack a large neocortex to do really complex memory-contingent calculations.

The following sections provide more detail on the main memory systems of the hippocampus and the amydala, including the other brain areas with which they interact.

The hippocampus

The limbic system was originally thought to be involved in generating and processing emotions. However, one brain structure within this system, the hippocampus, is now known to have a crucial function in the creation of memory. The hippocampus receives inputs from virtually the entire neocortex. Through specialized adjustable synaptic receptors called NMDA receptors, it can associate together virtually any constellation of properties that define an object and its context.

The amygdala

Just in front of the hippocampus is another memory structure that *is* primarily involved with emotional processing, the amygdala. The amygdala interacts with the prefrontal cortex to generate and process the major emotions of anger, happiness, disgust, surprise, sadness, and, particularly, fear. People who have sustained damage to their amygdalas have reduced abilities to react to and avoid situations that induce fear.

Orbitofrontal cortex

The amygdala and other structures of the limbic system interact with the part of the prefrontal cortex called the *orbitofrontal cortex.* This is the medial and inferior part of the prefrontal cortex (excluding the lateral areas called lateral prefrontal cortex). Suppose that, on some particular Friday evening while driving home, you're almost hit by another car at a particular intersection. It is very likely that, for a long time after that, when approaching that intersection, particularly on Fridays, you'll get a little twinge of fear or uneasiness. Your orbitofrontal cortex has stored the circumstances, and the amygdala has stored the fear.

The anterior cingulate cortex

The amygdala and other structures of the limbic system also interact with an area of mesocortex called the *anterior cingulate cortex,* which is the most anterior part of the cingulate gyrus (refer to Figure 2-4). This area of the brain seems to monitor the progress toward whatever goal you're pursuing and generates an "uh-oh" signal when things aren't working out to indicate a change in strategy may be in order.

The basal ganglia

Another group of subcortical brain structures that are crucial in planning, organizing, and executing movement (with the frontal lobes) are the *basal ganglia*. The basal ganglia consist of five major nuclei: the caudate, putamen, globus palladus, substantia nigra, and subthalamic nucleus (see Figure 2-5).

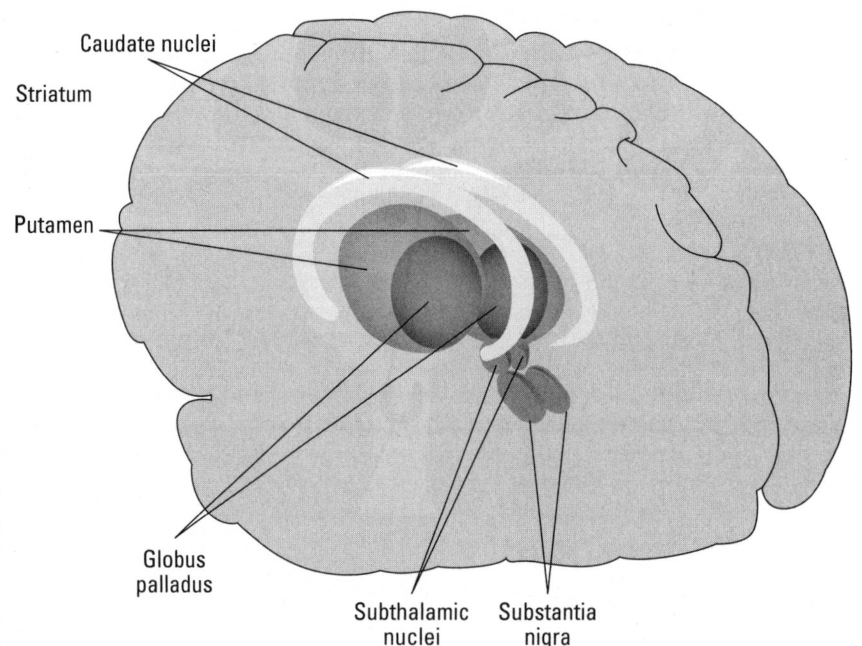

Figure 2-5: The basal ganglia and cerebellum.

These nuclei comprise a highly interconnected system that interacts with the thalamus and neocortex to control behavior. The following list explains each one:

- **The caudate:** Inputs to the basal ganglia from the neocortex primarily come into the caudate and another nucleus called the putamen. In fact, the caudate-putamen complex is now usually called the *striatum*). Anatomically, the caudate is a ring of cells that surrounds the basal ganglia complex except on the ventral side. The output of the caudate is to the globus palladus.

- **The putamen:** The putamen is a large nucleus within the basal ganglia complex with similar input-output connections as the caudate.
- **The globus palladus:** The globus palladus is the primary output nucleus of the basal ganglia via its inhibitory connections to the thalamus.
- **The substantia nigra:** The substantia nigra sends "enabling" connections to signals passing through the striatum. This nucleus is well known for its involvement in Parkinson's disease, which is caused by the death of dopaminergic neurons in this nucleus. Without these neurons, initiating voluntary movement becomes difficult.
- **The subthalamic nucleus:** The subthalamic nucleus also modulates the *globus palladus* output. This nucleus is frequently the target of deep brain stimulation surgical implants to relieve symptoms of Parkinson's disease.

Even basal ganglia experts readily admit the internal functional architecture of the basal ganglia is poorly understood at present. These nuclei are organized in a similar manner to the *cerebellum,* a phylogenetically old motor coordination center with which they interact. For more on the cerebellum, head to the later section "Coordinating movement: The cerebellum."

Transitioning between the brain and the spinal cord

The next three brain areas I cover — the midbrain, the pons and medulla, and the cerebellum — make the transition between the brain and spinal cord. These areas are phylogenetically *very* old, and they're found in all vertebrates. They control basic behaviors such as locomotion coordination, eye movements, and regulation of body homeostasis such as respiration, heart rate and temperature. The following sections have the details.

The midbrain

The most superficial of these three areas is called the *midbrain.* The midbrain contains low-level processors that control eye movements and help localize sound.

The major visual area in the midbrain is the *superior colliculus,* which controls eye movements called *saccades*. The superior colliculus receives projections from about one-tenth of all retinal ganglion cells in humans and a much higher percentage in other mammals. In non-mammalian vertebrates such as frogs, the superior colliculus (called the *tectum* in frogs and other non-mammals) is the major visual target of the retina, and it's the crucial coordinating area for most visually guided behavior.

Research at MIT in the late 1950s showed that certain ganglion cells in the frog eye responded only to small moving objects the size of bugs the frog preys on — bug detectors. Electrical stimulation in the frog's tectum caused frogs or toads to snap their tongues at imaginary bugs in the visual field area stimulated. In mammals such as humans, the superior colliculus primarily controls eye movements; it leaves the control of visually guided behavior, like hitting tennis balls, to the thalamic neocortical system, particularly in the parietal lobe.

Auditory fibers from the inner ear project to a processing area just below the superior colliculus, called, for a rather obvious reason, the *inferior colliculus.* Neurons in this structure project to neurons in the auditory part of the thalamus, which in turn projects to the auditory part of the neocortex on the superior aspect of the temporal lobe.

The midbrain also includes one of the basal ganglia, the *substantia nigra* (refer to Figure 2-5). Other nuclei in the midbrain, such as the locus coeruleus, Raphe, and ventral tegmental areas, make extensive but diffuse modulatory projections throughout the neocortex. This means they make only a few out of the thousands of synapses on specific target cells, meaning they contribute small, modulatory excitation or inhibition to large numbers of neurons in many brain areas rather than being the major cause of the firing in those target neurons.

The reticular formation

An important brain area that courses throughout much of the subcortex including the midbrain is the *reticular formation.* The word *reticular* is derived from the Latin word for "net." The reticular formation is not so much a defined structure as it is a continuous network that extends through and interacts with numerous brain areas. The reticular formation extends through the midbrain and pons and medulla and is continuous with reticular "zones" above the midbrain and below the medulla into the spinal cord. The reticular formation controls all life functions, such as heart rate, respiration, temperature, and even wakefulness. Damage to the reticular formation typically results in coma or death.

Processing the basics: The pons and medulla

The *pons*, below the midbrain, contains nuclei that mediate several auditory and balance functions. Auditory functions include neural comparisons of the loudness and timing between the two ears for auditory localization (in cooperation with the inferior colliculus in the midbrain). Some pons nuclei receive inputs from the vestibular system in the inner ear (the semicircular canals) and communicate with the cerebellum for balance. Damage to these pons nuclei or to the vestibular system can produce not only chronic dizziness but even the inability to stand up. The pons receives sensory input from the face and sends motor neurons to facial muscles to control both voluntary and involuntary facial expressions.

Below the pons is the most ventral area of the brain called the *medulla.* The lower border of the medulla is the top of the spinal cord. Much of the medulla consists of neural fiber tracts (called *pyramids,* because of their shape in cross section) that contain sensory information from the skin, muscles, and tendons going up to the brain, and motor command information from the primary motor cortex in the frontal lobe going down into the spinal cord to control the body's muscles. Two areas in the medulla, called the *gracile* and *cuneate nuclei,* integrate the ascending and descending information. Like the pons, some sensory inputs from the mouth (including neurons from the tongue containing taste information), face, and throat also synapse in the medulla.

Coordinating movement: The cerebellum

The cerebellum ("little brain") is a complex motor-coordination structure that, by some estimates, contains as many neurons as the entire rest of the nervous system. Like the neocortex, the cerebellum is heavily convoluted with gyri and sulci. It is one of the phylogenetically oldest parts of the brain, so all vertebrates have a cerebellum, even including cartilaginous fish such as sharks. The cerebellum helps learn and control the timing of motor sequences. Damage to the cerebellum produces slow, clumsy, robot-like movement.

The cerebellum extends out from the pons, even though most of its connections are to motor cortex via the thalamus. Structurally, the cerebellum is organized with an outer cortex, internal white matter (axon tracts), and four pairs of nuclei deep within it. Most inputs and outputs of the cerebellum occur through the deep nuclei. This activity is modulated by the connections between these nuclei and outer or cortical processing areas. Most output of the cerebellum inhibits the thalamus, where it modifies, via sensory feedback, cortically generated ongoing motor control to achieve smooth, fine, feedback-adjusted movements.

The structure of the cerebellum is similar to the structure of and interactions between the deep thalamus and more superficial and extensive neocortex in the larger part of the brain. Refer to the earlier section "The neocortex: Controlling the controllers" for more information about the neocortex.

Looking at differences: Size, structure, and other variations

Up to this point, I've been talking about the human brain as though all human brains were identical. But they're not. So the question then becomes how much variation is there between people, and is this variation important? The following sections delve into some of the differences.

Mine's bigger that yours!

One obvious difference in brains is *size*. An average size brain is about 1,300 cc. Among adults, a 10 percent variation around that norm is common, which means the brains of many people are up to 10 percent larger or smaller than the standard 1,300 cc.

Brain size tends to scale with body size, and thus large men tend to have larger brains than small women. How important is this variation? Not very, apparently. There are many documented cases of geniuses with rather small brains, and, of course, we've all encountered lots of dimwits with rather large brains (or at least, large heads, which is what we can observe).

It's what's inside that counts: Looking at differences in structural organization

What about the structural organization of the brain? Everyone has more or less the same major gyri and sulci, but there are variations. In fact, if you look carefully at any real brain, you can see that the two hemispheres are rarely exact mirror images. (That doesn't mean, however, that you can tell whether the person is right or left-handed by looking at the two hemispheres.)

In fact, this amount of variation between hemispheres is common between brains as well. This is one of the reasons why epilepsy surgeons functionally map brain areas around the area they're targeting: they need to make sure they're not removing crucial areas, such as those that control language. (The other reason for functional mapping is that the pathology itself in the epileptic brain may have caused some displacement of function away from the normal areas.)

What is not observable at the level of the overall structure of the brain is its internal wiring.

Considering gender-based brain differences

Although the overall structure and detailed connectivity of male and female brains are nearly identical, some recent research has revealed male versus female differences in what might be called "styles" of processing, with men preferring big picture, map-like ways of organizing knowledge and woman preferring a more linear, procedural system of knowledge. Ask a man for directions, for example, and he tends to tell you the general layout of where you are in relation to where you are trying to go, whereas a woman tends to tell you the sequence of turns you need to get there. Yes, Virginia, we really do think and speak in different languages, sometimes.

The following section discusses some of the controversial hypotheses related to differences between male and female brains.

Which part of your brain do you use? Lateralization differences

One apparent (but controversial) difference that has been suggested is that the amount of specialization between the two hemispheres (lateralization) is different between male and female brains. Brain scans tend to show that women use more, and more symmetrical, brain areas than men do for the same tasks. Women also tend to have a higher percentage of connecting fibers in the corpus callosum, the fiber tract that links the two hemispheres together.

The general idea is that men's brains are more lateralized than females, meaning that functions tend to be more restricted to one hemisphere in men than in women.

The correlate of this hypothesis is that men's brains are therefore more specialized, particularly with respect to spatial processing in the right hemisphere. The upside for men, according to this idea, is that they may be able to process spatial, right-hemisphere tasks in some sense more deeply because of lateral specialization. The downside is that they tend to be more disabled by strokes or lesions that affect only one side of the brain because there is less redundancy in the representation for that ability in the mirror image location in the other hemisphere.

It has also been suggested that not only is autism much more common in males that females, but that autism itself can be regarded as a kind of "ultra-male" brain configuration with savant skills in certain kinds of complex processing accompanied by poor language and social development.

The lateralization theory explains women's general superiority in language skills compared to men in this way: less specialization allows both hemisphere's in females' brains to process language.

The issue is far from resolved. Some propose that the observed male/female differences in brain activation and skills are entirely culturally based and have nothing to do with any intrinsic differences in brain structure of organization. Others — usually those who argue more for biologically based differences —suggest that male versus female brains are differentially *susceptible* during development to lateralization of language versus spatial skills.

How brain development plays into the debate

One thing making it even more difficult to evaluate any of these hypotheses is that experience clearly changes brain development. Therefore, gender-differentiating cultural experience may modify development to produce true male/female biological differences in adult brains. As you can imagine, this little controversy has generated an industry of Ph.D. dissertations and journal papers that keeps many cognitive scientists on both sides of the issue gainfully productive.

What mechanisms could produce gender-based lateralization differences during development? There appear to be developmental and operational effects on the brain from the hormones testosterone, estrogen, and oxytocin. Artificial testosterone injections tend to produce male-like behavior and proclivities in females, whereas estrogen does the opposite in males. Oxytocin, a hormone released during the birth process in females, has been suggested to mediate a kind of "mother bear" behavior, with high nurturing towards those within the clan but high aggression against those outside.

The Spinal Cord: The Intermediary between Nervous Systems

The spinal cord is a key part of the central nervous system. Specifically, within the spinal cord are the connections between the central nervous system (such as motor neurons from primary motor cortex) and the peripheral nervous system (skin, muscle, and tendon receptors going to the spinal cord, and *alpha* motor neurons relaying motor commands from the spinal cord to the actual muscles). The sensory and alpha motor neurons constitute the peripheral nervous system, which mediate voluntary behavior and sensing.

The spinal cord also has integration and coordination functions, although it is close to the lowest level of the hierarchy of controllers. It mediates feedback between sensory and motor pathways for each limb and coordinates limb movements for locomotion.

Finding central pattern generators

Some of the earliest neurological experiments by scientists like Sir Charles Scott Sherrington (1857 – 1952, Oxford) demonstrated the existence of *central pattern generators* in the spinal cords of four legged animals like cats. The central pattern generator can control alternation of left and right hind or forelimbs and proper phasing between forelimbs and hind limbs.

Forelimb/hind limb phasing is different for different gaits. In normal four-legged walking, for example, the two forelimbs are in antiphase with each other, as are the two hind limbs, and the fore- and hind limbs on each side are antiphased, so that as the left forelimb goes back, the left hindlimb goes forward. Other gaits, like the trot, gallop, or canter have different phase relationships.

By severing the connections between the brain and spinal cord in experimental animals, Sherrington showed that these gaits could be organized by the spinal cord itself in the absence of any brain control, when the animals were placed on a treadmill. The brain is needed, of course, to decide whether to walk, trot, run, hop or turn in normal behavior.

Each level of the spinal cord contains a local processing module for the area of the body controlled from that segment, plus connections to other spinal cord segments and to and from the brain (through the medulla all the way to neocortex). The top of the spinal cord deals with muscles and sensory information from the neck, while the bottom spinal cord segments deal with the toes. The spinal cord segments are designated as 8 cervical, 12 thoracic, 5 lumbar, 5 sacral, and 3 coccygeal. These are numbered from the top down, so the highest cervical spinal segment is C1, while the lowest sacral segment is S5.

Looking at the spinal reflex

The fundamental unit of coordinated action mediated by the spinal cord is the spinal reflex; you can see the anatomy of this reflex in Figure 2-6. Striking the kneecap stretches the quadriceps muscle, as though one's knees were buckling. Muscle spindle receptors in the quadriceps project to the spinal cord, and, through interneurons, activate alpha motor neurons that contract the muscle to maintain upright posture.

Alpha motor neurons are the neurons that innervate muscle cells, which in turn move the limbs. The motor neurons have their cell bodies in the ventral (toward the stomach side of the body) part of the spinal cord. The axons that drive the muscles exit the spinal cord via a tract called the *ventral root.*

On the dorsal side of the spinal cord are the axons of sensory cells, such as stretch receptors, that enter via the dorsal root.

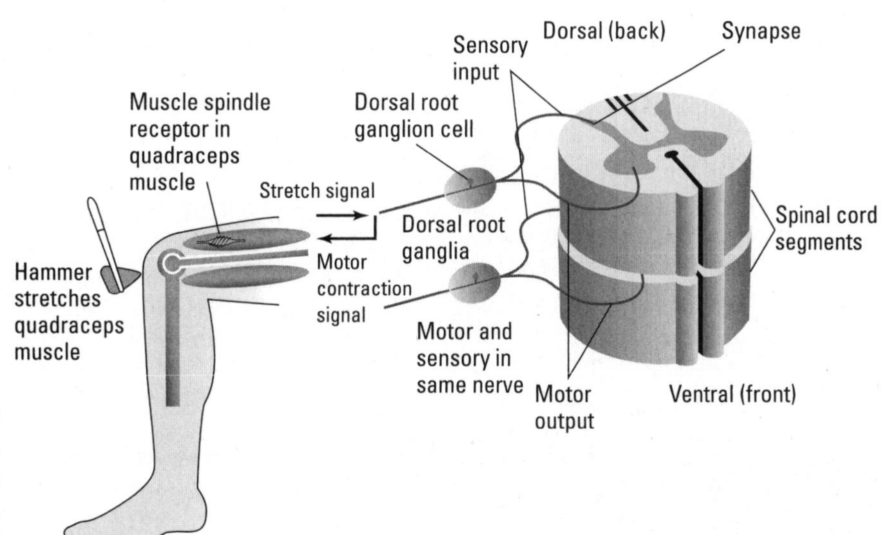

Figure 2-6: The spinal reflex.

When the doctor taps just below your knee cap with a rubber hammer, the tap slightly stretches the tendon from your knee to your foot, stretching the quadriceps muscle and spindle, which causes your leg to extend in compensation. This test checks the integrity of your peripheral nervous system (sensory and motor nerves acting through the spinal cord).

What doesn't Figure 2-6 show? It doesn't show several pathways connecting a spinal cord segment to other segments, and to the brain:

- **The original message gets sent to the brain.** The sensory input coming in the dorsal root synapses on spinal neurons that relay the sensory message all the way to the brain (somatosensory cortex, just posterior to the central sulcus).
- **Command messages come down from the brain.** Cortical neurons from the primary motor cortex come all the way down the spinal cord and synapse on the same alpha motor neurons that innervate muscles for the stretch reflex to allow you to voluntarily extend your leg.
- **Communication occurs between segments.** Motor events occurring in one segment send messages to other segments to coordinate body actions. If, for example, you were standing while your left leg started to buckle at the knee, your right leg would probably stiffen in compensation, and you would probably extend your left hand up, among other things. These connections also are involved in the central pattern generator control of gait mentioned previously.

Getting your muscles moving

How do muscles work? Muscles are groups of cells connected to each other in long parallel chains the ends of which are connected to bones by tendons. The muscle cells contain chains of proteins called *actin* and *myosin*. When these proteins are stimulated, they slide over each other, causing the cell to contract along its length.

Alpha motor neurons provide the stimulation by releasing the motor neuron neurotransmitter acetylcholine. When the acetylcholine is received by the receptors on the muscle cells, it causes a muscle cell action potential (an electrical pulse in the cell that causes it to momentarily contract) that causes the actin-myosin to slide, which in turn causes the contraction.

Fighting or Fleeing: The Autonomic Nervous System

The preceding sections outline the key components of the central nervous system (brain and spinal cord) and explained how the peripheral nervous system (motor and sensory nerves) gets in on the act. But when it comes to nervous systems, your body has more than just these two. There is another nervous system outside the central nervous system that controls not voluntary muscles as the peripheral nervous system does, but the heart, glands, and organs with smooth muscles (not under voluntary control) such as the intestines. This is the autonomic nervous system.

The autonomic nervous system consists of two divisions, the sympathetic and parasympathetic that often act in opposition to each other. The sympathetic system prepares the body for immediate action (fight or flight), at the expense of body regulating functions such as digestion. These two branches use two different neurotransmitters — norepinephrine (noradrenaline) for the sympathetic branch and acetylcholine for the parasympathetic branch — that tend to have opposite effects on the target organs. For example, norepinephrine speeds up the heart while acetylcholine slows it.

Norepinephrine also dilates the pupil and lung bronchi, decreases digestive functions, and inhibits bladder contraction and blood flow to the genitals. Many of these actions are mediated through the adrenal medulla. However, overstimulation of the sympathetic system from excessive stress is hard on the body and tends to be associated with heart disease and other stress-related chronic illnesses. Because social conflicts can also trigger the sympathetic system, the chronic stress of modern life may result in sympathetic overstimulation and its associated long term effects.

How We Know What We Know about Neural Activity

How do we know what we know about how the brain works? It's not obvious at all upon looking at the brain that it is the seat of thought, consciousness, and body control. Some very prominent ancient scientists thought the brain's purpose was to cool the body through the skull, while the *heart* was the seat of intelligence! Well, scientists have come a long way since then. The following sections outline some of the methods that have been — and continue to be — used to reveal the secrets of the brain and the nervous system.

Examining problems caused by brain injuries

Probably the first clues about the function of the brain came from dysfunction associated with head injuries. In the early 20th century, the scientific study of the brain began advancing significantly with the study of *cytoarchitectonics* (using stains to evaluate cell versus tract structure in various parts of the brain), and *tract tracing*, determining the connections between one area of the brain and another by tracing the axonal processes of cells (I explain axons in Chapter 3). Early tract tracing often involved seeing where secondary degeneration occurred after damage to an area the secondary area was connected to. Silver stains such as the Golgi stain allowed researchers to see the structure of individual neural cells in great detail.

Using technology to image the brain: From early EEGs to today

On the physiological front, electroencephalographic (EEG) recordings in the early 20th century revealed that the brain was constantly producing electrical oscillations that could be recorded from the surface of the skull. These oscillations changed with external stimulation and when different thought patterns occurred inside the brain. The EEG has a few deficiencies, though, which make it less than ideal for careful brain study.

- The EEG averages activity over large areas so localizing the source of the recorded activity in the brain is hard. The averaging also tends to miss much of the complexity of brain activity.
- The intrinsic ongoing activity in the brain tends to be larger than — and tends to swamp — the smaller, transient signal elicited by particular stimuli, such as showing someone a picture, which makes the EEG a poor method to study the effect of stimuli on brain activity

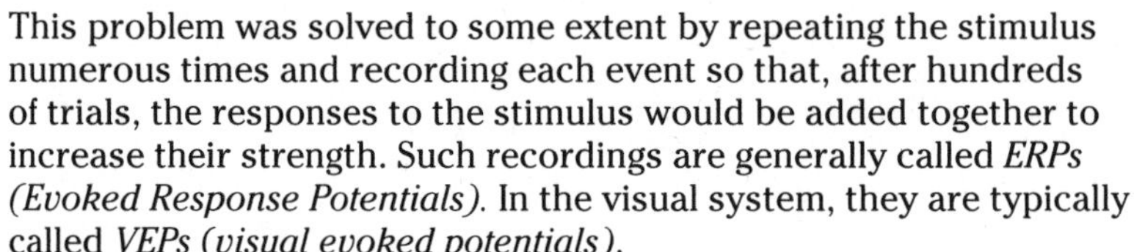

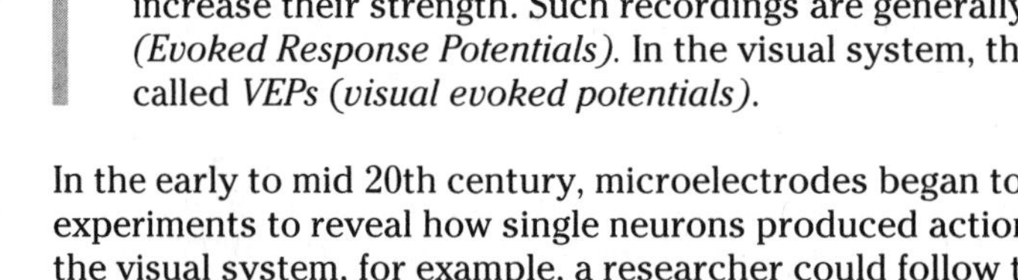

This problem was solved to some extent by repeating the stimulus numerous times and recording each event so that, after hundreds of trials, the responses to the stimulus would be added together to increase their strength. Such recordings are generally called *ERPs (Evoked Response Potentials).* In the visual system, they are typically called *VEPs (visual evoked potentials).*

In the early to mid 20th century, microelectrodes began to be used in animal experiments to reveal how single neurons produced action potentials. In the visual system, for example, a researcher could follow the signal from the photoreceptor's electrical response to light, through the retina, an on to the thalamus and visual cortex. Similarly, on the motor side, muscle action potentials, alpha motor neuron spikes, and motor cortex firing were recorded with microelectrodes. However, because microelectrode recording is invasive (you have to insert microelectrode needles near the neurons to get a recording), it has had little or no applicability in humans.

After the middle of the 20th century new imaging techniques, explained in the following sections, began to be commonly used.

PET and SPECT

PET stands for *Positron Emission Tomography.* In this technique, short-term weakly radioactive oxygen or sugar molecules are introduced into patients (or volunteers) while they perform some cognitive task. The most active neurons take up the radioactive substance because of their higher metabolism, and the locus of radioactivity in the brain is ascertained by a complex scanner and sophisticated software that detects the emission of positrons (anti-electrons).

Typically a scan is done at "rest" — before the cognitive task is performed — and then compared with another scan done while the subject is doing the task. In this way, the difference between the two scans is assumed to be task associated. The PET technique has a spatial resolution of millimeters, but because it takes only one snapshot of brain activity, it can't resolve brain activity changing in time.

SPECT (Single Photon Emission Computed Tomography), is similar to PET except the ingested radioactive emitter releases gamma rays that are detected by the scanner.

fMRI

The *fMRI* (functional Magnetic Resonance Imaging) is a derivative of MRI. The MRI imaging technique uses a high magnetic field and radio frequency to detect transitions in the spin of protons (usually in water molecules) in the brain, and was formerly referred to as *nuclear magnetic resonance* (NMR).

Reconstructed images have good contrast between areas of the brain with high neural cell body density (gray matter) versus areas consisting mostly of fiber tracts (white matter), at millimeter resolution. In fMRI, blood flow or blood oxygenation/deoxygenation is detected dynamically with a time resolution of several seconds (future instruments may even be able to do this even faster). Blood flow changes and blood deoxygenation are metabolic measures believe to reflect real neural activity.

Although the spatial resolution of fMRI is about an order of magnitude worse than MRI in the same instrument, typically a structural MRI scan is done first and then the fMRI functional scan is superimposed on the structural scan to locate areas of differential activity between task and rest. Because it uses no radioactive isotopes, fMRI is considered safe and protocols can be repeated many times in a single session to improve signal to noise.

MEG

MEG (magnetoencephalography) is based on the physics of electrical and magnetic fields. Active neurons generate electric currents that flow within the neurons and in the surrounding extracellular space. The EEG detects these currents after they have become diffused across extracellular space, including the skull between the electrodes and the brain, leading to poor spatial resolution.

A fundamental property of electric currents, however, is that they produce magnetic fields, which aren't shunted by the intervening volume electrical conductor. The neural magnetic fields are extremely small, though, so MEG technology requires heavily shielded recording rooms and exotic low strength magnetic field detectors called *SQUIDs* (*super-conducting quantum interference devices*). Current research and medical instruments use many of these SQUIDs over the entire brain to image a large region of the brain simultaneously. MEG has temporal resolution nearly as fast as EEGs, and spatial resolution on the order of fMRI.

DTI

DTI (diffusion tensor imaging) is a variant of magnetic resonance imaging, and is sometimes called *diffusion MRI.* It measures the movement (diffusion) of water in the brain. Because water moves most easily down the cytoplasm of long, cylindrical neural axons, DTI effectively measures axon tracts. It is an important clinical instrument because axon tracts may be destroyed by strokes or other brain disorders.

Optical imaging

Optical imaging is a promising future technology that involves either the changes in absorption or scattering of light by neurons due to their electrical activity, or the use of reporter (typically fluorescent) dyes that respond to changes in the concentration of certain ions (like calcium) entering the neuron during activity . Optical techniques have been used extensively in experiments in isolated tissue obtained from animals. Changes in scattering or absorption of infrared light have been used to a lesser extent in humans where optical access to tissue has existed, such as retina or cortex during surgery. There are also techniques to use infrared light obtain EEG-like recordings optically through the skull.

Chapter 3

Understanding How Neurons Work

In This Chapter

- Understanding what neurons do and how they work
- Communicating between neurons: The role of neurotransmitters
- Looking at the role of glial cells
- Examining the various techniques used to study neural activity

Everything you think, everything you do, and, in fact, everything you *are*, is the result of the actions of about 100 billion neurons. But what are neurons, and how do they enable minds?

Neurons, sometimes referred to as *marbles,* are specialized cells that process information. Some neurons are receptors that convert sensory stimuli into messages processed by the brain, while other neurons stimulate muscles and glands. Most neurons are connected to other neurons and perform computations that allow an organism — you — to behave in complex ways, based on what the body senses internally and externally, now, in the experienced past, and in the expected future. This chapter explains how neurons function, how we know what we do about how the nervous system operates, and what new techniques are extending our knowledge of brain function.

Neuron Basics: Not Just Another Cell in the Body

First of all, neurons are cells. Neurons, like other cells, consist of a bag of cytoplasm with a membrane separating their cytoplasm from the outside world. Within this cytoplasm are organelles, such as mitochondria that provide energy for the cell, and structures, such as the Golgi apparatus and the endoplasmic reticulum, that manufacture and distribute protein. The nucleus contains the DNA that codes for the identity of both the organism itself and the cell within the organism.

Here's what makes neurons different from other cells (if you're unfamiliar with general cell structure and components, you can find that information in most basic biology books, such as *Biology for Dummies,* 2nd Edition [Wiley]):

- A membrane with selective ion channels that can be opened by electrical events or binding by neurotransmitters
- Dendrites, branching extensions from the cell body that receive inputs from other neurons
- An axon, a single extension from the cell body that travels long distances (up to several feet) to contact other neurons

Figure 3-1 shows the main parts of a typical neuron. There are obvious structural differences between neurons and most other cells. While most non-neuronal cells resemble squashed spheroids, neurons typically have a "dendritic tree" of branches (or processes) arising from the cell body (or *soma)*, plus a single process called an *axon* that also emanates from the cell body but runs for large distances (sometimes even up to several feet) before it branches. Not so evident from Figure 3-1 is the fact that the membrane that encloses a neuron's cell body, axon, and dendrites is very different from the membrane that encloses most other cells.

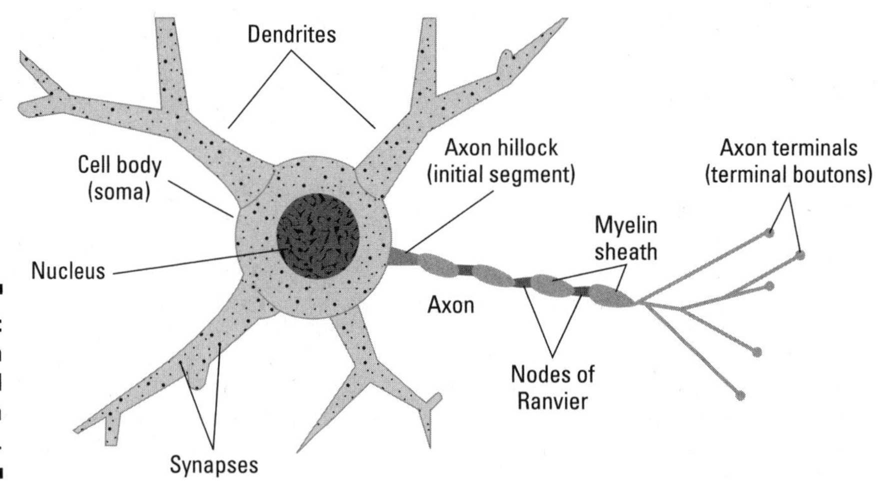

Figure 3-1: The main structural parts of a neuron.

Sending and receiving info between neurons: Synaptic receptors

What are the dendrites and axons for? Let's start with the dendrites, or *dendritic tree.*

Think about what the branches of a normal tree are for: They're the structures that make places for the leaves, whose purpose is to gather sunlight for energy and to build organic compounds that the tree needs. A neuron's dendritic tree does essentially the same thing. It's the structure that creates places for *synaptic receptors* that gather information from other neurons. And as trees may have thousands of leaves, neurons have thousands of these receptors.

While the dendrites receive inputs from other cells, the axon sends the output of the cell *to* other cells. The axon often looks like another dendrite, except that it typically doesn't branch very near the cell body like most dendrites.

Looking at the receptors: Pre- and post synaptic

What are the receptors for, you ask? Well, not for sunlight. Neural synaptic receptors receive information via little squirts of neurotransmitters from other neurons. Each little squirt of neurotransmitter tells the *postsynaptic receptor* (the receptor on the receiving neuron's dendritic tree) a message from the *presynaptic* cell (the one doing the squirting).

For the receiving neuron, it's like listening to hundreds of brokers simultaneously on the floor of a stock exchange. Some of the brokers are squirting happy messages about something or other and make the postsynaptic neuron happy, or *excited,* while others are sending low-key messages that make the postsynaptic neuron less excited. Millisecond by millisecond, the excitement level of the postsynaptic neuron is determined by which inputs, among the thousands it receives, are active and how active they are.

The meaning of the activity (or excitement level) of a neuron is a function of who is talking to it and who it is talking to. If our particular stockbroker neuron, for example, were listening to the output of other neurons talking about things like electrical wiring, certain mines in Chile, and production of pennies, our stockbroker neuron might embody information about the price of copper.

These connections become meaningful as a result of *learning.* Learning occurs when experiences modify the strength and identity of the interconnections between neurons and thus create memory. Suppose, for example, that your neuron receives inputs from some neurons that indicate you're seeing the color gray; other neurons that indicate it's big; others that indicate that it has a trunk; and others that respond to the image of tusks. Your postsynaptic neuron is going to be an elephant detector. The postsynaptic targets of this neuron might get other neuronal messages that the ears are big and that lions and giraffes are also nearby, in which case its activity would mean "We're in Africa, and this is an African elephant."

Firing off spikes to other neurons

When the postsynaptic neuron is excited, it fires off electrical pulses, called *spikes,* or *action potentials,* that travel away from its cell body and along its axon until they reach the places where the axon branches and the individual branches end in what are called *axon terminals* (also called *terminal boutons).* These axon terminals are next to other, postsynaptic neurons. These thousands of axon terminals are the places where your presynaptic neuron squirts its neurotransmitter message onto other neurons. And so it goes, times 100 billion, of course.

The junction between the *presynaptic* neurotransmitter squirting axon and the *postsynaptic* neurotransmitter-receiving dendrite is called a *synapse*, the most fundamental computing element in the nervous system. The gap between the pre- and postsynaptic neurons is called the *synaptic cleft.* The squirted neurotransmitter diffuses across the synaptic cleft from the presynaptic releasing site to receptors on the postsynaptic dendrite.

Many neuroscientists believe that if you are going to compare brains to digital computers, the transistor comparison should be at the level of the synapse. Because each neuron has about 10,000 synapses, and there area 100 billion neurons, the folks at Intel are going to have to burn a lot more midnight oil if they want their CPUs to catch up to the brain's computing power.

Receiving input from the environment: Specialized receptors

Some neurons receive inputs not as neurotransmitter squirts from other neurons, but as energy from the environment:

- In the eye, specialized neurons called *photoreceptors* catch photons of light and convert those into neurotransmitter release.
- In the ear, cells called *auditory hair cells* bend in response to sound pressure in the inner ear.
- In the skin, *somatosensory receptors* respond to pressure and pain.
- In the nose, olfactory neurons respond to odorants. Think of the odorants from the world outside the body as neurotransmitters themselves, with different olfactory receptors responding to different odorants.
- On the tongue, sweet, sour, bitter, and salty molecules excite different receptors in the taste buds.

The nervous system also has outputs such as the motor system. Motor neurons output their neurotransmitter onto muscle cells that contract when receptors in the muscle cells receive it. This action generally occurs after the brain has done a lot of planning that you're mostly not aware of about what you're going to do and which muscles in which patterns are going to contract for you to do that.

Ionotropic versus metabotropic receptors

Just as advice may fall on willing or deaf ears, the effects of a presynaptically released neurotransmitter on the postsynaptic neuron actually is mostly up to the *neurotransmitter receptor* on the postsynaptic cell. Although receptors are highly specific for a single neurotransmitter, each neurotransmitter has many different types of receptors. Some receptors act faster than others, some are longer or shorter lasting, and some have multiple binding sites that might include one for a particular fast neurotransmitter plus additional sites that bind and respond to one or more neuromodulators.

One important difference between receptors types depends on a crucial difference between their structures. The left image in Figure 3-2 shows a typical neurotransmitter receptor sitting in the cell membrane. Most neurotransmitter receptors have a binding site for the neurotransmitter on the outside of the membrane, which, when the neurotransmitter is bound, causes the receptor protein to refold in such a way that a pore opens in the receptor complex that allows a particular ion to go through the otherwise impervious cell membrane. These are thus called *neurotransmitter-gated* (also *ligand-gated*) receptors. The receptor type with the ion pore in the same receptor structure as the binding site is called an *ionotropic* receptor. Most fast excitatory and inhibitory receptors are of this type.

There is another type of receptor structure, common for neuromodulators. In this structure, the ion channel pore is not in the same place as the binding site for the neurotransmitter (refer to the right image in Figure 3-2). Instead, the neurotransmitter binding on the outside of the membrane causes part of the channel protein to change shape on the inside of the membrane such that it releases some internal, intracellular neural modulator. These modulators function as internal messengers inside the cell that often involve biochemical cascades that amplify their actions.

Neurotransmitter binding to these kinds of receptors can cause multiple effects, from opening several nearby channel pores from the inside to changing the expression of DNA in the nucleus. These types of effects tend to be much slower but longer lasting compared to ionotropic receptors. The receptor types without their own ion channels that have these indirect effects are called *metabotropic* receptors.

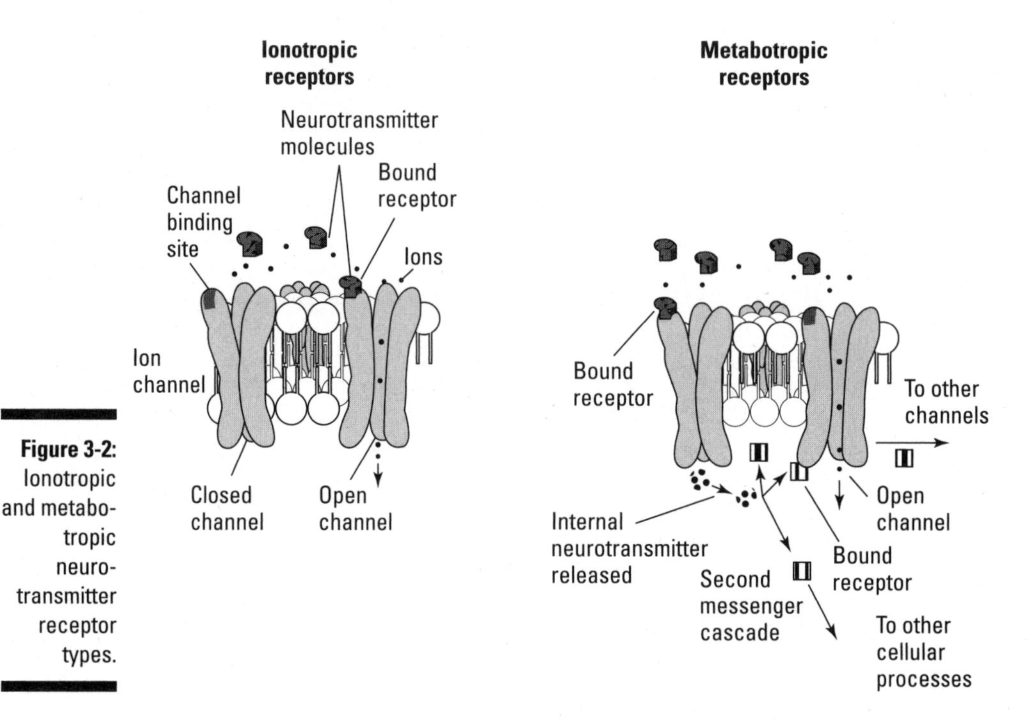

Figure 3-2: Ionotropic and metabotropic neurotransmitter receptor types.

The three major functional classes of neurotransmitters

The neurotransmitter "juice" squirted from the presynaptic to the postsynaptic neuron consists of neurotransmitter molecules. A variety of different types of neurotransmitters exist, each varying in its type of effect and time course on the postsynaptic cell. Neurotransmitters are grouped into three major functional classes:

- **Fast, excitatory neurotransmitters:** The most important neurotransmitters are the fast, excitatory neurotransmitters glutamate and acetylcholine. These communicate strong, immediate excitation from the presynaptic to the postsynaptic cell. Most long distance neural projections in the brain are made by axons that release glutamate. All voluntary muscle contractions are caused by the release of acetylcholine by motor neurons onto muscle cells.
- **Fast, inhibitory neurotransmitters:** Because large interconnected systems made only of excitatory connections tend to be unstable and subject to *seizures* (strong, uncontrolled, sustained, and recurrent excitation), subtlety and balance in the nervous system is guaranteed

by having inhibitory connections, using the fast *inhibitory* neurotransmitters gamma amino butyric acid (GABA) and glycine. Inhibitory neurotransmitters allow for computations like winner-take-all. For example, if most of the neurons are voting that you're looking at an elephant but a couple here and there are saying, "If it has four legs, it could also be a donkey," the inhibitory neurotransmitters suppress the "it could be a donkey" message.

- **Slow neuromodulators that include both excitatory and inhibitory types:** Most of the fast neurotransmitters are amino acids (other than acetylcholine), but most neuromodulators are small proteins like *somatostatin*, *substance P*, *enkephalins*, or organic chemicals called *catecholamines*, such as *epinephrine* and *norepinephrine*. The actions of these transmitters are more hormone-like, producing slow but long lasting changes. Think of their action as setting the "mood" within the brain as a background for what the faster neurotransmitters are doing at shorter time scales. For example, you might be resting quietly while digesting a good meal (which is a good idea) when a lightning storm strikes, triggering the release of norepinephrine, the fight or flight neurotransmitter. Its action shuts down digestion and other internal homeostatic processes in favor of increasing your pulse and blood pressure and diverting blood flow to the muscles in preparation for moving quickly to safety.

Some neurotransmitters are thought to have originated evolutionarily as hormones. Hormones are typically secreted into the blood stream where they influence cells that are receptive to them all over the body. Neurotransmitters, released by axons, appear to have been a way nervous systems evolved such that one cell could talk specifically to just a few other cells — something like whispering a secret to a few of your best friends rather than shouting news to an entire auditorium.

How Shocking! Neurons as Electrical Signaling Devices

A crucial aspect of neuronal function is their unique use of electricity. They use electricity both to secrete their neurotransmitters and for computation. Here's how. The membrane of the neuron contains a special kind of ion channel that is actually an active pump for ions. *Ions* are the charged atoms that are dissolved in fluids inside and outside the cell. The most important ions are the chloride salts of sodium (NaCl), potassium (KCl), calcium (CaCl2), and magnesium (MgCl2), which dissociate when dissolved into the cations Na^{+}, K^{+}, Ca^{++} and Mg^{++}, and the anion Cl^{-}. ***Note:*** A *cation* is positively charged (attracted to a negative cathode); an *anion* is negatively charged (attracted to a positive anode).

The fluid outside cells and in our blood actually has about the same ionic constituency as seawater, that is, mostly Na+, much less K+, and even less Ca++ and Mg++, with Cl– being the major anion. However, the ion pumps in neural membranes, called *sodium-potassium ATPase pumps,* actively pump sodium out of the cell and potassium in, in this ratio: 3 Na+ out to 2 K+ in (see Figure 3-3).

These pumps run all the time, creating a situation in which very little sodium and an excess of potassium exist inside the cell. However, because not as much potassium goes out as sodium comes in, there is a *net negative charge inside the cell* due to the deficiency of positive ions. This charge is about –70 millivolts and can be measured by inserting a probe attached to a sensitive voltmeter into the cell. Because this voltage is measured across the neuron's membrane (from inside the neuron to outside), it is called the *membrane potential.*

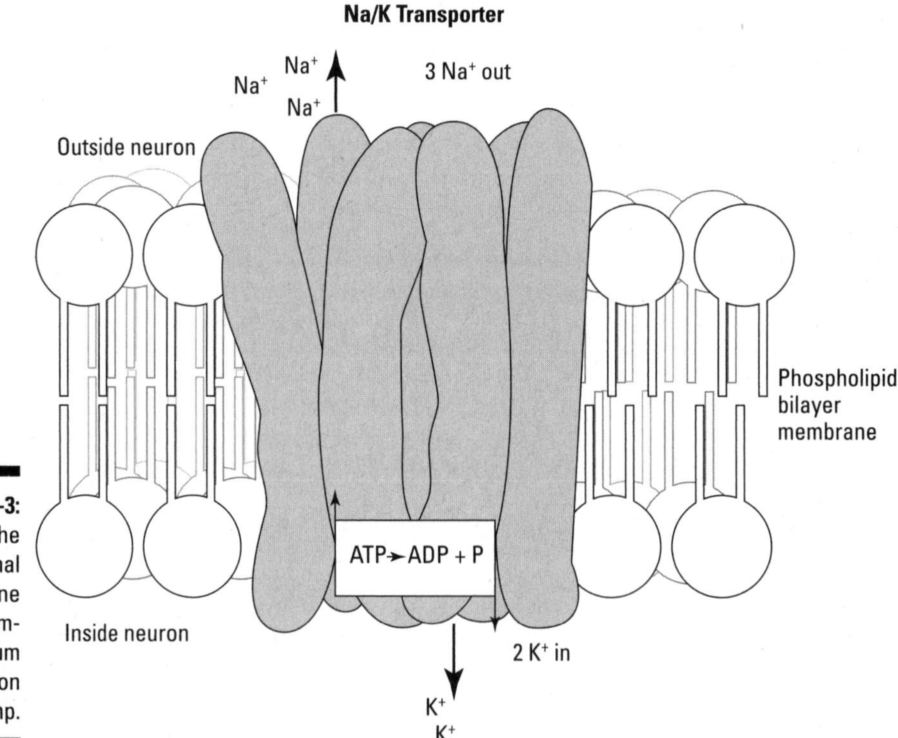

Figure 3-3: The neuronal membrane sodium-potassium ATPase ion pump.

The membrane potential of a neuron can be changed by ion channels in its membrane. This is where the ionic concentration imbalances matter because when neurotransmitter receptors open channels for particular ions, the ions flow through these receptors in the direction of restoring the imbalance set up by the sodium-potassium pumps. The flow of ions is a *current* that causes the *voltage* across the membrane (membrane potential) to change accordingly. The fast, excitatory neurotransmitters glutamate and acetylcholine, for example, bind to receptors that allow sodium ions to go through the channel pore. This entry of positive ions can locally reduce and even reverse the polarity of the negative potential inside the cell. This transient polarity change is called *depolarization*.

Inhibitory receptors are those that bind either GABA or glycine and typically open channels permeable to either Cl– or K+. If Cl–, which is negative, enters the cell, it can make the interior more negative, which is called *hyperpolarization*. When ion channels for K+ are opened, K+ actually goes out of the cell because of its higher concentration inside than outside the cell, and the cell also hyperpolarizes.

Now you can see why a single neuron can be so complicated. Some 10,000 excitatory and inhibitory synapses are opening and closing constantly in response to all the various presynaptic inputs, leading to a constantly changing net potential across the cell membrane that is the summation of all this activity.

Yikes, spikes — The action potential

You can think of the computation done by the neuron as the summation at the cell body of the continuously varying excitatory and inhibitory inputs impinging all over the dendritic tree (some inputs can be on the cell body, also). These inputs produce a net membrane potential at the cell body. What does the neuron do with this computation represented by this net membrane potential? As mentioned previously, neurons communicate their computation to other neurons by sending a series of voltage pulses down the axon. This section explains these voltage pulses and how they move along the axon.

The problems with getting from here to there

In your primary motor cortex are neurons that drive muscles in your toes. You might assume that axons act like wires — that is, the membrane potential from the neuronal dendrites sums at the motor neuron's cell body and this voltage travels to the end of the axon as if it were a wire. But that idea doesn't work. Neuronal membranes, including those of axons, are not totally impervious to ion flow outside of channels and are thus continuously leaky. Also, an electrical property of the neuronal membrane, *capacitance,* shunts (dampens and slows down) rapidly changing signals over the distances that axons need to communicate.

Helping the signal along: Voltage-dependent sodium channels

The answer to this problem is akin to the solution used to solve a similar problem in transatlantic communication cables. In transatlantic cables, devices called *repeaters* are installed at intervals to boost and clean up the signal, which becomes degraded as it travels from the shore or the last repeater. Neurons use a particular type of ion channel for the same function. This channel is called a *voltage-dependent sodium channel,* and its action is the basis of all long-distance neural communication by action potentials, colloquially called *spikes* (because they represent a sudden, brief change in membrane potential).

The voltage-dependent sodium channel does not open by binding a neurotransmitter, like the ligand-gated channels (discussed in the earlier section "Ionotropic versus metabotropic receptors"). Rather, this channel opens when the membrane depolarization created by other channels crosses a threshold.

Here's how it works: A particularly large number of these channels exist at the beginning of the axon at the cell body, called the *initial segment* (or sometimes, *axon hillock*); refer to Figure 3-1. When the summated action of all excitatory and inhibitory synapses on a cell is sufficiently depolarized, the voltage-dependent sodium channels in the initial segment (and often in the cell body as well) open and allow even more $Na+$ to flow through the membrane into the neuron. This action further depolarizes the membrane so that all the voltage-dependent sodium channels open in that region. These channels typically only stay open for about one millisecond and then close on their own for several milliseconds.

The pulse created by the voltage-dependent sodium channels at the initial segment is the action potential (spike). The voltage fluctuation created by this spike travels down the axon for some distance but is degraded by the shunting effects mentioned earlier. However, a little distance down the axon is another cluster of these voltage-dependent sodium channels. This distance is such that the degraded spike is still above threshold for these channels, so they produce another full spike at this new location. This spike travels further down the axon, gets degraded, reaches another cluster of voltage-dependent channels, receives the needed boost, travels down the axon some more, gets degraded, and, well, you get the idea.

Jumping from node to node

Many axons are sheathed between the clusters of voltage-dependent sodium channels by glial cell processes that provide extra insulation. These are called *myelinated axons* (refer to Figure 3-1). The gaps between the myelin wrappings where the spike is repeated are called *nodes of Ranvier,* and the "jumping" of the action potential from one node to the next is called *saltatory conduction.* Myelinated axons have fast conduction velocities in which the spikes travel at several hundred meters per second. Many smaller axons in the nervous system are unmyelinated and conduct action potentials more slowly. (One manifestation of the disease multiple sclerosis is the degradation of the axonal myelin coatings, resulting in slower, erratic, and eventually failed spike conduction along axons.)

Closing the loop: From action potential to neurotransmitter release

One last basic detail to consider in the picture of neuronal communication is how to get from action potential to the release of the neurotransmitter. This happens when the spike reaches the thousands of axon terminals after it branches (see Figure 3-4).

Here's what happens (refer to at Figure 3-4 for a visual of these steps):

- **Steps 1 through 3: The action potential arrives.** Within the axon terminal is another type of voltage-dependent channel, but this channel allows calcium (Ca++) rather than sodium to flow in. The spike-induced opening of these voltage-dependent calcium channels causes calcium to enter the axon terminal. Within the axon terminal, the neurotransmitters molecules are packaged in little membrane-enclosed spheroids called *synaptic vesicles*, a few of which are always near the membrane on the presynaptic side of the synapse.
- **Step 4: The vesicles bind with the membrane.** When calcium enters, it causes the membrane of several of these vesicles to fuse with the cell membrane.
- **Step 5: The transmitter is released.** When vesicles fuse with the cell membrane, the neurotransmitter molecules get dumped into the synaptic cleft.
- **Step 6: The transmitter binds with the receptors.** The neurotransmitter molecules that were dumped in the cleft quickly (within about a millisecond) diffuse across the cleft and bind receptors on the postsynaptic cell.

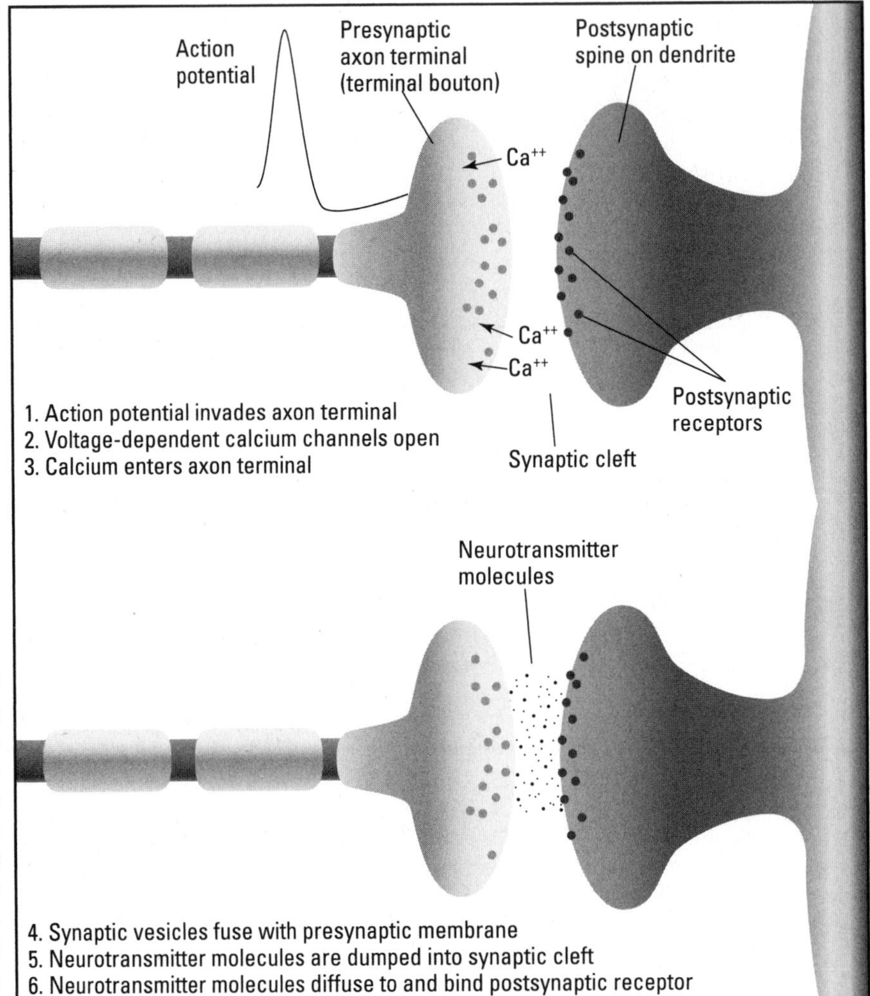
Action potential
Presynaptic axon terminal (terminal bouton)
Postsynaptic spine on dendrite
Ca++
Ca++
Ca++
1. Action potential invades axon terminal
2. Voltage-dependent calcium channels open
3. Calcium enters axon terminal
Synaptic cleft
Postsynaptic receptors
Neurotransmitter molecules
4. Synaptic vesicles fuse with presynaptic membrane
5. Neurotransmitter molecules are dumped into synaptic cleft
6. Neurotransmitter molecules diffuse to and bind postsynaptic receptor

Figure 3-4: Synaptic transmission from the presynaptic action potential to the postsynaptic receptor.

TIP

Here's a shorthand that some neuroscientists use when they think of this circuit: Around a given neuron, the extracellular concentration of various neurotransmitters is converted to membrane potential, which is converted to the rate of action potentials in the axon, which is converted to calcium concentration at the axon terminal, which is finally converted to neurotransmitter concentration outside the next cell.

Moving Around with Motor Neurons

Neurons originally evolved to coordinate muscle activity. Large, multi-celled animals can only move efficiently if muscles throughout the animal move in coordination. Coordinated muscle movement is achieved when neurons, embedded in a system that receives sensory input, can activate muscles in such a way as to produce specific muscle contraction sequences — which is precisely what the neuromuscular system does.

The neuromuscular system has, as its output, motor neuron axon terminals synapsing on muscle cells within a muscle, one axon terminal per muscle cell (although one motor neuron may have hundreds of axon terminals). The motor neuron axon terminal releases acetylcholine as the neurotransmitter (refer to the earlier section "The three major functional classes of neurotransmitters" for more on acetylcholine). Muscle cells have an excitatory, ionotropic receptor for acetylcholine that opens sodium channels in the muscle cell membrane. (To read more about ionotropic receptors see the earlier section "Ionotropic versus metabotropic receptors.")

Each action potential reaching a given motor neuron axon terminal releases a packet of acetylcholine that causes an action potential in the muscle cell. This muscle cell action potential has a much longer time course than the one millisecond action potential in most neurons. The effect of the muscle cell action potential is to cause actin and myosin filaments within the muscle cell to slide across each other, pulling the ends of the cell together (contracting it longitudinally).

A muscle is a set of chains of these muscle cells. The more each muscle cell is contracted and the more muscle cells in a chain are contracted, the shorter the muscle gets. The more chains that are contracted, the more force the muscle applies. These parameters are controlled by the number of activated motor neurons and the rate of firing in those neurons.

Non-neuronal Cells: Glial cells

Although the figure of 100 billion neurons in the brain is certainly impressive, within that same volume are at least ten times as many non-neuronal cells called *glia*. Glial cells fall into three major types — astrocytes, oligodendrocytes and Schwann cells, and microglia — each with a function, as the following sections explain.

Astrocytes

Astrocytes are glial cells that form much of the structure of the brain in which the neurons reside. Astrocytes regulate the brain environment and form the blood-brain barrier. In most of the body, the capillaries of the blood system are permeable to many substances so that oxygen, glucose, and amino acids move from the blood to the tissue, while carbon dioxide and other wastes go the other way. In the brain, the astrocytes form an additional barrier that is much more selective, creating a more finely controlled environment for the complex operations taking place there. The astrocytes do this by lining the blood vessels and only allowing traffic of substances between the capillaries and the brain that they themselves control.

There are several downsides to the relative chemical isolation of the brain created by the astrocytes, however. One is that many drugs that could potentially help treat brain dysfunctions can't pass this astrocyte barrier; therefore, treatments can't be done by injecting these drugs in the bloodstream. Another downside of brain isolation is that brain cancers are not readily attacked by the immune system because antibody cells also have a hard time getting from the blood to brain tumors. Most cancer treatment protocols effectively involve using toxic chemicals and radiation to wipe out the vast majority of dividing cancer cells, hoping the immune system can mop up the last few percent. But because the antibody mop-up operation in the brain is so inefficient, brain cancers typically have very poor prognoses.

Oligodendrocytes and Schwann cells

The second class of glial cells are called oligodendrocytes and Schwann cells. The function of both of these cells are to do the myelin wrapping of axons, discussed in the earlier section "Jumping from node to node," for saltatory conduction, wherein the nodes of Ranvier with high voltage-gated sodium channels form "repeaters" allowing long distance spike propagation. Oligodendrocytes perform this function in the central nervous system, while Schwann cells perform the same function in the peripheral nervous system.

An area of intense research now concerns that fact that peripheral nerves regenerate their axons, while central neurons usually do not. You can, for example, gash your finger and cut the nerves so severely that all sensory and motor control functions are lost in the finger tip. But wait a month or two, and full function generally returns. The same injury in the spinal cord, however, paralyzes you for life. Some differences between oligodendrocytes and Schwann cells in response to injury are thought to be important for regeneration.

Microglial cells

The last class of glial cells are microglial cells. These are scavenger cells that migrate through the brain when some area is injured and remove (scavenge) debris. This is just one aspect of the many housekeeping functions that different glial cells do, which include maintaining the proper ionic constituencies in the extracellular space, interacting with blood vessels during brain injuries, and providing the structural framework in which neurons initially grow during development.

Recording Techniques

The earliest brain recordings — electroencephalograms, or EEGs — use surface electrodes on the scalp to record ongoing brain potentials from large areas of the brain. Most of what researchers know about individual neuronal function (neurophysiology) began in about the middle of the 20th century with the invention and use of microelectrodes, which could sample the activity of single neurons, and oscilloscopes, which could display events lasting milliseconds or less. As the following sections show, several advances have been made in the kinds of devices used to record the workings of the neurons.

Single extracellular microelectrodes

Extracellular microelectrodes are very thin, needle-like wires that are electrically insulated except at their very tip. The tips of these are on the order of the size of a neuron's cell body (typically 20 micrometers across, depending on the neuronal type). The neuron is inserted into the neural tissue until the tip happens to be near one particular cell, so that the voltage the microelectrode detects is almost entirely due to the action potential currents from the ion channels opening in that one nearby cell soma. These electrodes also sometimes record action potentials from nearby axons. Extracellular microelectrodes have also been used to stimulate neurons to fire by generating pulses of current through the electrode and shocking the neuron to fire artificially.

Microelectrode arrays

Microelectrode arrays are clusters of microelectrodes that sample up to hundreds of cells simultaneously. Some of these have been implanted in paralyzed humans so that, for example, when the person thought of making some movement in his brain and activated his motor neurons, the electronically detected signal could be used to bypass the damaged spinal cord and activate muscles directly by sending shocking pulses to them.

Sharp intracellular electrodes

Intracellular recordings in mammalian neurons were first done with sharp intracellular microelectrodes, also called *glass pipettes*. These are glass tubes that are heated in the middle and then pulled apart so that the tube necks down and breaks in the middle. The tip at the break can have a diameter of less than one micrometer and still be hollow! These glass microelectrodes are sufficiently small compared to a 20-micrometer diameter cell that they can be inserted inside cells.

Intracellular electrodes inserted through the cell's membrane allow researchers to sample the electrical activity inside the cell from its synaptic inputs. Sharp intracellular electrodes are filled with a conductive saline solution to make an electrical connection between their open tip inside the cell and electronic amplifiers and displays. They yield more information about what's going on inside the cell, but because they inevitably damage the cell from the penetration, they can't be implanted permanently as would be needed for neural prostheses that would, for example, record motor cortex commands to artificially drive muscles in people who are paralyzed.

Patch-clamp electrodes

Patch-clamp microelectrodes are made like glass microelectrodes except that, instead of being inserted into the cell, they are placed against the cell membrane so that the glass makes a chemical bond, called a *gigaseal,* with the cell membrane. In the intact membrane patch, the electrode can monitor currents passing through ion channels in the membrane within the gigaseal area. In another configuration, the membrane within the gigaseal is ruptured by negative pressure, but the gigaseal along the perimeter of the pipette opening remains intact, so that, like a glass intracellular microelectrode, there is now electrical continuity between the interior of the cell through the saline solution in the pipette to the recording apparatus.

Optical imaging devices

Optical imaging advances in the late 20th century and the development of reporting dyes led to the use of optical recording techniques for monitoring neural activity. Here are the three main optical techniques for recording neural activity:

- **Fluorescent dye-mediated monitoring of ionic concentration changes:** This technique uses fluorescent dyes (dyes that absorb and then re-emit light) that change their fluorescence in response to the presence of ions like calcium, magnesium, or sodium. The most common dyes monitor calcium concentration, which is normally very low inside neurons but typically increases when the neuron is active due to calcium flux through cation channels that are not completely selective for sodium, and through voltage-dependent calcium channels that are sometimes common in neuronal dendritic trees as well as at the axon terminal. This means that activity within neural dendritic trees can sometimes be directly observed optically. Optical imaging also allows researchers to view the activity in multiple cells through a microscope.
- **Fluorescent dye-mediated monitoring of membrane potential:** *Potentiometric dyes* are dyes that bind neuronal cell membranes and change either their fluorescence or absorption of light in response to the level of depolarization of the membrane. The advantage of these dyes is that they give a direct reading of the electrical potential across the membrane so that membrane changes can be observed whether or not there happen to be ion channels there that flux calcium. On the other hand, the signal from these dyes is typically an order of magnitude less than that of calcium indicator dyes. These smaller signals are harder to detect.
- **Intrinsic optical changes in excited neural tissue, such as light scattering:** Intrinsic optical changes occur in neuronal tissue when cells are electrically active. The origin of these changes is unclear at the time of this writing but includes changes in light scattering due to transient cell swelling or rearrangement of intracellular organelle structures associated with electrical activation. One advantage of intrinsic optical techniques is that they require no dyes and are therefore less invasive. Intrinsic optical recordings similar to electroencephalograms (EEGs) are routinely carried out in humans using infrared light.

Part II

Translating the Internal and External World through Your Senses

"My senses are overwhelmed. The flowers look wonderful, the music sounds dreamy, the food tastes delicious, and the photographer makes my skin crawl."

In this part . . .

We humans are what we are and can do what we do because of our senses. We live in a world in which we can see, hear, and smell things external to us. We can also detect things that touch our skin or that we touch (intentionally or not), and we can taste what we eat. All these things are possible because we have specialized neurons, called receptors, that respond to light, sound, odors, touch, and taste.

What the world is to us is a reflection of what our receptors let us detect. We don't know (without special cameras), for example, that flowers have complex ultraviolet color patterns because our eyes don't detect ultraviolet light. In this part, I show how each of our senses receives information from the world and processes that information to produce our image of the world.

Chapter 4

Feeling Your Way: The Skin Senses

In This Chapter

- Getting messages from the sensory neurons
- Sending signals to the brain
- Looking at pain: Its causes, its effects, and ways we lessen it
- Discovering disorders related to the sense of touch

The skin is the boundary between us and what is not us. The touch boundary at the skin is so fundamental to our image of self that we take it for granted. How often, for example, do you say to yourself, "I am separate from the chair I am sitting in?" But this boundary is not impervious to sensation, so the question becomes, how is it that, when we touch things or are touched, we feel it?

The simple answer is that we feel things through our sense of touch. The sense of touch is called, technically, *somatosensory perception.* Our skin is sensitive not only to a variety of kinds of touch (tickle, pressure, and movement, for example), but we also can perceive other kinds of sensation with our skin, such as temperature and pain. These different kinds of perceptions are made when different kinds of receptors in the skin get activated.

This chapter covers the receptors that mediate touch and the brain processes that allow touch perception. It also covers different disorders related to somatosensory perception.

How Do You Feel? The Lowdown on the Skin and Its Sensory Neurons

The skin is an organ, one of the largest organs in the body in terms of area, and it has a number of important functions. It forms a protective barrier between you and the rest of the world, keeping what's you, inside, and what's not you, including bacteria, dirt, and parasites, outside. The skin is relatively impervious to water, and it also insulates, keeping you warmer on the inside than the outside when outside is cold and cooler on the inside than outside when the outside is hotter than body temperature.

General properties of the skin

To perform all these functions, the skin has several layers with different properties.

The dermis and epidermis

The outermost layer of the skin is called the *epidermis* (*epi* means "on" or "above," and *dermis* means "skin"). The epidermis is actually several layers of dead cell ghosts that provide mechanical protection from the outside. Because these cells are dead and have no pain receptors, dragging your fingernail lightly along your skin in order to remove a layer or two of these dead cells doesn't hurt (much). Cells at the bottom of the epidermis, where it meets the dermis, are constantly dividing, migrating outward, and dying to replace the dead layers as they wear off.

Below the epidermis is the *dermis,* the living layer of the skin that includes the bulk of the somatosensory receptors.

Somatosensory receptors

Throughout the skin are a variety of receptors for touch, temperature, and pain. Skin receptors allow us to have both passive and active senses of touch:

- **Passive sense of touch:** The passive touch sense occurs when something brushes our skin and we register the fact of the touch before we know what has done the touching.
- **Active sense of touch:** The active sense of touch is mostly done with our hands and fingertips. We can hold an egg or an apple or a pinecone in our hands and know what we're holding without looking. These different kinds of perceptions occur when different kinds of receptors in the skin are activated.

The sense of touch, or *somatosensory perception,* for most of the body (below the head) is relayed through the spinal cord, to the thalamus, and then to a strip in the parietal lobe where a "touch" map of the body exists.

The epidermis has very few receptors. The bulk of the somatosensory receptors are within the dermis. The next few sections explain the different types of somatosensory receptors.

Sensing touch: The mechanoreceptors

Inside the dermis are four distinctive types of touch receptors, termed *mechanoreceptors,* shown in Figure 4-1. In the next sections, I describe each of these mechanoreceptors.

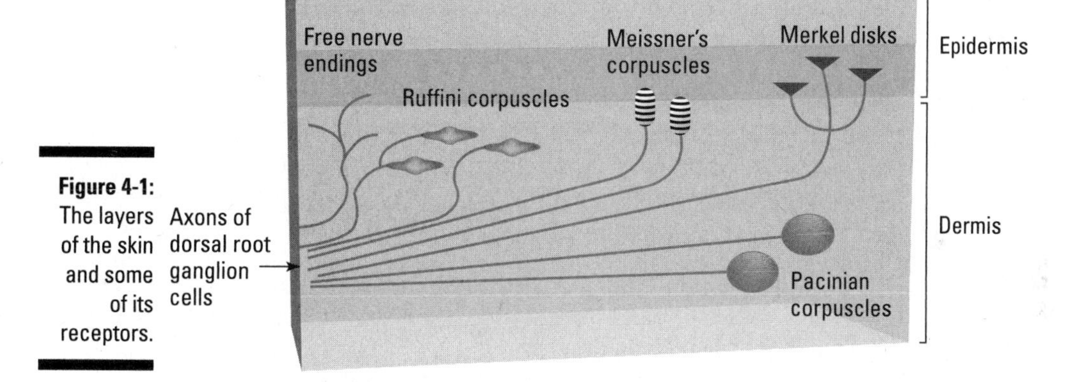

Figure 4-1: The layers of the skin and some of its receptors.

Merkel disks

Merkel disk receptors, as their name implies, are disk-shaped receptors located close to the border between the dermis and epidermis; sometimes they extend into the epidermis.

Merkel disks are receptors for pressure, meaning they're activated in the areas of your skin pressing against the chair in which you are sitting, for example. If someone were to sit in your lap, additional receptors in the skin over your thigh muscles would be recruited.

These receptors respond to relatively constant pressure over small areas of the skin, giving you the perception of the amount of *force* being exerted against different areas of your skin.

Meissner's corpuscles

Meissner corpuscles, like Merkel disks, also respond to pressure, but they can respond to more rapid changes in pressure than Merkel disks can, such as those generated by shearing forces. The sensation evoked when the Meissner corpuscles are stimulated is usually called *flutter.*

Like Merkel disks, Meissner corpuscles also have small areas of sensitivity, called their *receptive fields.* Small receptive fields are usually associated with shallow receptors (those that are near or in the epidermis), whereas receptors (deep in the dermis) typically have large receptive fields. Pressure at any point on the skin compresses the skin at greater distances deep in the skin than more shallowly.

Ruffini corpuscles

Ruffini corpuscles respond to skin stretch (think of the way your skin pulls when it's being dragged across a surface). These receptors, which serve an important function in protecting the skin from tearing, have large receptive fields.

Pacinian corpuscles

Pacinian corpuscles, which also have large receptive fields and tend to be deep within the dermis, are the fastest responding of all the touch mechanoreceptors. These receptors have a myelin wrapping similar to the glial wrapping around axons (refer to Chapter 3). However, in Pacinian corpuscles, the function of the wrapping isn't to allow the action potential to jump from one node to the next. Instead, it allows the receptor to respond to rapid changes in pressure. Through the Pacinian corpuscles, you are able to perceive vibrations. If you drag your fingertips across a coarse surface like sandpaper, your can assess its level coarseness, thanks to Pacinian corpuscles.

How mechanoreceptors work

Somatosensory neurons, the tips of which form the mechanoreceptors discussed, have an unusual morphology, or structure. This morphology is crucial to their function.

Their morphological classification is called *pseudounipolar.* Although this rather unwieldy name isn't particularly illuminating about their function, understanding their structure explains some aspects about how these receptors work. Figure 4-2 shows a diagram of a typical mechanoreceptor neuron.

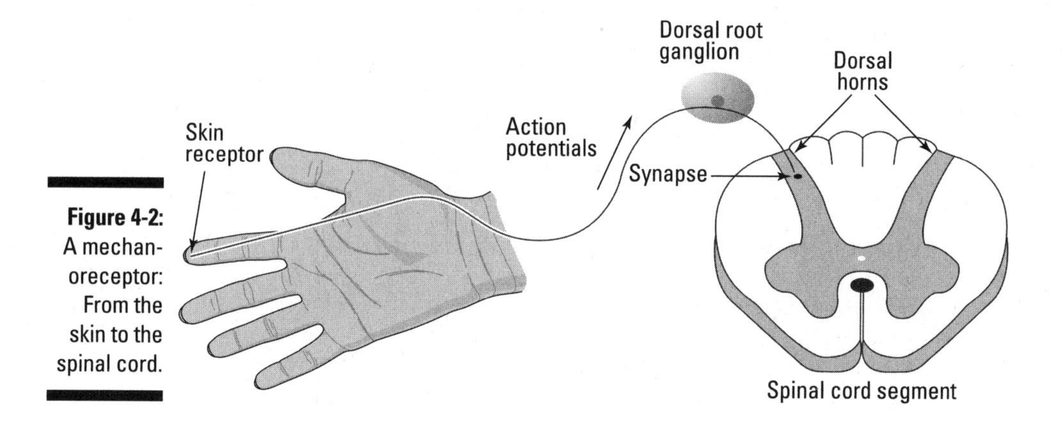

Figure 4-2: A mechanoreceptor: From the skin to the spinal cord.

The cell bodies of somatosensory receptor neurons for most of our skin (below the head and neck) are located in a series of what are called *ganglia* (concentrations of neural cell bodies) just outside the dorsal root of the spinal cord. They are thus called *dorsal root ganglia.* These ganglia and the neurons they contain are part of the peripheral nervous system.

The cell bodies of the somatosensory neurons have no dendrites. Instead, a single axon leaves the cell body and then *bifurcates,* or separates into two paths, a short distance away.

One end of the axon enters the spinal cord at the dorsal root and makes conventional synapses on spinal interneurons, enabling the stretch reflex and the relaying somatosensory information to other spinal cord segments and up the spinal cord to the brain.

Here's the interesting part: The other end of the axon goes away from the dorsal root ganglion in a bundle with other axons and ends up in the skin, where it forms one of the receptor types discussed in the preceding section: Merkel disk, Meissner corpuscle, Ruffini corpuscle, or Pacinian corpuscle. The mechanoreceptors are activated directly when a mechanical force stimulates an axonal ending of one of these neurons. This activation occurs through a special ion channel that responds to the stretching of the membrane (in a typical neurotransmitter receptor, the activation is triggered by voltage or ligand binding). Here's what happens:

1. The stretching causes action potentials to originate in the axonal ending and then proceed *toward* the cell body in the dorsal root ganglion (most axons conduct action potentials *away* from the cell body).

2. The action potential continues past the axonal bifurcation point near the cell body into the spinal cord, where it reaches axon terminals and makes conventional synapses onto spinal interneurons.
3. The interneurons connect the receptor neuron to motor neurons for reflexes and also send messages about the receptor activation to other spinal cord segments and up to the brain.

Sensing temperature and pain

I'm sure you are aware that you can detect more than just various kinds of pressure on your skin. Two other skin senses are temperature and pain. These receptors have similar structures, or, really, lack of structure. All the mechanoreceptors discussed in the preceding section consist of an axon terminal with ion channel stretch receptors embedded in some sort of structure, such as a corpuscle, disk, or myelin wrapping, that gives the receptor its particular responsiveness to different stimulation frequencies.

Receptors for temperature and pain look like the axon terminals without any other structure around them (see Figure 4-3). They are typically called *free nerve endings*. Free nerve endings for temperature have ion channels that respond to particular temperatures, while other free nerve endings generate action potentials in response to extreme force on the skin or other potentially damaging stimuli that is felt as pain. Some receptors — those having what are called *transient receptor potential* (TRP) channels — respond to both.

Different temperature receptors respond best to particular temperatures. Warmth receptors respond best to particular temperatures above body temperature (98.6 degrees Fahrenheit), while cold receptors respond best to particular temperatures below body temperature. You judge a wide range of temperatures (cool, damp, chilly, cold, warm, humid, hot, and so on) by sensing the unique ratio of activation of the different receptors activated at any particular temperature.

Extreme heat, cold or skin pressure, however, activates receptors that are interpreted as pain. Although different types of pain receptors work by different mechanisms, what they have in common is that the sense of pain signals impending damage to the skin. Pain receptors also exist that respond to chemical damage from acids or bases, and other types of damage such as that caused by a cut. For more information on pain, head to the later section "Understanding the Complex Aspects of Pain."

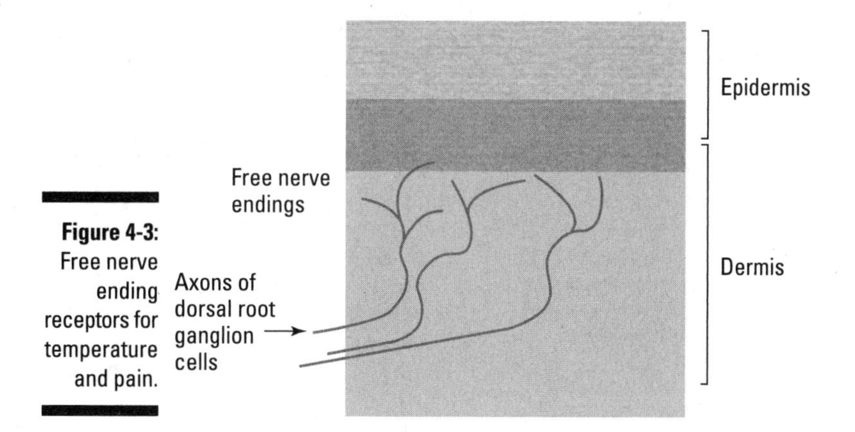

Figure 4-3: Free nerve ending receptors for temperature and pain.

Sensing position and movement: Proprioception and kinesthesis

Although not located in the skin, receptors mediating *proprioception* (position sense) and *kinesthesis* (movement sense), are either free nerve endings or structures similar to mechanoreceptors like Ruffini corpuscles (refer to Figure 4-1) and have similar layouts as the cell bodies in the dorsal root ganglia (refer to Figure 4-2).

These receptors are embedded in muscles, tendons, and ligaments around joints. The receptors in muscles and tendons that have relatively sustained responses called *proprioreceptors* signal muscle force and joint position. Similar receptors with more short-lived, or *transient,* responses signal when the joint is moving, allowing us to have the movement sense of kinesthesis. For example, proprioreceptors allow you to touch your nose with your eyes closed. Transient, kinesthetic receptors allow you to reach out quickly and then stop your hand in the right place to grab a thrown ball.

Skin Receptors, Local Spinal Circuits, and Projections to the Brain

Skin receptors allow you to respond to things that contact your skin and to be aware of what those things are. The touch, temperature, and pain messages these receptors encode get passed onto other neurons within the same spinal cord segment, as well as onto other nearby spinal cord segments and to the brain, where the perception occurs.

Somatosensory receptor outputs

The output of most somatosensory receptors participates in at least three different kinds of neural circuits:

- **Local reflexes** are those that primarily involve contraction of a single muscle, such as a flexor like the biceps that contracts when, for example, you touch something hot. The circuit for this action consists of the neurons in your fingertip that contain the temperature sensor for heat contacting the spinal cord interneurons in the dorsal root area of the spinal cord (refer to Figure 4-2). This activates motor neurons in the same spinal segments, which cause contraction of muscles that withdraw your finger.
- **Coordinated movement** involves receptor connections through interneurons to other spinal cord segments. When you throw something, mechanoreceptors in the skin of the hand work with proprioceptive and kinesthesis receptors associated with muscles in your fingers, hand, arm, and shoulder. Even your leg muscles are involved. Locomotor activities like walking also require coordination between spinal segments so that you do not, for example, try to move one foot before the other has hit the ground. Receptor output from any one segment in the spinal cord can project up or down to other spinal cord segments for coordinated activity of multiple muscles.
- **Messages from skin receptors** are also passed to the brain where you become conscious of them. There are two major pathways, the *lemniscal* pathway and the *spinothalamic* pathway. Both of these pathways lead to the ventral posterior nucleus of the thalamus on the opposite side of the body from the skin receptors, following the nearly universal principle that the right side of the brain deals mostly with the left side of the body, and vice versa. (For more on the interaction between the right and left sides of the brain, refer to Chapter 2.)

Kicking it up at the doctor's office

The spinal reflex is the most elementary unit of behavior through which a sensation causes an action. In a standard neurological exam, while you sit, the doctor uses a little triangular rubber hammer to tap your patellar tendon just below the kneecap. This tap causes the tendon to stretch. This stretching simulates what would happen if you were standing and your knees started to buckle. You have an automatic mechanism by which the output of the proprioceptors in that tendon cause the quadriceps muscles to contract and extend your knee. When you're sitting, this action results in the kicking motion. The exam thus tests the integrity of the entire circuit from sensory receptor to muscle contraction, and everything in between.

Locating the sensation: Specialized cortical sensory areas

Reflexes (like the spinal reflex you can read about in the sidebar "Kicking it up at the doctor's office) occur in a local circuit, faster than you are aware of it. Awareness has to wait until the signal reaches the brain. This section discusses how the signals from the skin go to the brain, where in the brain they go, and how brain activity is related to your consciousness of skin sensation. In this section, I also give details about the neural processing in somatosensory pathways (for a refresher on brain organization and processing, refer to Chapter 2).

The somatosensory part of the thalamus (the ventral posterior nucleus) projects to a narrow strip of cortex just posterior to the central sulcus, making it the most anterior part of the parietal lobe (see Figure 4-4).

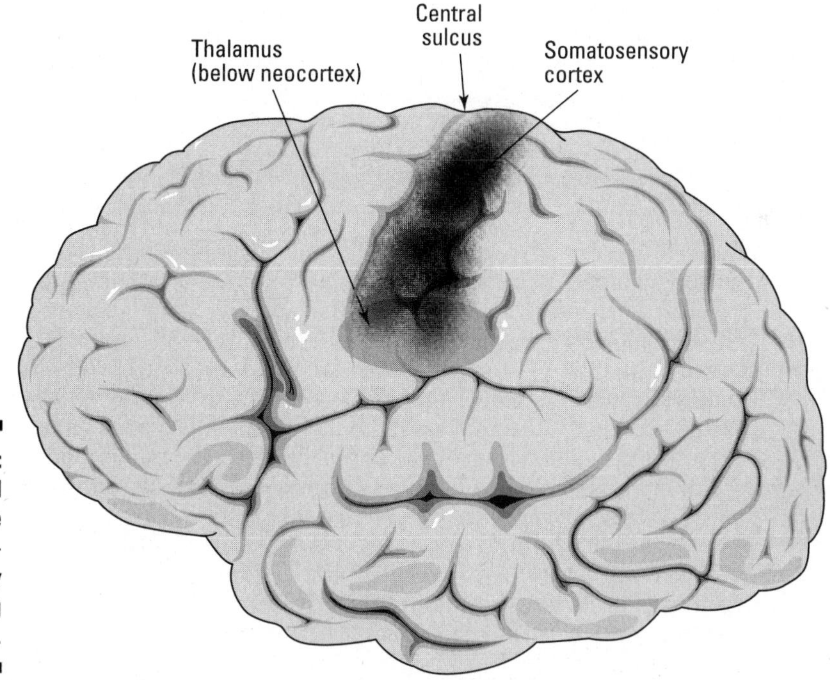

Figure 4-4: The parietal lobe: the somatosensory receiving area.

Mapping skin receptors to specific brain areas: Cortical maps

The mapping of skin receptors to a specific area of neocortex illustrates one of the most fundamental principles of brain organization, *cortical maps.* The projection from the thalamus is orderly in the sense that receptors on nearby parts of the skin project to nearby cortical neurons. Figure 4-5 shows a representation of the skin map on the somatosensory cortex.

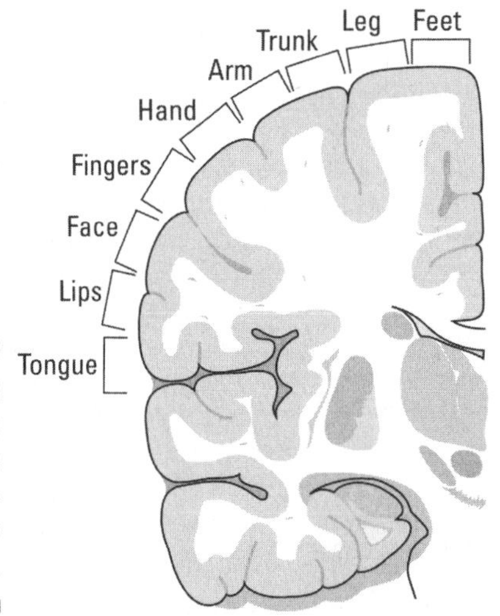

Figure 4-5: The skin map on the somatosensory cortex.

The fundamental idea about a given area of cortex being devoted to receptors in a given skin area is that activity in this area of cortex is necessary for the perception of the skin sense (along with other parts of the brain). We perceive activity in that bit of cortex area as a skin sensation not because that area has some special skin perceiving neurons, but because it receives inputs from the skin and has outputs that connect to memories of previous skin sensation and other associated sensations. In other words, the perception produced by activity in this and other areas of cortex is a function of what neural input goes to it and where the outputs of that area of cortex go.

This skin map on the cortex is called a *homunculus,* which means "little man." However, the surface area of the somatosensory cortex onto which the skin receptors project is not really a miniature picture of the body; it is more like a band or strip, as established in the studies by Canadian neurosurgeon Wilder Penfield (1891 – 1976). Because of the difficulty in mapping a three dimensional surface onto a two dimensional sheet (think of how two-dimensional maps of the earth compare to the three-dimensional globes), the image is distorted, depending on choices that the "map-maker" makes about what is relatively more important to represent accurately versus what is less.

Also note that some areas of the body, such as the hands and fingers, are located in the cortex map close to areas such as the face that are actually quite distant, body-wise. Some researchers suggest that the phantom limb feelings (including pain) that sometime occur after an amputation may occur because some neural projections from the face invade the part of the cortex that was being stimulated by the limb and cause sensations to be perceived as being located there, even when the limb is gone.

Receptor densities

Equal areas of the skin do not map to equal areas of cortex because different areas of the skin have very different receptor densities. The skin in the fingertip, for example, has many more receptors per area than the same areas in the skin on the stomach or back. The higher density of receptors in the fingertip allow fine *two-point discrimination,* which is the distance over which one can distinguish a single point pressing on the skin from two points close together. This distance is millimeters on the fingertip, lips, and forehead, for example, but centimeters on the stomach, back, and legs. You need high two-point discrimination in your fingertips to manipulate objects precisely and in the skin on your face to make precise facial expressions.

A given area of cortex processes inputs from about the same absolute number of receptors, so that skin areas with high receptor density, such as the fingertips, get proportionally more cortical area compared to the skin area, than areas with low receptor density. You can look at Figure 4-5 and see where skin receptor density is relatively high or low.

Understanding the Complex Aspects of Pain

Although pain, like other skin senses, is usually associated with something happening at some particular part of your body that you need to do something about, it also has other causes and effects. Pain can produce long lasting changes in mood and can arise from causes that are not themselves perceived and not well localized, including depression not associated with any physical source. Moreover, unlike the topographic but distorted mechanoreceptor somatosensory map in the anterior parietal cortex (refer to Figure 4-5), no similar map exists for pain. Although you can localize pain to some part of your skin, the perception of pain has a more general effect on mood, similar to pain that cannot be localized. Pain pathways to the central nervous system are diffuse, have multiple causes (*multimodal*), and impact general mood often in long lasting ways.

Not only does pain affect general mood, but it goes the other way, too: Mood can affect the feeling of pain. A person's mental state and attitude can cause him to ignore or not even feel pain, while anxiety can intensify the pain associated with any particular physical stimulus.

Reducing — or overlooking — pain

For thousands of years, people have been aware that it is possible to cognitively mitigate the sensation of pain. Think of soldiers ignoring significant wounds, yogis lying on beds of nails, and subjects ignoring pain after hypnosis. Drugs have also been found that specifically reduce pain without causing loss of consciousness or other sensations. Many of these phenomena arise because the pain system in the brain uses a particular set of neurotransmitters.

Neurotransmitters that reduce or block pain

Here's a mystery that puzzled researchers for a long time: Why does a substance produced by a poppy plant (morphine) relieve pain? Most psychoactive drugs mimic the action of known neurotransmitters, but until a few years ago, there was no known neurotransmitter that mediated the vgeneral effects of pain.

This all changed with the discovery of discovery of endogenous opioids (that is, opioids that are developed naturally within the body). Of these morphine-like substances, the most common are the endorphins (a term which is an abbreviation of *endogenous morphines*). Common situations in which endorphins are produced include childbirth and running the last few miles of a marathon.

Opiates like morphine and heroin reduce the feeling of pain because they mimic the action of substances the body produces on its own to control pain. These drugs bind these same receptors and, at low doses, produce similar effects. However, when injected in large doses, these drugs produce the opposite of pain — a "high" — and are addictive. The drug naloxone antagonizes the effects of these opioids and is often given to addicts to reverse the effects of heroin they have injected.

The existence of endorphins also explains another mystery of pain management, the *placebo effect.* The placebo effect occurs when patients are given a substance that itself has no pain-blocking potential but, because the patients believe they have been given a real drug that will alleviate the pain, find that the pain is actually alleviated.

Although the placebo effect is robust and common, those in the medical community tend to dismiss it as being "psychological," that is, not based on any medically demonstrable or quantifiable basis. However, it turns out that the drug naloxone not only reduces the effects of opioids, such as heroin, but it also reduces the placebo effect. What this means is that the placebo effect isn't just psychological; it actually has a physiological component, involving the cognitive stimulation, from belief, of the body's internal endorphin production that objectively and measurably reduces pain by binding the endorphin receptors.

Using distraction to alleviate pain: The gate theory

Another mystery about the sense of pain is that it is often reduced by cognitive distraction. Numerous well-documented cases exist of people enduring in some survival situation despite being seriously wounded and not even realizing it. A hypothesis put forward to explain this effect is called the *Melzack and Wall gate theory.*

According to the gate theory, messages from pain receptors in the skin mingle within the central nervous system with messages from ordinary mechanoreceptors. The operation of the neural circuit is such that when only pain receptors are activated they pass through the gate and reach the brain. However, if other mechanoreceptors are sufficiently activated, even in other parts of the body, they can block the neural gate and suppress the pain signal to the brain. As it turns out, even cognitive activity can be sufficiently distracting to close the pain gate at high spinal or even brain levels.

Pain-free and hating it: Peripheral neuropathy

In the preceding sections, I discuss several ways in which the sense of pain can be reduced, including the body's own production of endorphins. It has occurred to many people to wonder why we have a sense of pain at all. Feeling pain is, well, painful. Wouldn't we be better off if we could just eliminate pain?

The answer to the question of whether we would be better off without a sense of pain is a resounding *no.* We know this because this situation occurs in some people. One of these is a condition called *peripheral neuropathy,* in which many neurons such as pain receptors in the peripheral nervous system die or become inactive due, for example, to vascular problems associated with diabetes. Loss of pain sense in parts of the body can also be the result of certain strokes and types of brain damage.

People with peripheral neuropathy tend to injure themselves without knowing: They burn themselves while cooking, break bones during routine physical activity, and develop asymptomatic skin lesions that are ignored until they become serious infections. The sense of pain is necessary to prevent harm to the body.

The loss of feeling in a limb is so disabling that people with sensory peripheral neuropathy are effectively paralyzed in that limb, refusing to use it, even if the motor neuron circuitry is actually intact. A recent technique called *constraint induced therapy* (CIT) holds promise for reversing this type of disability. In this therapy the patient is forced to use his "paralyzed" arm because therapists temporarily constrain the other "good" arm in a sling. In this situation not only will the patient begin using the affected arm, but, with practice, can often achieve the considerable dexterity necessary for common two-armed life tasks such as tying shoes and opening jars.

Chronic pain and individual differences in pain perception

Although pain is a necessary function for preventing damage to the body, in some cases, pain itself becomes disabling. Chronic pain can occur in disease conditions such as cancer, in which case the normal function of pain that forces you to rest, protect, or not use some injured part of the body until it heals is simply inappropriate in a disease state in which destruction is occurring from the cancer all over the body that cannot be healed from rest. Pain can also arise from psychological factors or from factors that cannot be medically identified and are assumed to be psychological. Examples include some types of chronic pain and depression.

Pain from both medically identified sources and that which is psychological (or cognitive) appears to activate a brain area called the *anterior cingulate cortex*. The anterior cingulate cortex is the anterior portion of an area of the mesocortex, just above the corpus callosum. (Refer to Chapter 2 for discussion on the different areas of the brain; for more on the development and function of the mesocortex, head to Chapter 12.)

The anterior cingulate appears to be a high-level cortical monitoring center. It tends to be activated by pain, anticipation of pain, and failure in goal-seeking activity. Its function seems to be to arbitrate between taking different strategies in response to experience. At a low level, after placing your hand on a hot stove burner, it may make you cautious when you're around the stove. At a higher level, getting reprimanded for sending a flaming e-mail at work may make you wary of doing so again.

Considerable individual differences with respect to pain tolerance exist, just as there can be differences in tolerance in different situations for a particular person. Men are reported to be less tolerant of chronic pain than women, though they are more tolerant of acute pain. Pain tolerance generally increases with age, based on tests for pain tolerance such as the total time one can stand to have one's arm immersed in ice water. It is not clear whether the increase in tolerance with age is based on psychological or physical factors. Athletic training and strong motivation to obtain some goal can significantly reduce the disabling effects of pain.

Suggestions that different cultures or ethnic groups have intrinsically different pain thresholds — in other words, there's a physiological difference among cultures about pain tolerance — have almost always been shown to be the effect of at what point the perceived stimulus is reported as painful or unbearably painful, not whether the pain itself is perceived. Cultures that encourage expression of emotions in general tend to be associated with lower pain-reporting tolerance.

Chapter 5

Looking at Vision

In This Chapter

- Seeing the role the eye and its components play in vision
- Examining the vision centers of the brain
- Looking at how we see color, depth, and shapes
- Uncovering the causes of visual impairment and the secrets behind optical illusions

How do you see? Most people think that, when we look at things, the light coming off those things enters the eye and sends a camera-like image of what we're looking at to the brain. However, the retina itself is an extension of the brain and already modifies the camera-like image that it receives, and this image will be further modified by the rest of the brain.

Vision occurs when retinal photoreceptors capture photons and the retina and brain perform a complex analysis of this information. Visual input is processed in parallel pathways through the retina, thalamus, and occipital lobe of the cortex. Specific classes of neural cells in the retina sense different aspects of that image, such as the colors present, whether something is moving, and the location of edges. These neurons act like little agents that "shout out" the presence of these features in the image so that you can recognize objects and determine their distance from you.

This chapter covers the cellular agents involved in sight. Early in the visual system, these agents are very simple-minded and react to qualities like color and intensity. Higher in the system, these agents become very sophisticated and very picky, some responding only to certain faces. This is their story. Oh — and after you've studied all this visual processing neural circuitry, think about the fact that blind people, without any input from the eyes, still can visualize.

The Eyes Have It: A Quick Glance at Your Eyes

You may have heard that the eye is like a camera. As Figure 5-1 shows, this comparison springs from the fact that the eye has evolved tissues that act like the optical elements of a camera, namely, the lens, which acts like a camera's lens, and the pupil, the opening through which the light enters, acts like an aperture.

When light enters the eye, the cornea, which is the outermost clear layer at the front of the eye, first focuses it, and then the lens focuses it further, and the pupil, located between the cornea and lens, opens and closes to let in more or less light, like a camera aperture. The image formed by the cornea, lens, and pupil is projected onto the *retina*, the neural lining inside the eye. The retina is where the real action in vision takes place. The following sections outline what happens to this image once it hits the retina.

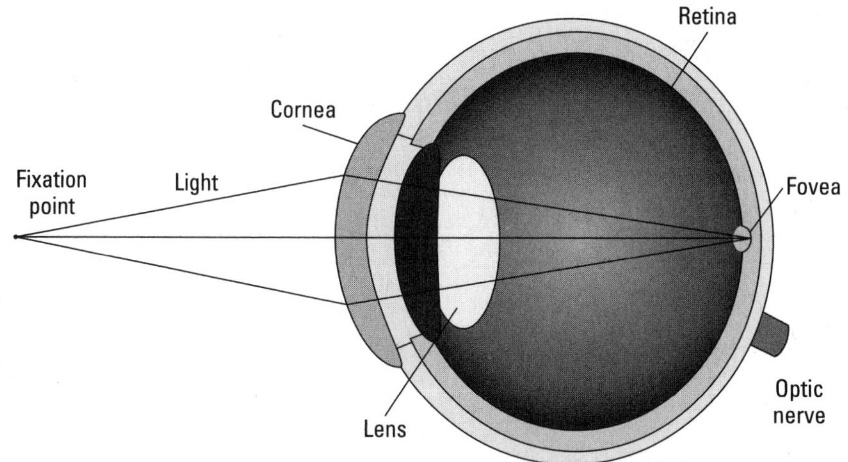

Figure 5-1: Light entering the eye.

Although you may think that the eye's lens functions the same as the camera lens, this isn't quite right. Instead, the cornea, which is at the air/tissue interface, does about 70 percent of the focusing. The lens of the eye, being slightly more dense than the surrounding tissue, does the remaining 30 percent rest. The lens also changes shape to adjust the focus for near and far objects, a process called *accommodation*.

The retina: Converting photons to electrical signals

The smallest units of light are called *photons.* When photons hit the retina, they are absorbed by *photoreceptors*, specialized neural cells in the retina that convert light into electric current that modulates the release of a neurotransmitter (glutamate). This whole process — from photons to electrical current to neurotransmitter release — is called *phototransduction.*

Two main types of photoreceptors exist:

- **Rods,** which work in very dim light for night vision.
- **Cones,** which function only in bright daylight. Humans have three different types of cones that allow us to perceive color.

At night, when only your rods are absorbing enough photons to generate signals, you have no color vision because the signal from a rod contains no information about the wavelength of the photon absorbed. During the day, however, three different cone types are active (red, green, and blue). Individually, cones don't signal wavelength either, but the brain can deduce wavelength from the ratio of activity of different cones. In very blue light, for example, the blue cones are relatively more activated than the green and red cones.

Because you have three types of cones, simulation of color in television and computer screens requires that they have three different colored-light emitters: red, green, and blue. If you had only two cones, like many other animals, display screens could get by with two different colored-light emitters. Similarly, if humans, like some fish, had four cones, a TV with only three colored-light emitters couldn't simulate color accurately, just like a printer with one color cartridge out of ink can't simulate all colors.

Catching photons: Light and phototranduction

When a photoreceptor absorbs a photon of light, a cascade of events occurs that result in a message being sent to other neurons in the retina:

1. **A chemical reaction occurs.**

The molecule rhodopsin (in rods; similar molecules exist in cones), which absorbs photons, starts out in a kinked form, called *11-cis retinal.* When this molecule absorbs a photon of light , a molecular bond in the middle of the 11-cis retinal flips from a kinked to a straight configuration, converting it to what's called *all-trans retinal.*

The all-trans retinal is a *stereoisomer* of 11-cis retinal, which means it has the same chemical composition but a different structure.

2. **All-trans retinal reduces the concentration of cyclic GMP (cGMP).**

   cGMP is an intracellular messenger inside the photoreceptor that keeps depolarizing ion channels in the cell membrane open.

   These depolarizing ion channels are like the channels of metabotropic receptors (refer to Chapter 3) except that they're triggered by light absorption rather than a neurotransmitter binding to a remote receptor. When light reduces the internal concentration of cGMP in the photoreceptor, it reduces the number of these channels that are open, *hyperpolarizing* the receptor.

   In all *vertebrates* (animals with backbones, like humans), photoreceptors hyperpolarize to light by a similar mechanism, using similar photochemistry. Some non-vertebrates, like barnacles and squids, have photoreceptors that use different photochemistry to depolarize to light.

3. **The hyperpolarization of the photoreceptor causes a structure at its base, called the *pedicle*, to release less *glutamate,* the photoreceptor neurotransmitter.**

   The photoreceptor pedicle is very similar to a conventional axon terminal except that, instead of individual action potentials releasing puffs of neurotransmitter, light absorption continuously modulates the release of neurotransmitter.

4. **The modulation of glutamate release drives other cells in the retina.**

   The outputs of photoreceptors drive two main types of cells called bipolar and horizontal cells. These cells are discussed in the next section.

Photoreceptors do not send an image of the world directly to the brain. Instead, they communicate with other retinal neurons that extract specific information about the image to send to higher brain centers. The following sections explore that communication.

Getting the message to the brain

Why doesn't the eye just send the electrical signal from all the rod and cone photoreceptors directly to the brain? The main reason is that there are well over 100 million rods and cones but only a million axon transmission lines available to go to the brain (refer to Chapter 3 for more about axons). Even worse, these transmission lines work by sending a few action potentials per second along the axon, further limiting what information each line can send.

The retina gets around these limitations on transmission capacity in a few interesting ways, as the following sections explain.

The most common misconception about the retina is that it sends some sort of raw image to the brain. In reality, the retina processes the image and sends information extracted from the image to at least 15 different brain areas by as many pathways.

More photoreceptors in the center of the eye

The retina has more photoreceptors and other retinal cells in the center of the eye (called the *fovea;* refer to Figure 5-1) than it does in the periphery. You notice things happening in your periphery, but to see and identify something clearly, you have to look directly at it to place the image on the high-resolution fovea.

Modulating responses around average light level

The retina uses *adaptation* in which the photoreceptors respond briefly whenever the light level changes but then settle down and reduce their output after a few seconds. Adaptation saves energy and spikes if cells don't have to keep telling the brain, "Yes, the light level is still the same as it was a few seconds ago." Photoreceptors change their dynamics so that their basal neurotransmitter release modulates their responses around the average current light level. This type of temporal adaptation is cellular.

Minimizing information across space

Adaptation occurs in another way in the retina (and the brain) via neural circuit interactions. This type of adaptation minimizes information across space. A process called *lateral inhibition* reduces how much information is transmitted to the brain because photoreceptors communicate the difference between the light they receive and the surrounding light rather than the absolute level of the light they receive. The next sections explain the neural circuitry in the retina that makes lateral inhibition possible.

Processing signals from the photoreceptors: Horizontal and bipolar cells

As mentioned previously, lateral inhibition is one way your nervous system overcomes the limitations on how much information can be transmitted from the retina to the brain. In lateral inhibition, photoreceptors don't convey the absolute level of the light they receive; instead, they communicate the difference between the light they receive and the surrounding light. This section explains the neural circuitry that makes lateral inhibition possible.

Photoreceptors connect to two neural cell classes: horizontal cells and bipolar cells (see Figure 5-2). Horizontal cells mediate lateral inhibition, and bipolar cells pass the photoreceptor signal, which has been modified by the horizontal cells, on toward the brain. The next sections explain how this process works.

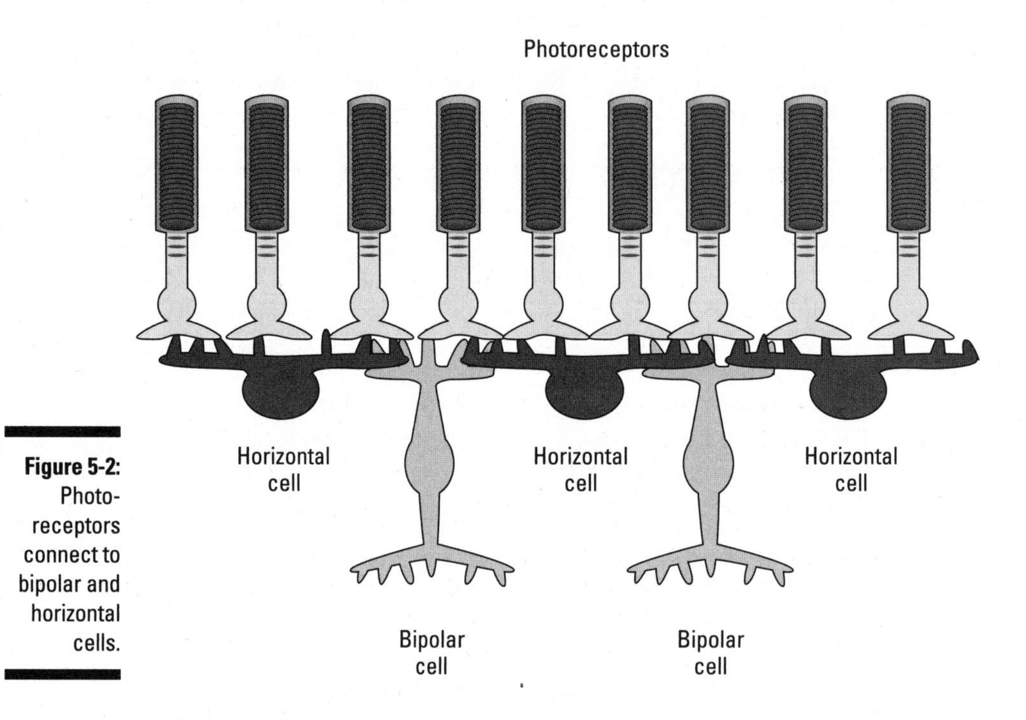

Figure 5-2: Photoreceptors connect to bipolar and horizontal cells.

Step 1: Reducing redundant signals (horizontal cells and lateral inhibition)

Suppose you're staring at a stop sign. You don't need all the cells responding to different parts of the sign to report with high precision that exactly the same shade of red everywhere occurs over the entire sign. The retina can avoid sending redundant spatial information because *lateral inhibition* uses horizontal cells to allow photoreceptors to communicate the difference between the light they receive and the surrounding light.

Here's how it works: Horizontal cells receive excitation from surrounding photoreceptors and subtract a percentage of this excitation from the output of the central photoreceptor. This action allows each photoreceptor to report the difference between the light intensity and color it receives and the average intensity and color nearby. The photoreceptor can then signal small differences in intensity or color from those of nearby areas. These highly precise signals go to the next cells, the bipolar cells.

Step 2: On to bipolar cells and to the processing layers beyond

The signals from the photoreceptors that have been modified by the horizontal cells are then sent to the *bipolar cells* (refer to Figure 5-2). The bipolar cells then carry these signals to the next retinal processing layer. Bipolar cells come in two major varieties:

- **Depolarizing bipolar cells**, which are excited by light areas of the image.
- **Hyperpolarizing bipolar cells**, which are excited by dark areas.

How do depolarizing and hyperpolarizing bipolar cells have the opposite responses to light? Both types of bipolar cells receive glutamate from the photoreceptors, but each has a different receptor for glutamate. The glutamate receptor on depolarizing bipolar cells is an unusually fast metabotropic receptor that is inhibitory. As mentioned previously, photoreceptors are hyperpolarized by light so that, when illuminated, they release less glutamate. This reduction of glutamate is a reduction of the inhibition of depolarizing bipolar cells (in other words, dis-inhibition = more excitation). The glutamate receptors on the hyperpolarizing bipolar cells respond in a conventional manner, that is, reduction of glutamate results in less excitation.

As mentioned, bipolar cells carry the signal forward toward the next layer of retinal processing before the brain. These cells terminate in a second synaptic layer in the retina where they connect to two kinds of post-synaptic cells:

- **Retinal ganglion cells:** These cells send the final output of the retina to the brain.
- **Amacrine cells:** These cells mediate lateral interactions, something like horizontal cells do with photoreceptors (refer to the preceding section).

I cover both of these cells types in more detail in the next section. (Stick with me through the next section: we're almost to the brain!)

Sending out and shaping the message: Ganglion and amacrine cells

The visual image that the photoreceptors capture and that the horizontal and bipolar cells modify passes to another group of neurons, the *ganglion cells,* where yet another set of lateral interactions, mediated by amacrine cells, occurs.

Ganglion cells are the output of the retina. Think of them as train depots: The information from the eye is finally onboard and ready for the ride to the brain. The destination? The ganglion cells send their axons to at least 15 retinal recipient zones in the brain. Why so many receiving zones, and exactly what kind of information does each tell the brain? The next sections explain.

Converting analog to digital signals to go the distance

As I explain in the previous sections, the connections within the retina are between cells much closer than one millimeter to each other. But the messages going from your eye to your brain have to travel many centimeters. A few centimeters may not sound like a lot to you, but when you're a cell, it's a marathon! Traveling this distance requires axons that conduct action potentials, by which ganglion cells convert their analog bipolar cell input into a digital pulse code for transmission to the brain. (See Chapter 3 for a discussion of action potentials.)

Breaking down into ganglion cell types and classes

The depolarizing bipolar cells, which are excited by light, are connected to matching ganglion cells called *on-center.* The hyperpolarizing bipolar cells, which are inhibited by light, are connected to matching ganglion cells called *off-center* ganglion cells.

In addition to performing other functions, amacrine cells modulate signals from bipolar cells to ganglion cells much the same way horizontal cells modulate signals from photoreceptors before sending them to the bipolar cells. That is, amacrine cells conduct inhibitory signals from the surrounding bipolar cells so that the ganglion cell responds to the difference between the illumination in its area and the surrounding areas, rather than to the absolute level of illumination (as coded by its bipolar cell inputs). This action reduces the amount of redundant information that must be transmitted over the limited number of ganglion cell axon transmission lines.

Despite this similarity between amacrine and horizontal cell function, amacrine cells come in more varieties and are more complicated. As a result, the same bipolar cell inputs create different ganglion cell classes. The two most important ganglion cell classes are

- **Parvocellular (small cells):** These are selective for color and fine detail.
- **Magnocellular (large cells):** These are selective for motion and low contrast.

Parvocellular ganglion cells are by far the most numerous ganglion cells in the retina. Both of these ganglion cell classes have on-center and off-center varieties.

Other types of amacrine cells produce ganglion cell classes that respond only to specific features from the visual input and project to particular areas of the brain. For example, some ganglion cells respond only to motion in a certain direction and help you track moving objects or keep your balance. Other ganglion cells sense only certain colors, helping you tell ripe from unripe fruit, or red stop lights from green lights. Still others indicate the presence of edges in the scene.

Ganglion cells report specific features of the visual world "upstream" so that the retina doesn't have to send an overwhelming amount of millisecond by millisecond "pixel" information to your brain.

From the Eyes to the Vision Centers of the Brain

The previous section explored how the retina converts light into ganglion cell pulses that signal different things about the visual image. These pulses, called *action potentials,* can travel the centimeter distances to the brain over the ganglion cell axons. In this section, I finally get to the heart of the (gray) matter and discuss where these pulses go in the brain and what happens after they get there.

Destination: Thalamus

The main output of the retina is to an area of the brain called the thalamus (refer to Chapter 2 for a general description of the thalamus). The visual sub-region of the thalamus is called the *dorsal lateral geniculate nucleus* (dLGN). Both the parvocellular and magnocellular ganglion cell classes — refer to the earlier section "Breaking down into ganglion cell types and classes" — project to the dLGN.

Figure 5-3 shows that a bundle of axons leaves each eye and that the two bundles join a few centimeters later. These bundles of axons are called the *optic nerves* (the term *nerve* is a general term for a bundle of axons). The junction point where the optic nerves first meet is called the *optic chiasm,* which means "optic crossing."

Crossing to the other side: The left-right sorting of images

A very interesting thing happens at the optic chiasm. Some of the ganglion cell axons from each eye cross at the chiasm and go to the other side of the brain, and some don't. Which do and which don't, you may wonder, and why.

Look carefully at the right eye in Figure 5-3. The part of the right retina closest to the nose (called the *nasal* retina) receives images from the world on the right side (right visual field), while the part of the right retina farthest from the nose (called the *temporal* retina) receives input from the left side of the world. In the left eye, the right visual field falls on its temporal retina. What happens at the optic chiasm is that axons sort themselves so that the information received from the right visual field is dealt with by the left side of the brain, while the right side of the brain deals with the left side of visual space (that the left brain deals with right side should be a familiar theme by now).

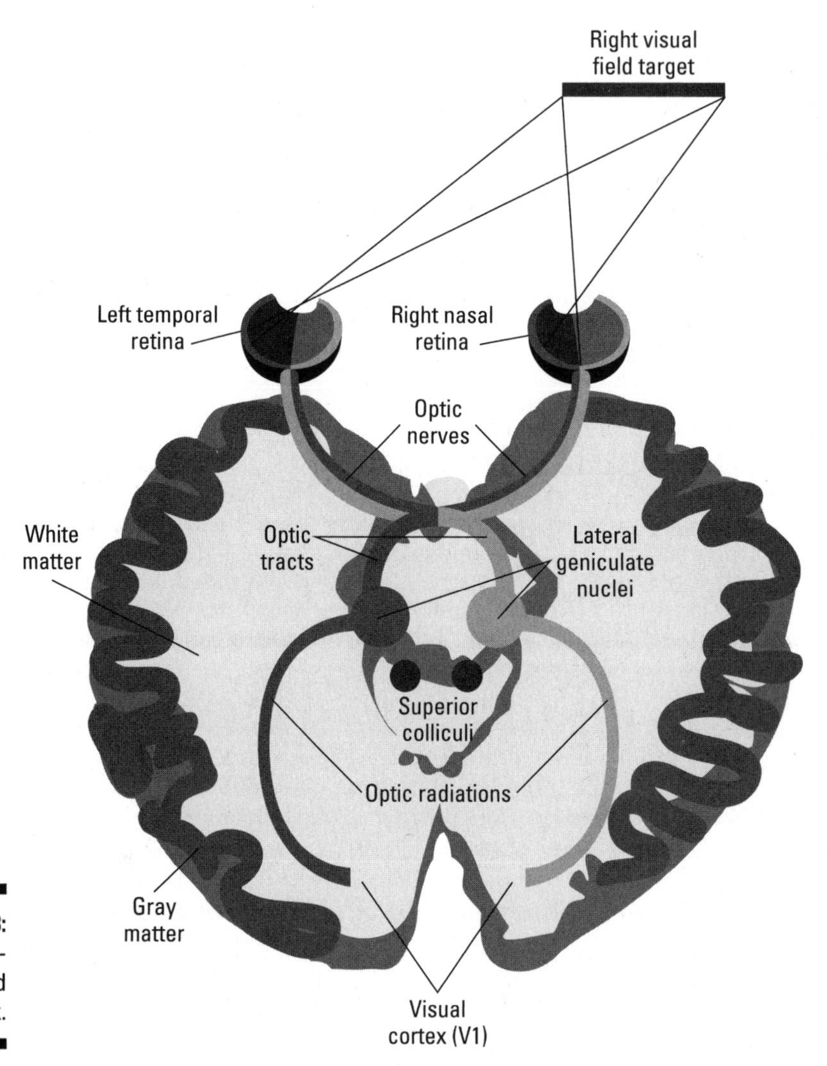

Figure 5-3: The thalamus and neocortex.

This left-right sorting occurs because axons from nasal retinal ganglion cells (that see the visual world on the same side as the eye) cross at the optic chiasm and go to the opposite side of the brain, while the axons of temporal retinal ganglion cells do not cross.

So the nerves after the optic chiasm have different axons than the optic nerves going to the chiasm; they also have a different name, the *optic tracts*. The left optic tract has axons from both eyes that see the right visual field, while the right optic tract has axons from both eyes that see the left visual field. This means that there are cells in the cortex that are driven by the same visual field location in both eyes, in addition to the left visual cortex dealing with the right visual field, and vice versa.

Looking at the visual signal in the thalamus

What happens to the visual signal in the thalamus? Twenty years ago, most researchers would have answered very little because ganglion cell axons *synapse* on cells in the thalamus that have very similar response properties to their ganglion cell inputs. Each thalamic relay cell receives inputs from one or a few similar ganglion cells; for this reason, they're often referred to as *relay cells*. For example, some layers of the dLGN receive inputs from only parvocellular ganglion cells, and within those layers, a few on-center parvocellular ganglion cells drive an on-center parvocellular type relay cell. Similarly, off-center parvocellular ganglion cells drive off-center parvocellular relay cells and a corresponding situation exists for on-and off-center magnocellular relay cells in other layers.

Why do ganglion cells make this relay stop then? Is it because the ganglion cell axons simply can't grow far enough? This explanation seems unlikely because all mammals have this visual relay through the thalamus, despite large differences in brain sizes (consider the distances involved in elephant versus mice brains). The more likely explanation is that, although the relay cells in the thalamus seem to respond very much like their parvocellular or magnocellular ganglion cell inputs, other inputs to the dLGN from other parts of the brain allow gating functions associated with attention. (A *gating function* is a modulation of the strength of a neuron's responses to any particular stimulus based on the context of importance of that stimulus.)

How does attention use a gating function in the thalamus? Imagine you're meeting someone you've never seen before and you've been told this person is wearing a red sweater. As you scan the crowd, you orient to and attend to people wearing red. This task is accomplished at numerous places in your visual system, including your thalamus, because cells that respond to red things in your thalamus have their responses enhanced by your attention. If you got a text message that your party took off the red sweater because the plane was too hot and was wearing a green shirt, you could switch your attention to green, with green-responding cells' outputs enhanced.

The thalamic relay cells in turn send their axons to the visual area of the neocortex at the back of the head, called the *occipital lobe* (refer to Chapter 2). This fiber tract of axons is called the *optic radiation* due to the appearance of the axons "fanning out" from a bundle. I discuss cortical processing in later sections starting with "From the Thalamus to the Occipital Lobe."

Other destinations

Some ganglion cells project to retinal recipient zones other than the thalamus. These zones, explained in the following sections, carry specific information extracted from images for functions like the control of eye movements, pupillary reflexes, and circadian rhythms.

The superior colliculus: Controlling eye movement

This zone receives axons from almost all ganglion cell classes except the parvocellular cells. The superior colliculus controls eye movements. Our eyes are almost never still; instead, they jump from fixation to fixation about three to four times per second. These large rapid eye moments are called *saccades*. Saccades can be voluntary, such as when you're visually searching for something, or involuntary, as is the case when something appears or moves in your peripheral vision that draws your attention and your gaze (like red sweaters or green shirts).

The accessory optic and pretectal nuclei

Several accessory optic and pretectal nuclei receive inputs from ganglion cells that detect self movement. These visual nuclei are essential for balance and enable you to maintain fixation on a particular object while you or your head moves. They project to motor areas of the brain that control eye muscles so that no *retinal slip* (or movement of the image across the retina) occurs despite your movement or movement of the object of your attention.

One important function of this pathway is for *visual tracking*, the ability to follow, for example, the flight of a bird across the sky while keeping the bird image centered on your high acuity fovea. You can do this not only when you are standing still, but also while you are running — a handy skill if you are a wide receiver running to catch a forward pass.

The suprachiasmatic nucleus

Suprachiasmatic means "above the optic chiasm." This area regulates *circadian rhythms*, the body's intrinsic day-night cycle, which includes being awake and sleeping. Humans are built to be active during daylight hours and to sleep at night.

This natural cycle is activated by a class of ganglion cells that are intrinsically light-sensitive; that is, they have their own photoreceptor molecules and respond to light directly, as opposed to being driven by the photoreceptor-bipolar cell sequence (explained in the earlier section "Processing signals from the photoreceptors: Horizontal and bipolar cells"). These *intrinsically photoreceptive* cells, as they're called, send information about day versus night light levels to the area of the brain that controls your circadian rhythms. (See Chapter 11 for more about what happens during sleep.)

The Edinger-Westphal nucleus

Like the suprachiasmatic nucleus (refer to the preceding section), the Edinger-Westphal nucleus receives inputs from intrinsically photoreceptive ganglion cells that inform it of the current overall light level. This nucleus controls your pupil's level of dilation.

From the thalamus to the occipital lobe

The cells in the dLGN of the thalamus that receive projections from the retina project to the occipital lobe of the cerebral cortex at the back of your brain. This is the pathway that mediates almost all of the vision you're conscious of. (Contrast this with vision functions, such as pupil contraction and dilation, that you're neither conscious of nor able to voluntarily control.) The area of the occipital lobe that receives this thalamic input is called *V1* (meaning, "visual area 1").

In addition to being called Vi, this area goes by other names, such as *area 17* (according to the general number system of cortical areas from Korbinian Brodmann, a 19th century German anatomist who assigned numbers to every area of neocortex) and *striate cortex* (referring to a dense stripe through this area that appears in histological stains for cell bodies unique to V1). To keep things simple, throughout this book, I refer to this area as area V1.

Neurons in area V1 project to other areas of cortex and these to other areas still so that virtually all the occipital lobe and most of the parietal lobe and inferior temporal lobe have cells that respond to certain types of visual inputs. What all these different visual areas (more than 30 at last count) are doing is responding to and analyzing different features of the image on the retinas, enabling you to recognize and interact with objects out there in the world. The way these visual areas accomplish this feat is through neurons in different visual areas that respond to discrete features of the visual input.

What happens in V1 and other visual areas

Take a brief look at the numbers. A little over one million retinal ganglion cells project to about the same number of relay neurons in the dLGN of the thalamus. However, each thalamic relay neuron projects to over 100 V1 neurons. In other words, the tiny area of the visual image subserved by a few retinal ganglion and thalamic cells drives hundreds of V1 neurons.

What the hundreds of V1 neurons are doing with the output of a much smaller number of ganglion cells is extracting local features that exist across several of their inputs.

As David Hubel and Torsten Weisel of Harvard University famously showed, V1 cells are almost all sensitive to the orientation of the stimulus that excites them. This means that these cells don't fire action potentials unless a line or edge exists in the image, which is represented by several ganglion cells in a line in some direction being activated.

All stimulus orientations (vertical, horizontal, and everything in between) are represented in V1 so that some small group of ganglion cells that respond to local light or dark in some area of the image gives rise to a much larger group of V1 cortical cells that respond only to a particular orientation of an edge in that area.

Other V1 neurons only respond to certain directions of motion, as though a particular sequence of ganglion cells has to be stimulated in a certain order. As with orientation, all directions are encoded, each by a particular cell or small set of cells. Other V1 cells are sensitive to the relative displacement of image components between the two eyes due to their slightly different viewing position (called *binocular disparity*).

Cells in V1 are specific not only for position in space as projected onto the retina but also to *specific features* such as orientation and movement direction. Any given pattern on the retina stimulates a majority of the ganglion cells there but only a minority of V1 cells that receive from that area. But the firing of the selective V1 cells codes specific visual information.

As mentioned earlier, area V1 is at the posterior pole of the occipital lobe. Just anterior to V1 is (you may have guessed) area V2. Anterior to that is V3. Neurons in these areas tend to have relatively similar response properties. For now, just think of V1 – V3 as a complex from which projections to other areas arise (yes, I know this is a gross oversimplification of the undoubtedly important differences in their functions that additional research will make clear).

Looking at the dorsal and ventral streams

Understanding the immensely complex visual processing network that takes up nearly half of all the neocortex is one of the most challenging areas of research in neuroscience today. One of the most important organizing principles we currently have is that there is a structural and functional division in the visual processing hierarchy.

The V1–3 complex gives rise to two major pathways: the dorsal stream and the ventral stream. The response properties of the neurons and the visual deficits after damage to these two streams support the idea that these areas have important functional differences.

The dorsal stream

The dorsal stream is the projection into the parietal lobe. Cortical areas in the dorsal stream, such as areas called *MT* (middle temporal) and *MST* (medial superior temporal) are dominated by cells that respond best to image movement. In MST particularly, there are cells that respond best to the types of visual images that would be produced by self movement, such as rotation of the entire visual field, and *optic flow* (the motion pattern generated by translation through the world, with low speeds around the direction to which you are heading, but high speeds off to the side). In addition, *motion parallax,* in which close objects appear to shift more than distant ones when you move your head from side to side, is encoded by motion-selective cells in the dorsal pathway.

The dorsal stream has been colloquially referred to as the "where" pathway, although more recently many neuroscientists have preferred the phrase the "how to" pathway. Lesion studies show that this pathway is necessary for visually guided behavior such as catching a ball, running through the woods without bumping into trees, and even putting a letter in a mail slot. Damage to this area results in deficits called *apraxis*, the inability to skillfully execute tasks requiring visual guidance.

Life in a virtual disco

A classic (but unfortunate) case exists in the clinical literature of a woman who suffered bilateral damage to her left and right side MT areas. Although this woman's vision is normal as assessed by eye charts and object recognition tests, she is severely disabled because she has no ability to judge movement. For example, she cannot cross the street because she cannot gauge when oncoming cars will reach her position. She routinely overfills a cup when pouring tea because she cannot tell when the liquid will reach the top. This woman lives in what is essentially a perpetual strobe-light disco, with no ability to assess or deal with continuous motion.

The ventral stream

The ventral stream goes to areas along the inferior aspect of the temporal lobe (the so-called *infero-temporal cortex).* The ventral pathway is often called the "what" pathway. Cortical areas in this stream have neurons that are not generally motion selective but that prefer particular patterns or colors (almost all neurons in ventral stream area V4, for example, are color selective).

As you move from posterior to anterior along the inferior temporal lobe, you find cells that respond only to increasingly complex patterns such as the shape of a hand. Near the pole of the temporal lobe is an area called the *fusiform face area,* with cells that respond only to faces. Damage to this area has resulted in patients with normal visual acuity but who cannot recognize *any* faces, including their own.

Crosstalk between the dorsal and ventral streams

Despite the clear segregation of functions between the dorsal and ventral streams, they clearly exist in a network in which there is crosstalk. For example, in *structure from motion* experiments, researchers put reflector spots on various body parts of actors who wore black suits and filmed their movements in very low light so that only the dots were visible on the film. Anyone seeing these films can tell, once the film is set in motion, that the dots are on the bodies of people, what the people are doing, and even their gender. In this case, motion-detecting neurons from the dorsal pathway must communicate with object detecting neurons in the ventral pathway.

Another example of dorsal-ventral pathway crosstalk is depth perception. The visual system estimates depth or distance to various objects in the environment in a number of ways. Some cues, such as *pictorial depth cues,* can be represented in pictures and photographs. These include near objects overlapping objects that are farther away and relative size (nearer objects are larger than more distant ones — a tiny car in a picture must be farther away than a large person). Ventral pathway pattern-based cues must work with dorsal pathway motion-based cues to give a unified judgment of depth.

Impaired Vision and Visual Illusions

We tend to believe that we see "what is really out there" when, in reality, what we "see" is a construct from a combination of the current images on our retinas and our past experience. If you have a color-vision defect, for example, you may show up for work with a pair of socks that you think match but that your coworkers see as being different (to which your best response might be "Funny, but I have another pair at home just like these!"). There are also things — such as optical illusions — that none of us see as they really are. Visual deficits and illusions both tell neuroscientists a lot about how our visual system is built and works.

Looks the same to me: Color blindness

As I note in the earlier section "The retina: Converting photons to action potentials," the three different cone types (red, green, and blue) enable you to see color. Take away any of those particular cone types, and you have color blindness.

The most common (by far) forms of color blindness results from the absence of one cone type in the retina. About 1 in 20 men and 1 in 400 women are missing red cones (a condition called *protanopia)* or green cones *(deuteranopia).* People with these conditions can't discriminate red from green colors.

Why the difference between genders? The genes for these pigments are on the X chromosome. Because men, who have an X and a Y chromosome, only get one copy of the gene, they are red-green color blind if that gene is defective. Women, on the other hand, have two X chromosomes and therefore two copies of the red and green genes. If they end up with one bad copy of the red-green gene, they still have red-green color vision. For women to be red-green color blind, both genes on both X chromosomes have to be bad. Red-green color blindness is thus much rarer in women.

Even rarer in both men and women is the loss of blue cones, called *tritanopia.* These folks can't distinguish blue-green colors.

An example of an acquired form of color blindness involves damage to cortical area V4 in the ventral stream, resulting in *achromatopsia. Achromatopsia* differs from retinal color blindness in that, in retinal color blindness, the person can't discriminate between certain hues. In achromatopsia, on the other hand, the different colors appear as different shades of gray but without color, enabling the person to discriminate between them.

Understanding blindness

People's dread of losing their vision has nearly the same intensity as their dread of contracting cancer. Although many blind people have led productive and very satisfying lives, loss of sight is considered one of the most disabling of all possible injuries. In this section, I discuss some of the most common causes of blindness.

Most blindness, at least in the developed world, originates in the retina. The most common forms of retinal blindness are retinopathies, such as retinitis pigmentosa, macular degeneration, and diabetic retinopathy, which cause photoreceptors to die, but numerous other causes of blindness occur as well:

- **Retinitis pigmentosa:** Retinitis pigmentosa involves a hereditary degeneration of retinal rod photoreceptors. This condition progresses from night blindness to loss of all peripheral vision (tunnel vision, sparing central vision because there are no rods in the fovea) to loss of all vision. Currently, unfortunately, there is no treatment for this disease, and the resulting blindness is irreversible.
- **Inherited metabolic disorders:** Macular degeneration and diabetic retinopathy involve death of retinal cells as a result of inherited metabolic disorders, usually also beginning with photoreceptor death followed by death of other retinal neurons.
- **Glaucoma:** Unlike the preceding retinopathies, glaucoma involves a primary death of retinal ganglion cells, most commonly due to inherited excessive pressure within the eye. One glaucoma subtype, *closed-angle glaucoma,* is treatable with laser surgery. The other form, *open-angle glaucoma,* can often be controlled with medication.
- **Cataracts:** In third-world countries, lens cataracts and corneal opacities cause a high percentage of blindness, but most of these conditions can be easily treated with modern surgical technology.
- **Eye and head injuries:** Eye injuries that cause retinal detachment can lead to retinal death. Severe injuries can cause the loss of an entire eye (or both eyes). Head injuries, tumors, or vascular lesions can impact the optic nerve or any of the visual processing areas of the cortex (explained in the earlier section "From the thalamus to the occipital lobe").

Amblyopia, often called "lazy eye," is one form of visual disability that was extremely puzzling to researchers and doctors — at least until they gained a better understanding of plasticity mechanisms in visual cortex (Chapter 16 has more on plasticity). It turns out that during development, the eyes compete for cortical synapses. If one eye is optically much worse than the other due to a cataract or extreme near- or far-sightedness, the ganglion cells in that eye fire much less vigorously than in the other, and the other eye "takes over" all the available synapses in V1. This condition is virtually irreversible in humans after about the age of 6, so even if the optical problem in the affected eye is remedied after that age, the disadvantaged eye is still blind, even though it is, in all other ways, normal. For this reason, children younger than 6 should have their eyes examined and any defects corrected.

Visual illusions

How is it that we sometimes see something that is not there? Some optical illusions, like mirages and rainbows, are due to optical properties of the atmosphere and can be photographed.

Other illusions, however, seem to be constructions of our brains, so that we perceive something that is not photographable. Typical examples of these include the Ponzo (railroad track) illusion, in which two identical lines appear to be different sizes when placed on parallel lines that converge in the distance, and the Necker cube, where the face of the cube seems to change, depending on which side the viewer focuses on. Another famous visual illusion is the Kanizsa triangle (see Figure 5-4), in which a solid white triangle seems to overlay a black triangle outline. The catch? There is no white triangle.

Each of these illusions can be explained similarly. Our visual system evolved to make sense of images projected onto our retinas resulting from real, three-dimensional objects in the real world. In other words, we see what we *expect* to see. The illusion image of the Kanizsa triangle, for example, is a very complicated two dimensional image, what with the three precisely spaced angles and the three precisely arranged circle segments. In the three dimensional world, such an image is possible only when a solid white triangle is present. Hence, that's what we see.

The important point about illusions like these is that they reveal interesting things about how the visual system works. As I discuss earlier in this chapter, the visual system does not convey the image on the retina to some special place where it is looked at by some brain entity. The visual system extracts information that allows us to identify objects and interact with them. To do this, the system "interprets" the visual input according to internal models that come from our own experience and from the evolution of our species in dealing with the real world. Compared to this, we have little evolutionary history dealing with black markings on white paper, something we have to learn to produce and interpret.

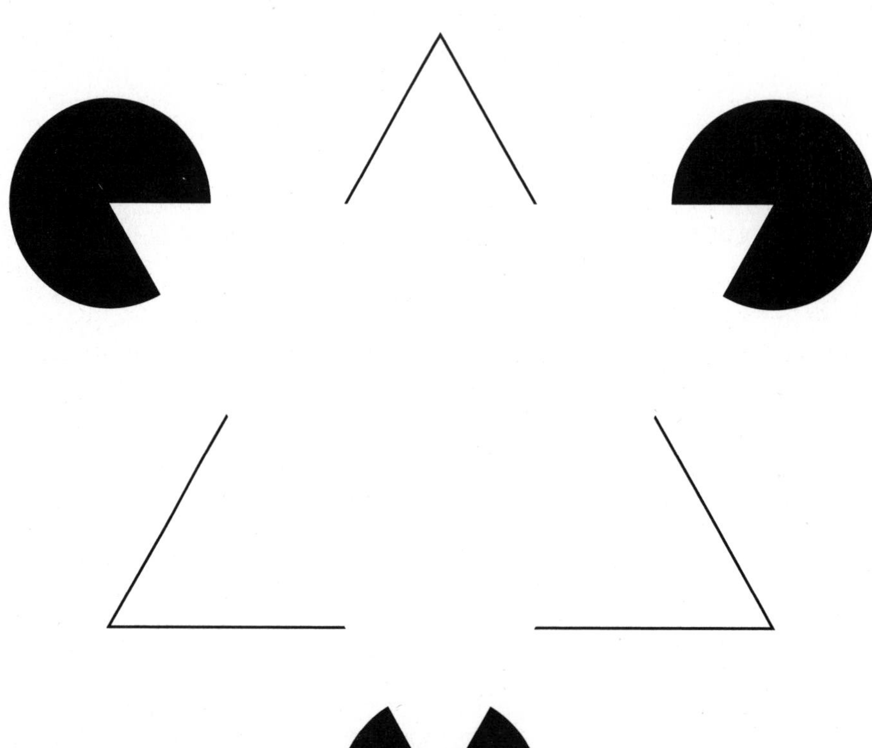

Figure 5-4: The Kanizsa triangle; there is no solid white triangle.

Chapter 6

Sounding Off: The Auditory System

In This Chapter

- Understanding the roles the outer, middle, and inner ear play in hearing
- Learning about sound processing in the brain
- Looking at common hearing impairments

The sense of hearing, like vision, captures energy from the environment to inform us about what is out there. What hearing detects is sound. Sound is our perception of air vibrations that are created by the vibration of objects in the world.

Think about a plucked guitar string. The string bows out first in one direction, making air molecules on that side slightly denser for a moment. These molecules then make the air denser a little farther away. The pulse continues traveling away from the string and spreading out in space. When the string bows the other way, it makes the immediate air region less dense. This low pressure pulse then spreads. The series of high and low pressure pulses form a wave whose frequency is the frequency of string vibrations.

The sound waves spreading out from the sound source form larger and larger spheres of waves, so the sound energy is spread out over a larger and larger area. This means that the sound's intensity per any receiving area declines with distance. Ultimately, the sound cannot be heard far away because the sound energy per area from the source is on the order of the energy of random air molecule movements. The auditory system's sensitivity to pressure waves in the air is very close to this absolute threshold; in other words, we can hear (detect) air molecule density changes close to those occurring randomly due to the air motion itself. You can't do much better than that, even in theory. Our sense of hearing allows us not only to detect that sound exists, but also what made it, what it means, and from where it came.

Through a complex series of interactions between the different parts of the auditory system — the topic of this chapter — we can not only recognize a sound's existence, its origin, and the direction from which it came, but we can also perform higher functions, like appreciating music and understanding language.

The Ear: Capturing and Decoding Sound Waves

The first stages of auditory processing are mechanical rather than neural. Just as the eye, like a camera, has a lens and pupil that focus and regulate light levels, so the auditory system physically transforms incoming sound waves. Figure 6-1 shows the anatomy of the ear.

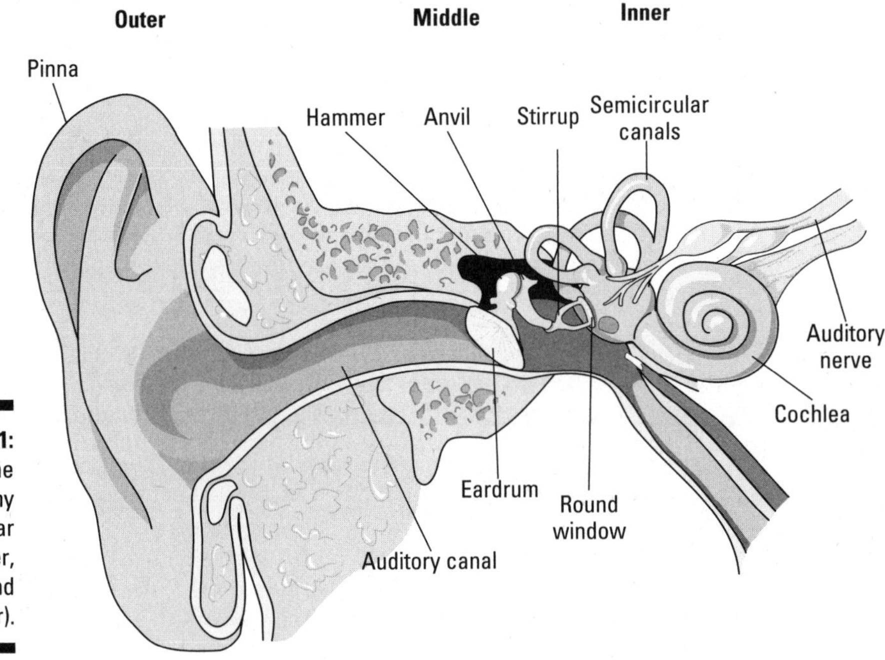

Figure 6-1: The anatomy of the ear (outer, middle and inner).

The first of these transformations are mechanical and involve the outer ear, the middle ear, and the inner ear. Very generally, here's what happens to a sound wave that enables you to hear:

- **The outer ear:** The outer ear has three parts: the pinna, the auditory canal, and the eardrum. The pinna "captures" and filters the sound waves and directs them via the auditory canal to the eardrum.
- **The middle ear:** Three little bones in the middle ear vibrations into pressure waves and send them to the inner ear.

- **The inner ear:** The pressure waves from the middle ear go to the oval window of the cochlea where auditory receptors (hair cells) are located. The mechanical structure of the spiral cochlea causes high frequency vibrations to be concentrated near the oval window, while low frequency vibrations travel to the end of the spiral. The pressure waves in the cochlea bend the auditory hair cells within it which generate action potentials in the auditory nerve. At this point, the mechanical transformations are complete, and the neurons take over, which you can read about in the later section "Making Sense of Sounds: Central Auditory Projections."

The following sections go into greater detail on the role each part of the ear plays in hearing.

Gathering sound: The outer ear

The pinna, auditory canal, and eardrum constitute the outer ear. This is the first structure involved in processing sound.

The pinna, the first part of the outer ear

The first part of the outer ear that sound waves hit is the *pinna.* Despite its technical name, you're very familiar with the pinna: It's the part of the auditory system you can see, tuck your hair behind, pierce, and, if you're particularly talented, wiggle. So why don't I just call it the ear? Because technically the ear is made up of several components, the pinna being only one of them, and in neuroscience — and several other disciplines — you need to be able to distinguish this structure from other auditory structures considered part of the ear.

The pinna does two things with incoming sound waves:

- It concentrates sound by reflecting sound from the larger pinna area to the smaller auditory canal opening.
- Because of its complex shape, the pinna changes the frequency content of the incoming sound based on the direction from which the sound came. This helps us in *localization,* or telling where the sound came from (you can read more about this in the later section "Locating sound").

Everyone's pinna is slightly different and changes as a person grows. What this means is that, during development, we must "learn" our pinna's directional frequency transformation. This learning almost certainly involves neural plasticity — that is, the ability of neural connections to reorganize themselves — in the auditory cortex.

Pitch, aging, and pulling a fast one on the teacher

The frequency of sound is measured in cycles per second, the term for which is Hertz (abbreviated Hz), named after a German scientist who studied alternating electrical phenomena. (When musicians talk about *pitch*, this is what they're referring to.)

Humans hear frequencies from about 20 to about 20,000 Hz, at least when they're young. As we age, we tend to lose high frequency sensitivity, even in the absence of any ear disease or damage. In fact, people over 60 rarely hear well above around 16,000 Hz. This loss tends to be relatively greater in men than women. High school students have been known to take advantage of this by setting the ring tones on their cell phones at high frequencies that their teachers cannot hear.

Sailing into the auditory canal

The pinna reflects the sound waves to the auditory canal, which, at its most basic level, connects the pinna to the eardrum. The auditory canal is slightly resonant at midrange frequencies (frequencies important for hearing human voices). Once the sound waves enter the canal, they are further transformed to deliver more energy from these important frequencies to the eardrum.

Banging the (ear)drum

The auditory canal ends at the eardrum. When the sound waves reflected by the pinna and transmitted along the auditory canal hit the eardrum, the eardrum begins to vibrate, which triggers responses in the middle ear.

The middle ear

On the other side of the eardrum are three tiny bones called the *hammer* (Latin name: *malleus), anvil* (*incus*), and *stirrup* (*stapes*) (*malleus, incus*, and *stapes* are the original Latin, medical terms). These constitute the middle ear.

The three middle ear bones are the smallest bones in the body, but they have a crucial function: Through them, the vibrations of the eardrum activate the oval window (part of the inner ear) at the entrance to the cochlea.

Think of these three bones as a lever, which in physics terms, is a rigid object that, when used with a fulcrum (or pivot point), can increase mechanical force at another point. In terms of the function of the three bones of the inner ear, here's how this idea plays out: The eardrum, which is attached to the hammer bone, vibrates in response to sound as pressure changes in air. At the other end of the three bones is the stirrup, which is attached to the oval window at the cochlea, which is filled with fluid. The ball-like head of the hammer where it connects to the anvil is like a fulcrum in this system, so that a larger, but weaker movement of the eardrum causes a smaller but stronger movement of the stirrup. The leverage of the bones is necessary because fluid in the cochlea is stiffer than air at the eardrum.

The leverage effect of the middle ear bones is on the order of 100 to 1, which means that the force is amplified a lot. A second mechanical advantage is that the area of the eardrum is larger than the area of the oval window, meaning that the force is amplified by this ratio as well.

Playing chords to the brain: The inner ear

After the oval window of the cochlea has been activated by the bones of the middle ear, the inner ear gets into the action. The inner ear consists of the cochlea and its contents and connections (see Figure 6-2). When the stirrup forcefully vibrates the oval window at the entrance to the cochlea, pressure waves are sent through the fluid that fills the cochlea. These pressure waves bend the transducers for sound, the auditory hair cells, located in the Organ of Corti within the cochlea. These hair cells are so sensitive to small movement that they can respond to deflections of the cilia by the widths of no more than an atom!

Interestingly, the cilia on the auditory hair cells in the cochlea don't just look like the kind of cilia that other, non-neuronal cells have to move things around in the extracellular fluid; they also are apparently derived evolutionarily from motor type cilia, such as used by some single-celled organisms to move.

Opening ion channels to fire action potentials

At the base of the auditory hair cell cilia are specialized channels that resemble mechanoreceptors (refer to Chapter 4) in that, when the cilia bend, they stretch the hair cell membrane, which in turn opens ion channels that depolarize the hair cell and causes the axon terminals of the auditory nerve fibers at other end of the hair cell to fire action potentials. These action potentials go down the auditory nerve to the cochlear nucleus.

The Organ of Corti actually contains two types of hair cells, called inner hair cells and the outer hair cells. Even though there are fewer inner hair cells than outer ones, the inner hair cells make more connections to the auditory nerve than the outer hair cells do. Still, outer hair cells have an important function: They control the stiffness of the membrane near the hair cell cilia. This outer hair cell motor function boosts the hair cell response to low amplitude sounds, particularly at high frequencies, but weakens it for excessively loud sounds to prevent hair cell damage.

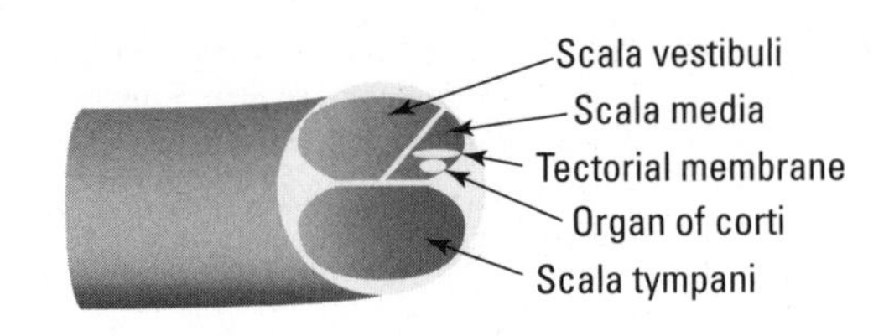

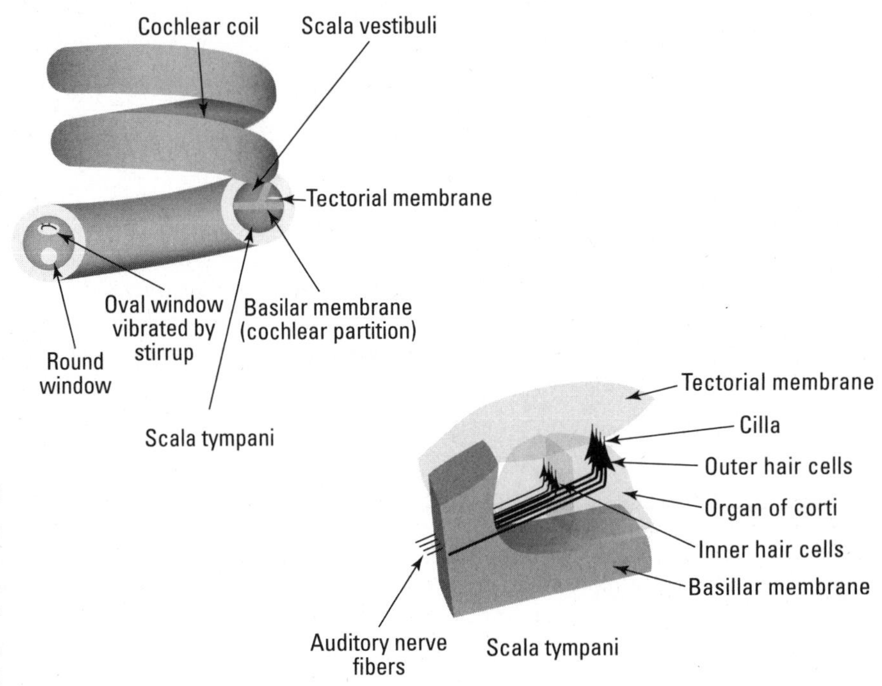

Figure 6-2: The cochlea and the Organ of Corti.

Sending information about frequency and amplitude

About 30,000 auditory nerve fibers travel from the cochlea to the cochlear nucleus (refer to Figure 6-2) in the brainstem. The structure of these neurons is something like that of mechanoreceptors in the dorsal root ganglia (refer to Chapter 4). In the auditory system, the cell body of the sensory neuron is in the cochlear nucleus. The cochlear neurons have two axons. One goes to the auditory hair cell in the cochlea, where bending of the cilia induce action potentials in the cochlear neuron axon. The other axon projects to the superior olive, carrying the auditory message.

The message the auditory fibers send the brain about sound in the environment consists of two basic properties of sound —frequency and amplitude — similar to the way the retina informs the brain about color and light intensity. However, whereas the retina represents space explicitly in terms of location on the retina, there is no similar link between the location of the sound and position in the cochlea. Instead, when it comes to the auditory sense, the cortex computes the location of the sound in space (you can read more about this in the later section "Locating sound"). In other words, the auditory nerve output doesn't explicitly include a representation of space (which is one reason the cochlea can mediate hearing even though it has only 30,000 auditory fibers —compare this to the over one million retinal ganglion cell axons that leave each eye!).

Although position in the cochlea doesn't reflect sound location, the cochlea is extended in one dimension, forming a spiral consisting of about 2.5 turns (refer to Figure 6-2). What is represented in this single dimension is frequency.

The mechanical structure of the cochlea (stiffness, size) is such that high frequencies only cause vibrations in the portion of the cochlea near the oval window, while low frequencies produce larger vibration amplitude toward the end of the cochlea farthest from the oval window (the center of the coil). The position along the cochlea where the vibration occurs codes frequency. Hair cell responses near the oval window indicate high frequencies; those at the farthest end of the cochlea indicate low frequencies. If position along the cochlea can encode frequency, then firing rate and number of cells firing can encode amplitude, thus generating the neural representation of the audio input. Simple enough, right?

Unfortunately, it's not as simple as place coding for frequency versus a firing rate code for amplitude. The reason has to do with a fundamental property of neurons, that is, their maximum firing rate (refer to Chapter 3). Because the action potential itself lasts about one millisecond, neurons can't fire more than 1,000 action potentials per second; in general, sustained rates are less than 500 per second.

For reasons having to do with sound localization (head to the later section "Locating sound" for more on that), auditory fibers usually fire just at the peak of the sound pressure wave. Here's what this means for different frequencies:

- **For frequencies below about 500 Hz:** Auditory fibers can fire every sound pressure wave peak, so that their firing rate actually directly codes frequency. In other words, at low frequencies, the fibers fire at the sound frequency.
- **For frequencies higher than 500 Hz:** Auditory fibers can't fire at the sound frequency, so they fire every other pressure peak, or every fifth peak, or tenth, or, you get the idea. In that case, the frequency is encoded by place along the cochlea.

Both low and high frequency sound amplitude is encoded by both the number of spikes that occur at each pressure pulse peak, and the number of active fibers (loud sounds will recruit more fibers to fire than fainter sounds).

Why hearing aids work

The one dimensional frequency structure of the cochlea has made auditory prostheses so successful. Most hearing loss is associated with death of the auditory hair cells. Successful auditory prostheses consist of linear electrode arrays that are slid into the cochlea. Electronic processing splits the incoming sound into frequency bands and generates stimulating current at the appropriate location of the prosthesis to directly drive the auditory nerve fibers. At the time of this writing, well over 50,000 auditory prostheses have been implanted in the United States, many allowing recipients to communicate verbally with little handicap. There is no current prosthesis with even remotely comparable success for vision loss.

Making Sense of Sounds: Central Auditory Projections

The 30,000 auditory fibers convey messages about sound to the cochlear nucleus and beyond in a mixed code. Each fiber is sensitive to a particular frequency band, and different fibers have different thresholds and firing rates that encode the magnitude of sound.

Stops before the thalamus

Auditory nerve fibers don't project directly to the thalamus; they relay twice before that. As mentioned previously, when the auditory hair cilia bend, the hair cell membrane stretches and opens up ion channels that depolarize the hair cell and causes the axon terminals of the auditory nerve fibers to fire action potentials.

The auditory nerve action potentials head to the cochlear nucleus where the neuron cell bodies for those axons are located and cause an action potential in the cell body, which in turn causes an action potential in a second axon that crosses the body midline and projects to a nucleus, called the *superior olivary nucleus,* on the opposite side (see Figure 6-3). In other words, action potentials from the right cochlear nucleus cross to the superior olivary nucleus on the left, and action potentials from the left project to the superior olivary nucleus on the right. The superior live nuclei are located in the pons (refer to Chapter 2 for general information about the areas on the brain).

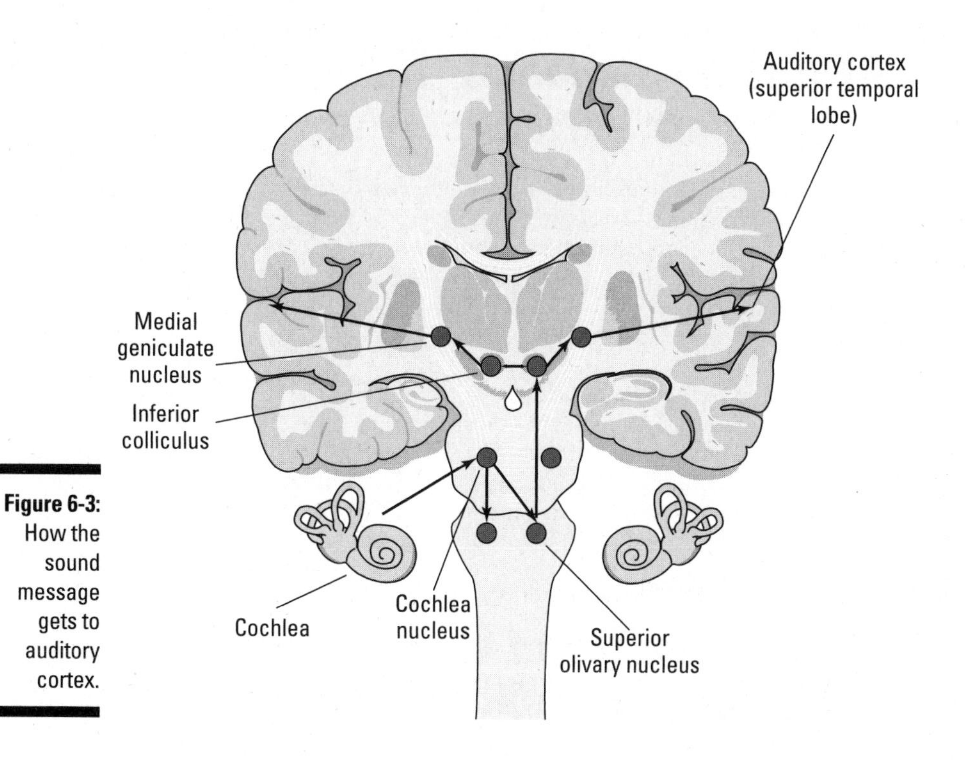

Figure 6-3: How the sound message gets to auditory cortex.

Relay cells in the superior olivary nucleus project to the *inferior colliculus,* located, as you may guess, just below the superior colliculus in the midbrain. Finally, inferior colliculus neurons project to the medial geniculate nucleus of the thalamus, which you can read about in the next section.

All the projections after the contralateral projection to the superior olivary nucleus are *ipsilateral*, meaning they stay on the same side. Therefore, the medial geniculate on the right side of the body receives mostly input from the left ear, and the medial geniculate on the left side of the body receives mostly input from the right ear.

Why does the auditory message pass through two processing stops before reaching the thalamus? The answer has to do with localizing sound. Both the superior olivary nucleus and inferior colliculus on each side of the brain receive not only contralateral but some ipsilateral projections. Neural comparisons between the outputs from each ear set up the neural coding for sound localization. In the auditory system, sound localization is mediated by a neural computation of the difference in sound volume and sound arrival time between the two ears. These computations require precise neural timing, which means they must occur very early in the auditory processing stream, as close to the two cochlear nuclei as possible. The later section "Locating sound" goes into the details of the mechanisms for sound localization.

Off to the thalamus: The medial geniculate nucleus

The medial geniculate nucleus on each side of the brain receives inputs primarily from the ear of the other side. The responses of medial geniculate neurons appear to be similar to those in the auditory nerve; that is, they are frequency codes so that each medial geniculate relay cell prefers a particular band of frequencies similar to its cochlear nucleus axon input, and louder sounds recruit more cells firing at higher rates.

In addition to these similarities, typical thalamic attentional gating functions almost certainly occur there, as well. For example, if you barely hear a suspicious noise, you can concentrate and attend to hearing the occurrence of another instance of that noise. The neural enhancement of *attended auditory input* — sounds you *deliberately* try to hear — is mediated in multiple brain areas that typically include the thalamus.

Processing sound in the brain: The superior temporal lobe

As Figure 6-3 shows, the medial geniculate nucleus of the thalamus conveys auditory information to the superior temporal lobe at an area slightly posterior to the middle of the superior sulcus. This area of the superior temporal lobe is called the *primary auditory cortex,* or *A1.* It is also referred to as *Heschl's gyrus.*

What varies with position within A1 is the frequency to which neurons respond. Such a frequency map is called *tonotopic.* Just as real images in the visual system are made of *patterns* of lines and edges, among other things, that are put together as shapes in the visual cortex, the auditory cortex represents patterns of frequencies whose combination indicates some meaning, such as the word "hello" or a door slamming.

Other aspects of sound, particularly localization, also appear to either be processed by neurons in A1 or depend on projections from A1, since lesions in A1 impede a person's ability to locate the direction a sound is coming from.

Figure 6-4 shows the primary auditory cortex and other areas important for discriminating between sounds.

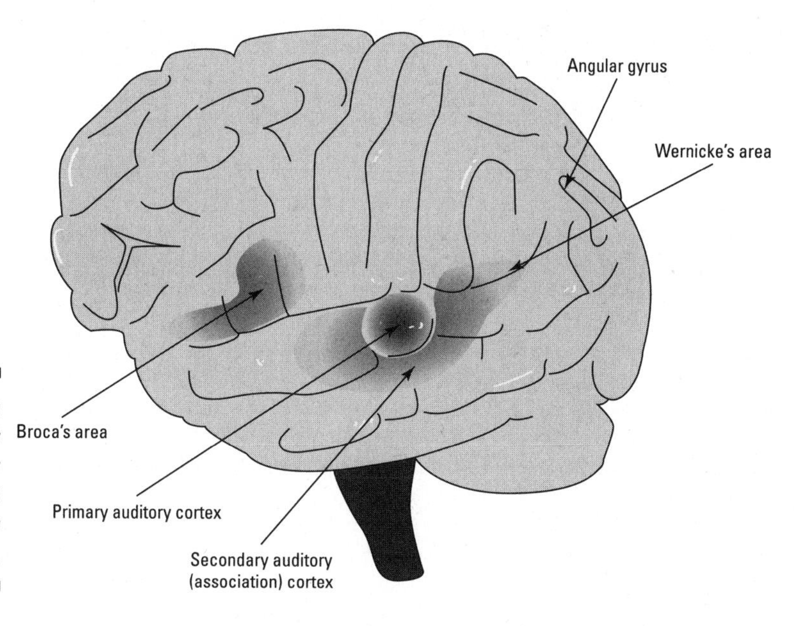

Figure 6-4: The primary auditory cortex and other key areas.

Handling complex auditory patterns

Around auditory area A1 are several higher order areas, which, because no standard naming convention for them exists, are usually referred to as *auditory association areas.*

Many cortical neurons outside A1 respond best to complex or environmentally relevant sounds rather than pure tones. An example of a complex sound is a chirp, where the frequency changes throughout the duration of the sound. Neurons have been recorded that respond well to other real sounds from the environment, such as keys jangling, doors closing, or a table being dragged across the floor, rather than to tones.

This cortical preference for higher-order sound patterns is clearly a common theme in cortical processing. At lower levels (those closer to the input) in the auditory system (and other sensory systems, for that matter), most neurons respond to most stimuli, with the neural representation projected by a relatively small number of axon transmission lines. In the cortex, neurons become pickier, preferring more complex patterns, so that, for any given real sound, only a minority of cortical neurons respond. But because billions of cortical neurons are available, your recognition and memory of particular sounds is linked to the firing of a relatively small percentage of particular neurons whose activity is specific to that stimulus. Much of the specificity in the neural responses is achieved by learning and experience.

Hearing with meaning: Specializations for language

Although the functional differentiation among the higher order auditory processing areas (after A1) remains somewhat unclear at present, one stream of auditory processing is well known because of its importance in speech processing. This is the projection to *Wernicke's area* (area 22), located at the border between the superior, posterior part of the temporal lobe and the parietal lobe. Patients with damage to Wernicke's area have difficulty understanding language. When they speak, although the word flow is fluent, it is often nonsensical, typically described as a "word salad."

Wernicke's area has extensive connections with another language area, Broca's area, located in the frontal lobe just anterior to the motor areas controlling the tongue, vocal cords and other language apparatus. Damage to Broca's area leads to difficulty in producing speech, but deficits in speech understanding are only really apparent for complex sentence constructions, like passive voice.

Both Wernicke's and Broca's areas, shown in Figure 6-4, are located on the left side of the brain in virtually all (approximately 95 percent) right-handed people. The situation is more complicated for left-handers. A small majority of left-handers also have language primarily on the left, but the rest have language on the right side of the brain.

Some research suggests that there are two sorts of left-handed people: *strong left-handers,* whose brains really are reversed compared to right-handers, and *mixed left-handers,* whose brain lateralization is like right-handers but who happen to be left-handed. Some evidence has suggested correlated differences in some traits between the two types, with strong left-handers writing in the "upside down" hooked hand position, while the mixed left-handers tend to write with the hand in a mirror symmetric position to typical right-handed writing.

The identification of left-sided language areas constituted one of the first instances of known functional lateralization in the brain. Given that damage to Wernicke's area on the left side causes profound language processing dysfunctions, the question arises as to what function is mediated by the mirror symmetric area on the right side of the brain. Recent evidence suggests that damage to the right side area 22 results in an inability to process *prosody* in language — the changes in tonality and rhythm — that conveys meaning. Patients with this damage, for example, have trouble distinguishing sarcastic versus questioning versus other tones of voice, and in particular, don't "get" jokes and other forms of humor.

Perceiving music: I've got rhythm

Music is another complex stimulus for audio processing that appears to rely more on the right than the left side of the brain, indicated by the fact that most people recognize melodies better with their left ear (right brain).

Interestingly, there is a clinical case of bilateral damage in higher order auditory areas that resulted in a very specific loss. Called *amusia,* this condition is one in which the sufferer was unable to recognize melodies. Despite the amusia, the patient had normal ability to comprehend speech and complex environmental sounds, and she knew the names of the songs on records she owned. She just couldn't recognize the songs when they were played.

Locating Sound

Localization of the source of a sound is very different from localizing it visually. The optics of the eye preserves topography between direction in the external world and position on the retina that is maintained in projections to higher visual centers. But the ears don't encode auditory direction in any map; instead, sound localization is computed neurally in the auditory system through neural comparisons between the two ears. This computation begins in the superior olivary nucleus and inferior colliculus and is represented in the firing of cells in the auditory cortex that respond best to sound emanating from particular elevation and azimuth (direction in the horizontal plane) ranges.

Computing azimuth (horizontal angle)

The auditory system uses two different methods to compute azimuth: interaural intensity difference and interaural time difference.

These two methods work best in complementary frequency ranges. Spike firing at the exact sound pressure peak works better at low frequencies because the axons can fire fast enough to follow more cycles and stay in what's called *phase lock*. On the other hand, the interaural intensity difference is larger for high rather than low frequencies because the head itself weakens high more than low frequencies.

Interaural intensity difference

A sound source that is not directly in front or behind the listener is louder in the closer ear than in the ear farther away. Some neurons in the superior olivary nucleus and inferior colliculus receive both ipsilateral and contralateral inputs. These neurons are arrayed so that some respond best when the inputs are equal between the two ears, while others prefer different percentages of left versus right ear strength. These neurons project in an ordered way to the auditory cortex where auditory location is then represented by position on the surface of the cortex.

Interaural time difference

Sound source position also affects the relative time that sounds reach the two ears. Most of us are familiar with counting seconds from the time we see lightning until we hear thunder, with a five second delay for each mile. In other words, the speed of sound is very roughly 1,000 feet per second, or one foot per millisecond. If the human head is (very roughly) about 6 inches (one half foot) wide, then the difference in arrival time between the two ears for a source directly to one side is about one half of a millisecond.

Given that action potentials last about one millisecond, you may think that the auditory system wouldn't be able to resolve a one half millisecond difference. But it can! The arrival time of the sound is encoded by auditory fibers because they fire at the exact peak of the sound pressure wave, which of course is delayed between the two ears.

The decoding process for interaural time difference is exquisite. Axons from the two cochlear nuclei take slightly different length paths from the left versus the right ear to the superior olivary nucleus and colliculus. The paths to neurons there that are equal result in the two signals arriving at the same time when the sound source is either directly in front or behind. Neurons that receive inputs with a slightly longer path from the left versus right ear respond best if the sound is proportionally closer to the left than right ears so that the spikes arrive at the same time. An array of neurons with different path lengths produces a linear map of neurons, each of which prefers a slightly different direction. This map is also projected to auditory cortex in an orderly way so that position represents horizontal direction.

Detecting elevation

As long as the head is vertical, no useful binaural signal difference exists between the two ears to enable you to estimate elevation. Instead, estimating elevation depends on the frequency reflection characteristics of the pinna, or outer ear (refer to the earlier section "The Ear: Capturing and Decoding Sound Waves" and Figure 6-1).

Auditory localization in barn owls

Barn owls have auditory localization so precise that, with only the sound of a mouse's movements to guide them, they can locate and catch the critters in total darkness. Because of this ability, barn owls have been used as a model in a number of labs to study neural sound localization mechanisms. Researchers have found that, among other things, the owl's two ears are not symmetrical but are arranged so that there is an intensity difference processed by the ears as a function of elevation as well as azimuth.

The complex shape of the pinna functions as a frequency filter when it reflects sound that is elevation dependent. Almost all sounds in the natural world consist of multiple frequencies in particular ratios that you have become familiar with. The degree to which your pinna distorts the sound tells your auditory cortex the sound's elevation.

I Can't Hear You: Deafness and Tinnitus

As with any system, things can go wrong. Glitches in the auditory system can range from the profound, like not being able to hear at all, to merely irritating, like periodic ringing in your ears. This section goes into a little detail on two conditions — hearing loss and tinnitus — both of which can range from mild to severe.

Hearing loss

Deafness or partial deafness can be congenital (existing at birth) or the result of disease or damage from loud noises. Congenital deafness is rare compared to acquired deafness, occurring in less than one in 1,000 births. About one quarter of congenital deafness is due to inner ear malformations; the other causes of congenital deafness are unknown.

The causes of acquired deafness are easier to pinpoint. Severe inner ear infections can cause deafness by destroying or compromising the three middle ear bones. Called *conductive hearing loss,* in this condition, the sound vibrations are not mechanically transmitted from the eardrum to the cochlea. Otosclerosis, the growth of spongy masses on the middle ear bones, is a hereditary conductive hearing loss with an onset typically between the ages of 15 and 40. It is somewhat more common in women than men. Maternal rubella (German measles) can cause the birth of deaf children from damage by the virus.

Some infections as well as other unknown causes can result in *Meniere's disease,* a partial loss of both hearing and vestibular (balance) function due to a common mechanism that affects both the cochlea and the semicircular canals near the cochlea that function in balance.

Besides the age-related decline in high frequency sensitivity (*presbycusis,* which occurs in almost everyone, particularly males), the most frequent cause of hearing loss is damage to the eardrum or auditory hair cells. Eardrums can be damaged mechanically by items entering the auditory canal or by high pressure such as diving into deep water. Loud noises can damage hearing permanently by destroying auditory hair cells. The louder the noise, the less time it takes for hearing loss to occur. Exposure to loud work environments, for example, can accumulate over decades to damage hearing, while hearing loss can occur from single exposures to very loud sounds, such as the levels experienced in rock concerts near the loudspeakers. Auditory hair cell damage is permanent, but auditory prostheses can be threaded into the cochlea to stimulate the cochlear fibers electrically to restore some hearing.

Oh those bells bells bells bells bells bells bells: Tinnitus

Tinnitus is the perception of ringing in the years. Chronic, severe tinnitus is one of the most psychologically debilitating diseases known, having reportedly led some patients to drug abuse and even suicide. Tinnitus has multiple causes: ear infections, allergies, reactions to certain drugs, and exposure to loud noise (many rock musicians have tinnitus).

So-called *objective tinnitus* that arises from muscle spasms in the ears can be detected by ear doctors with special instruments that detect the actual noise. *Subjective tinnitus* can arise from numerous causes but is often associated with hearing loss.

There are few universally effective treatments for tinnitus, but, in extreme cases, sometimes patients wear noise producing earphones that mask the tinnitus noise with electronically generated white noise, giving some relief. In extreme cases, some patients have opted for surgical treatments, in which the surgeon *lesions* (destroys) some of the auditory nerve for relief, but which also reduces normal hearing. Tinnitus can also sometimes appear suddenly and then disappear within a few months.

Chapter 7

Odors and Taste

In This Chapter

- Understanding the sense of smell
- Looking at the mechanisms for processing taste
- Recognizing the connection between smell and taste
- Learning how things go wrong with these senses

Let a dog outside, and you immediately observe that he lives in a world of smell. He runs around in a complex pattern, sniffing, moving, and sniffing again. After a minute, your dog knows what other creatures have been in your yard, what they were doing, and probably when they were doing it. Dogs and other mammals live in an olfactory world that is mostly invisible to humans. Dogs' noses contain on the order of one billion olfactory receptors, a number roughly equal to the total number of neurons in the rest of their brains. Humans must make due with about 10 million receptors — a hundredth of what dogs have — that project to a brain overwhelmed by other, mostly visual, information.

Whereas senses like vision and hearing are detected as energy (light photons and sound waves), smell *(olfaction)* involves detecting actual substances from the world. The other sense that involves substance detection is taste, which functions as a pleasure/pain gateway to the body, encouraging us to ingest substances that are sweet and sometimes salty, which generally are needed for nutrition, and discouraging us from eating bitter and many sour substances, which may be toxic or spoiled. Taste receptors occur in four major types, primarily on the tongue. But detecting complex tastes, referred to as *flavors,* requires simultaneously smelling what we are tasting. (If you hold your nose, for example, you'll find it difficult to distinguish substances like cherry, chocolate, or coffee based on taste alone.)

Taste, like other senses, relays through the thalamus to cortex, but olfaction is the one sense that projects directly to cortex, without going through the thalamus. The non-thalamic odor sense gives rise to behavior whose cause we do not consciously perceive, such as some aspects of sexual attraction. This chapter explains how both of these senses work.

What's That Smell?

Whether it's button, snub, Greek, or hawk, your nose has a role in both respiration and smell. It filters, warms, and humidifies the air you breathe and analyzes that air for odors.

This analysis begins with the olfactory receptors in the roof of your nose (see Figure 7-1). The olfactory receptor neurons have cilia that stick down into the mucus that lines the roof of the nose. This mucus layer is called the *olfactory mucosa,* which is Latin and more or less means "smelling mucus." The cilia have receptors that respond to different odors.

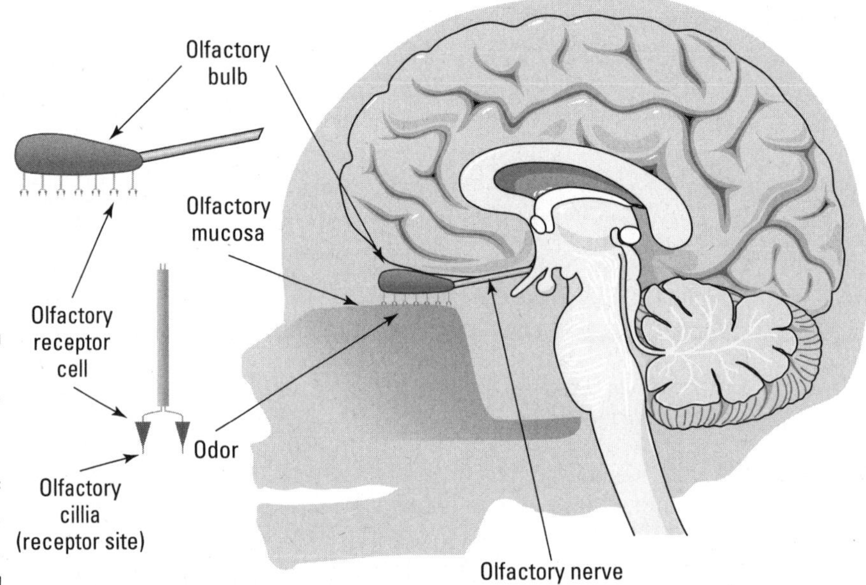

Figure 7-1: Olfactory receptors: The world as a stew of neurotransmitters.

Here's how synaptic transmission works in general: A presynaptic neuron releases a neurotransmitter that geometrically fits into a portion of a postsynaptic receptor. In metabotropic receptors, this binding causes a second messenger cascade inside the cell that results in the release of intracellular messengers that bind on the inside of other receptors and open other channels. (Refer to Chapter 3 for a detailed discussion of this process). With the sense of smell, the olfactory receptors function much like other metabotropic receptors, except that *a molecule from the world* binds to a receptor and gets the rest of the process going.

Smelling comes down to molecules from the world binding to particular types of receptors, followed by the brain analyzing the result. Each of the approximately 1,000 different olfactory receptor types responds to many different odors, and each odor binds to a somewhat different subset of these receptors. In this way, each odor has a unique *receptor activity signature.* The brain identifies the odor from this signature. In mammals that are less visually dominated than humans, much of the brain is dedicated to olfactory processing.

Sorting things out through the olfactory bulb

In humans, about 10 million olfactory receptors consisting of about 1,000 different receptor types project to the olfactory bulb where there are about 1,000 – 2,000 different recipient zones called *olfactory glomeruli* (singular, *glomerulus*). The fact that the number of glomeruli is about the same as the number of olfactory receptor types isn't just a coincidence: Most glomeruli receive inputs from mainly a single receptor type. If you were to take a snapshot of what goes on in the olfactory bulb when someone is smelling a particular odor, you'd see a pattern of activity across the glomeruli that constitutes the signature for that odor.

However, the process is a bit more complicated than receptor type A smelling roses, receptor type B smelling pie, receptor C smelling dirty socks. Real smells in the world are made of a complex mix of odors. The smell from your cup of coffee, for example, contains at least 100 different kinds of odor-producing molecules, which means that 100 different signatures are superposed across the olfactory glomeruli. In other words, your brain still needs to do a bit of processing before you can recognize what you're drinking.

Projecting along different paths

The olfactory bulb has several main projections. These projections are unique among the sensory systems in that they go directly to several areas of the cortex without relaying through the thalamus first. Figure 7-2 shows the overall olfactory projection scheme.

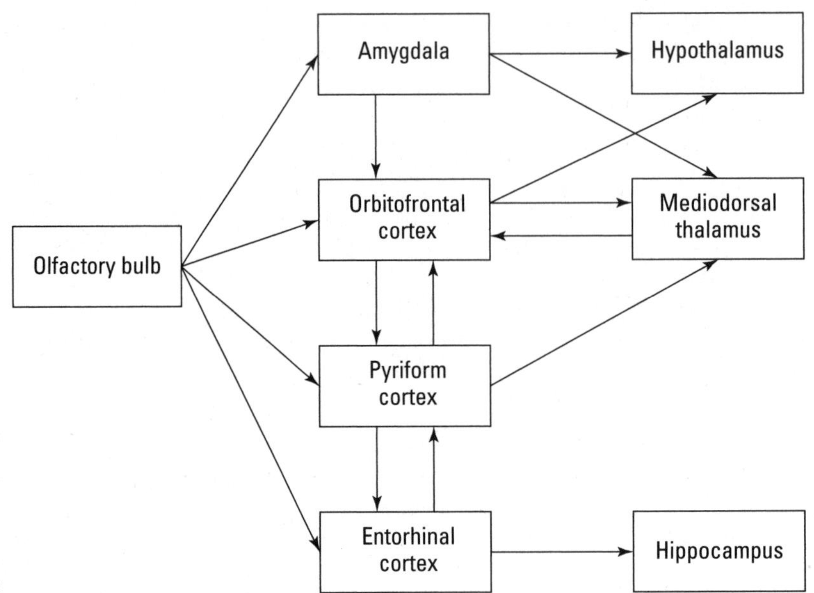

Figure 7-2: The olfactory system in the brain.

The olfactory bulb projects to these areas of the cortex:

- **The orbitofrontal cortex:** The orbitofrontal projection mediates conscious odor discrimination.

  The olfactory area of the thalamus (the *mediodorsal thalamus)* is reached by a relay projection from the orbitofrontal cortex rather than from the bulb directly. Because activity in the thalamus is normally necessary for consciousness of any sensory input, this indirect loop must have evolved to allow awareness of and attention to odors.
- **The pyriform cortex:** The projection to the pyriform cortex is an ancient pathway which makes unconscious and reactive odor discrimination possible.
- **The entorhinal cortex:** Projections to the entorhinal cortex, amygdala, and hippocampus allow you to remember odors that were part of an important experience. The projection from the amygdala to the hypothalamus mediates hormonal responses to odors.

The next sections have the details on some key pathways the projections take and what scientists think these pathways signify.

Looping between the orbitofrontal cortex and the mediodorsal thalamus

The loop between the mediodorsal thalamus and orbitofrontal cortex mediates attention and other conscious aspects of odor detection.

Many neuroscientists believe that the thalamus is one of the central controllers of brain function. It functions like a traffic cop in a town where all roads meet in the town center so that the most urgent messages can only get from one place in the brain to another at the cop's direction.

According to this idea, animals more primitive than mammals had sensory systems that were initially separate and highly specialized; visual circuits mediated visual behavior, auditory circuits mediated responses to various sounds, and so forth. Mammals, on the other hand, seem to have erected a new unified sensory system that always involved a loop between the thalamus and some area of neocortex. In this scheme, thalamic "association areas" control the traffic through various thalamic nuclei specific to each sense. This thalamic-neocortical system is akin to a new interstate that connects major hubs at high speed, bypassing the old routes and smaller towns.

The explanation for the difference between the other sensory systems and the olfactory system has to do with the fact that the olfactory system was already well-developed in early mammals. To extend the interstate highway metaphor, picture what happens when an interstate has to traverse a major city. The interstate structure is clean and regular outside the city where it's built to efficient specification, but inside the city, all kinds of exceptions and non-standard compromises have to be made. The age of the well-established olfactory system forced the brain to "hack" a thalamic pathway by projecting from olfactory cortex back to a switching area in the thalamus, which then projects back to cortex in a manner similar to the other senses.

The thalamic switching and integration mechanisms mean that when you encounter a fire, you see the flames, hear the crackling, and smell the smoke as a unified experience, which also includes remembering the taste of something burned and the pain of touching something hot.

Projecting to the pyriform cortex and the amygdala

Other significant projections from the olfactory bulb are to the pyriform cortex and the amygdala.

One aspect of smell that differentiates it from the other senses is that it so often intrinsically generates an emotional response. Although disturbing pictures and sounds generate emotional responses as well, this reaction constitutes a minority of our normal visual and auditory experience (I hope!). In contrast, a large percentage of smells evoke emotional reactions, being either immediately disgusting or displeasing (such as the smells emitted by stink bugs and skunks) or pleasing (such as the aroma of fresh peaches or chocolate chip cookies baking).

The direct projections from the olfactory bulb to the pyriform cortex and amygdala appear to be part of a neural circuit that mediates approach/withdrawal behavior for smells. Some of these behaviors are inborn (genetic). For example, from birth, sweet smells are pleasing, while sour and sulfurous smells are turned away from, as demonstrated in films of newborn facial expressions when the babies are exposed to such smells.

Other smells, like cheese, are learned. Children often turn away from the strong sour cheese smell initially but later learn to like it. The opposite occurs when spoiled food is ingested. Even a single incident can trigger lifelong aversion to that food, even when the food isn't spoiled. These latter, learned smell aversions involve projections to the amygdala and hippocampus, which are the main memory organization structures in the brain (refer to Chapter 2 for more on these brain areas).

Projecting to the entorhinal cortex and hippocampus

In all the other sensory systems besides the olfactory system, the peripheral receptors project to the thalamus and a succession of cortical areas that in turn project to memory structures such as the entorhinal cortex and hippocampus. The entorhinal cortex is a kind of "pre-sorter" front-end processor for inputs to the hippocampus, which has most of the actual associative synapses.

The olfactory system has, however, retained an older pattern in which the olfactory bulb projects directly to these structures. This pathway is concerned with memory associations of very universal smells such as something rotting, rather than sophisticated smells such as hazelnut coffee and dark chocolate, which require higher order cortical processing to establish their identity prior to being stored in memory.

Projecting straight to the amygdala

The amygdala lies just in front of the hippocampus and has a similar memory function, except that it is specialized to deal with emotionally salient memory formation. What this means is that (1) the amygdala receives inputs from structures such as the olfactory bulb because certain smells have intrinsically strong emotional aspects, such as human waste and spoiled food on the negative side and sexual odors on the positive side. It also means that (2) the amygdala communicates with areas of the frontal lobe such as the orbitofrontal cortex that are involved in the control and initiation of behaviors that are highly emotional.

Getting more specific in the orbitofrontal cortex

As mentioned previously, olfactory bulb neurons in each glomerulus tend to receive inputs from the same receptor type among the 1,000 different receptor types, but each of those receptors responds to many odors. The neurons in the orbitofrontal cortex, however, are much more specific: They respond well to far fewer odors than the olfactory receptors or glomeruli neurons do.

This odor feature specificity is similar to cortical feature specificity found in other sensory systems. For example, in the visual system, most cells in the retina and thalamus respond to just about any light increases and decreases in some area, but cortical cells respond only to edges having a particular orientation and, sometimes, only when moving in a particular direction. The visual cortex can be said to detect the coincidence of aligned, linear features. The olfactory cortex, on the other hand, detects the coincidence of co-occurring smells, such as the more than one hundred distinct smells thought to make up the aroma of any particular type of coffee.

The orbitofrontal cortex also plays a role in your perceptions of flavor and in learning:

- **Generating the perception of flavor:** The neurons in the orbitofrontal cortex pair smells with tastes to generate the perception of flavor. By combining these two senses, you can more easily distinguish between things like chocolate and coffee. Without the sense of smell, such as when your olfactory mucosa is messed up because you have a cold, food is tasteless and less palatable.
- **Learning which foods to avoid:** The orbitofrontal cortex also has a role in learning. Adults consume different foods that are not all equally palatable to children, as anyone who has tried to get their kid to eat broccoli or spinach knows. During development, humans learn through culinary experience to find many foods quite palatable that they didn't initially like. We may also learn, even from a single bad experience, to avoid certain foods that we may have once liked. The way this works is that the orbitofrontal cortex encodes a specific representation of the food that includes both its odor and taste, and that representation is paired in the amygdala with the unpleasant experience of a stomach-ache or worse. Future encounters with the offending food trigger the amygdala-orbitofrontal cortex memory representation that programs both a negative emotional response and avoidance.

The later section "Having Good Taste" explains the sense of taste in more detail.

Human pheromones: Myth or misconception?

It is well known that smell contributes importantly to sexual behavior. What is less well known is that much of the smell contribution occurs unconsciously. For example, smell induces an orienting toward the opposite sex, even though we're generally not aware that smell is playing any sort of role at all (although using perfume makes the connection between smell and attraction more obvious). It has long been known anecdotally and more recently proven by controlled research that women who live in close physical proximity tend to synchronize their menstrual cycles unconsciously through smell.

The nose contains another organ of smell near the vomer bone, the *vomeronasal organ.* This organ may have receptors for human sexual odors that are unconsciously processed in limbic system structures such as the amygdala and that influence sexual behaviors such as those previously mentioned.

The traditional biological term for an odor that influences species-specific inter-individual communication affecting behavior is a *pheromone.* There has been some (though diminishing) controversy about using this term for human odor communication. The first aspect of the controversy is biological. Until recently, many thought humans didn't have a vomeronasal organ (we do). The second reason is more philosophical or political: Should a term used to refer to mediating behavior in animals be used for conscious humans?

Regardless of the term used, odor communication occurs between humans, using odors similar to those used in other animals (which is why perfumes are often made of extracts from animal sexual scent glands), and these odors act on receptors in a similar structure, have similar central projections, and result in unconscious behavioral effects.

Having Good Taste

One reason for human's evolutionary success is that we are omnivores. Across the Earth, people can be found who eat just about anything that has nutritional value. To do this, humans must be able to discriminate between non-toxic and toxic plants, ripe versus unripe fruit, and fresh versus spoiled meat. Because you're tasting what's already in your mouth and are about to swallow, you need to accomplish this discrimination rapidly and accurately.

Taste, like the other senses, starts with a set of receptors. These receptors reside on the tongue in structures called *papillae* (see Figure 7-3). The papillae appear as little bumps on the tongue that you can actually see (and probably spent some time looking at when you were younger!).

Papillae come in four types: filiform, fungiform, foliate, and circumvallate. With the exception of the filiform, all the other papillae (collectively called your taste buds) contain the taste receptor cells; you can find these papillae at the tip, sides, and back of the tongue. Taste occurs in these areas. The central portion of the tongue has almost exclusively filiform papillae; taste does not occur there (the filiform papillae have a mechanical function in some species that use rough tongues to rasp ingested food).

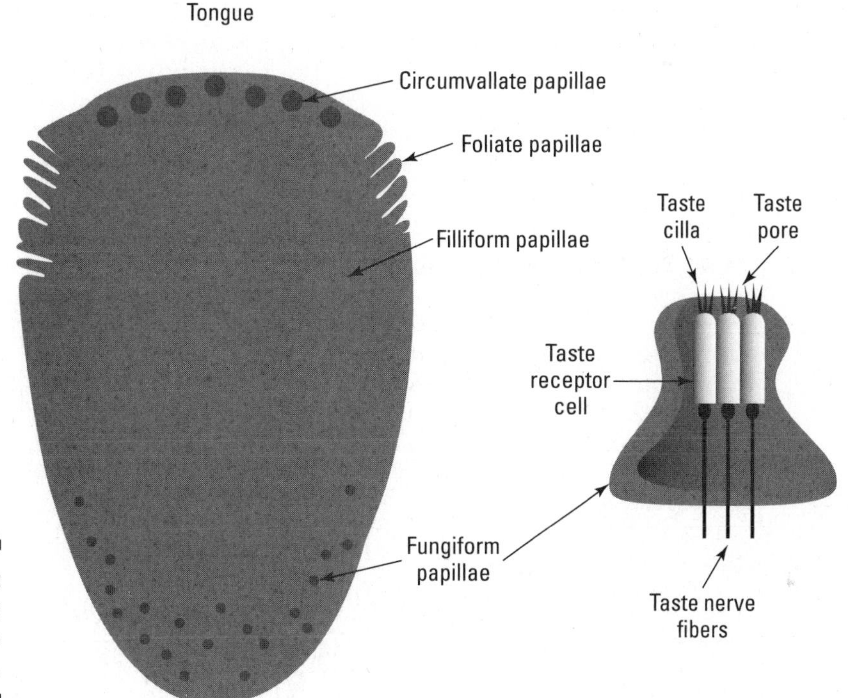

Figure 7-3: The tongue and its taste receptors.

The discriminating tongue: The four basic tastes

All the taste buds have about 10,000 taste cells, and most taste buds have receptors for several tastes. The "basic" taste sensation comprises these basic tastes: sweet, salt, bitter, and sour, plus a fifth basic taste, *umami,* which has recently been added by researchers who have located a specific receptor type for it. This receptor responds particularly well to the amino acid L-glutamate. (The word *umami* comes from the Japanese term meaning "savory and pleasant taste.") The experience of the umami is that of a meaty, MSG-like taste. There are receptors for umami, like the other four "classical" tastes, throughout the tongue, and represented in every one of the three taste bud morphologies (fungiform, foliate and circumvallate). There may also be CO_2 receptors that are activated by carbonated liquids.

The widespread idea of strong regional specialization of the tongue for specific tastes — that is, that you taste sweet things on one area of the tongue, sour things on another, salty things on yet another, and so on — came from a mistranslation of an article by the famous German psychophysicist Edwin G. Boring at the turn of the 20th century. In reality, as I mention earlier, most taste buds have receptors for several tastes. Nevertheless, on a more subtle scale, some evidence remains for *relatively* enhanced taste for sweet at the front of the tongue and bitter at the back, with considerable individual variation even in this tendency.

Sweet

The sweet taste is mediated by receptors that respond to sugars, the source of energy for metabolism. In human evolutionary history, obtaining enough energy was almost certainly the major food intake demand, and humans are genetically programmed to like the taste of sucrose and other sugars. In fact, many human taste receptors respond almost exclusively to sugar. Unfortunately, while eating as much of anything sweet as could be found probably suited most of our distant ancestors, plentiful supplies of sweet foods in modern times have forced most of us to use discipline and diets or risk becoming obese.

Salt

Another resource limited in the natural world is salt (NaCl), which constitutes the major part of the fluid throughout our bodies. As with sweet receptors, a large number of taste receptors respond almost exclusively to salt. However, other salts, such as potassium chloride (KCl) also activate these receptors. As with sweets, human's innate desire for salt combined with its ready availability today can easily result in excessive consumption, sometimes leading to hypertension and other health problems.

Sour

The sour taste is really an ability to detect acidity, so sour receptors react to H+ ions. Because sourness can be a characteristic of foods that are quite palatable (think sauerkraut and lemons) as well as a characteristic that indicates spoilage (think spoiled milk), the reaction to the sour taste is quite complex and depends on learning.

Bitter

Bitterness, like sourness, is sometimes but not always a sign of food toxicity or being unripe. The substance that best exemplifies the bitter taste is quinine. We learn through experience that some foods with a bitter taste should be avoided, while other foods may be deliberately spiced with a small amount of bitterness and are okay to eat. Some aspects of the spicy taste may be mediated by the slight burning response conveyed by activation of the trigeminal nerve.

Sending the taste message to the brain: Taste coding

As I explain earlier in this chapter, there were about 1,000 different receptor types for smell, and the olfactory receptors project to receptor type-specific regions in the olfactory bulb. The situation with taste is simpler in the sense that there are only five basic tastes. On the other hand, the projection to the brain by different taste receptor types doesn't result in an areal specialization by receptor type.

Projecting to the brain stem via the chorda tympani and the glossopharyngeal nerve

As described earlier, each taste bud has multiple (five to ten) taste receptor cells, with most taste buds having at least one of each of the five basic receptor types. The taste receptor cells project to an area of the brainstem called the *nucleus of the solitary tract* (see Figure 7-4) via two different nerves:

- **The chorda tympani** carries the taste message from most of the front of the tongue (mostly fungiform papillae).
- **The glossopharyngeal nerve** carries messages from foliate and circumvallate papillae, and a few rearward fungiform papillae.

No clear difference exists in the functions of these two pathways. Taste receptors probably evolved from more general somatosensory receptors that happened to project by different nerves from the front versus the rear of the tongue.

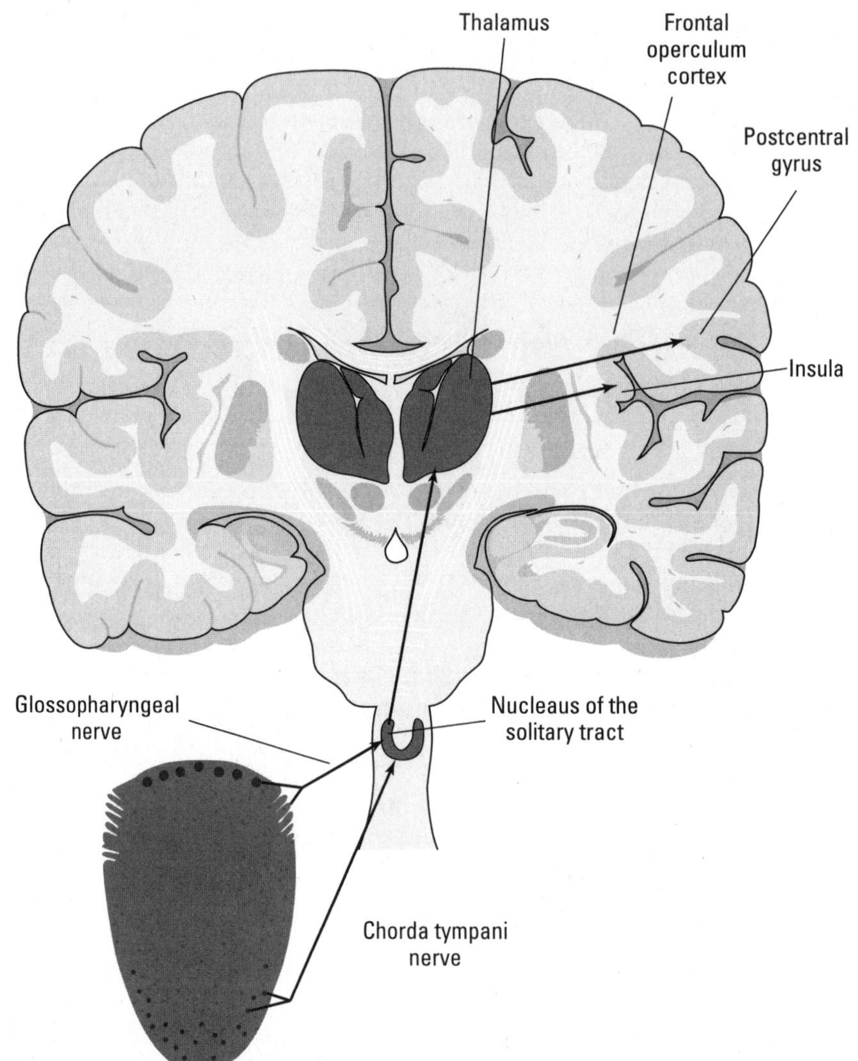

Figure 7-4: Central projections in the taste pathway.

Distributed versus labeled line coding

Each axon of the chorda tympani nerve receives inputs from several taste receptor cells. Nerve fibers that are very selective, such as some salt fibers, receive inputs almost exclusively from salt receptors. Other chorda tympani fibers receive a mix of inputs from different receptor cell types so that these fibers represent other tastes by a pattern of firing across fibers.

In other words, the messages conveyed by each chorda tympani fibers can be regarded as either *labeled lines* (they respond only to salt, for example) or *distributed* (they respond to multiple tastes spread across many fibers):

- **Distributed coding:** A large percentage of taste receptors respond to some degree to sweet, salt, sour, and bitter tastes, leaving it to the higher brain centers to identify taste from the ratio of receptor activation distributed across a number of receptors.
- **Labeled line coding:** Some receptors for salt and, to some extent, sweet appear to be highly specific, so our brains can consider activation of these labeled line receptors as due to sweet or salt.

The idea of a fiber-firing code revolves around what a cell in the brain can "know" about the taste from firing in any particular fiber. Some of the salt fibers, for example, are exclusive enough in their response to salt that you can interpret the action potentials firing rate as a labeled line message of the amount of NaCl (or similar salts such as KCl) present. But in a distributed code, the chorda tympany fibers respond broadly, and you can't interpret the firing as any particular taste without comparing the firing in one fiber to the firing that occurs in other fibers. The neurons that these fibers ultimately project to in the cortex perform this comparison and allow both subtle and complex taste identifications.

Other sensory systems use both of labeled line and distributed coding strategies. What's unique in taste is that the combination exists in the same nerve.

Identifying and remembering tastes

Taste cell signals go to the brain by several different routes (refer to Figure 7-4).

- Receptors in the front and sides of the tongue project via the chorda tympani nerve
- Those in the back of the tongue (mostly circumvallate and some foliate receptors) project via the glossopharyngeal nerve.
- A few taste receptors in the mouth and larynx travel via the vagus nerve. (***Note:*** There are so few of these taste receptors compared to those in the tongue that I don't discuss them in any detail.)

All three of these nerves project to the *nucleus of the solitary tract* (NST) located in the medulla. Taste neurons in the NST receive other inputs that modulate their firing, associated with satiety. For example, eating sweet substances will, after some time, reduce the responses of sweet-responding

NST neurons, which in turn will reduce how much pleasure you get out of continuing to eat sweet things. (In short, it's one of the reasons why you can eat only so many chocolate bars before your body says, "Enough!")

The NST projects to the taste portion of the thalamus, the *ventroposteromedial thalamus (VPM).* The reciprocal connections between this thalamic area and the cortical processing areas for taste mediate conscious awareness of the taste sensation. Lesions in the VPM can cause *ageusia,* the loss of the sense of taste, for example.

The VPM thalamus projects to two areas of the neocortex: the *insula* and *frontal operculum cortex* (refer to Figure 7-4). These are both considered "primary" primary cortical taste areas. The insular lobe is a distinct, small cortical lobe at the junction of the parietal, frontal, and temporal lobes. The opercular cortex (*operculum* means "lid") lies just above the insula. These two primary gustatory areas are evolutionarily old and embedded in a complex of cortical circuits that are also affected by olfaction and vision, and which appear to be involved in satiety, pain, and some homeostasis functions.

The insula and operculular cortex project to the orbitofrontal cortex where they are also combined with olfactory signals to determine *flavor,* the complex "taste" derived from a combination of taste and smell. They also project to the amygdala, which mediates memory for emotionally salient experiences. The amygdala projects back to the orbitofrontal cortex.

The cortical circuitry, in concert with the amygdala, allows you to associate the sight, smell, and taste of very specific foods with your experience before and after ingesting it. Your food experiences then program your food preferences and various triggers for hunger at the sight and smell of food.

The Role of Learning and Memory in Taste and Smell

So how is it that you learn to like that spicy chili that made you almost gag when you had the first spoonful? It's unlikely that the human species evolved over any long period of time to have a specific craving for Chef Pancho's Five Alarm Chili. Instead, although you may have found the taste overwhelming initially, after a few bites, your receptors adapted (or died!), and you found the flavor (combined taste and smell experience) to be quite enjoyable.

The cortex is adept at modifying neural activity from experience to enhance detection and discrimination. If scientists recorded neurons in your orbitofrontal cortex as you ate this chili (something that's been done in monkeys — with something other than chili, of course), they would see neurons whose firing increased to the paired associations between the specific taste and smell of Pancho's Five Alarm. Whereas a chili novice may not be able to differentiate Pancho's chili from another cook's chili, a regular Pancho's customer could probably tell the difference immediately.

Eating is not just a matter of smelling and tasting. Food texture, which is perceived by mechanoreceptors while chewing, is an important component of the enjoyment of eating. This is one reason why people who congenitally lack both taste and smell still enjoy eating, and why attention to and manipulation of texture is part of the fare of all really high-quality restaurants (refer to Chapter 4 for information on mechanoreception).

Lacking Taste and Smelling Badly

Often we tend to take the senses of taste and smell for granted. As mentioned earlier, when we have a head cold with congestion, our sense of smell may be severely compromised and with it our sense of flavor (the combination of taste and smell). However, our lack of appetite when we are sick would not be a good permanent situation because we need to eat regularly to stay alive. In a sense, we are all "addicted" to food, which is generally a good and necessary thing (when we're able to control ourselves at the buffet bar).

Smelling poorly or not at all

People vary in their ability to detect and discriminate odors. Women tend to have a better sense of smell than men. People with schizophrenia often have some deficiencies in the sense of smell (researchers think this is linked to generalized neural problems in the frontal lobe). Some people, however, have no sense of smell, called *anosmia.* Some are born without a sense of smell, and others may lose some or all smell sense from brain damage or damage to the olfactory mucosa. Continuous exposure to noxious substances that damage the tissue, for example, can have this effect.

There are also people who congenitally lack a sense of taste *(ageusia),* although total lack of the taste sense is very rare. More common is a very poor sense of taste *(hypogeusia),* which may also be associated with or even caused by damage to the olfactory mucosa (in which case anosmia leads to hypogeusia due to the involvement of smell with taste). Hypogeusia is also rare as a congenital condition but may occur as a result of nerve damage (such as to the chorda tympani nerve) from disease or injury. Temporary hypogeusia often accompanies some cancer chemotherapies. Aging, in general, tends to reduce sensitivity to bitter tastes. This gives children some leg to stand on in their complaints about some vegetables being strongly bitter to them when they are not so to their parents.

Transplanting olfactory neurons

Normally, olfactory receptors die, and new ones are regenerated in a continuous cycle, even in the adult nervous system. This is unusual when compared to the other sensory systems. In the sense of sight, for example, dead neurons in the retina and cochlea are not generally replaced, and the losses are permanent. Some research has been directed at transplanting olfactory mucosa neurons (mostly the olfactory ensheathing cells, a kind of stem cell in the olfactory system) into damaged areas of the brain or spinal cord in the hope that the olfactory cells will not only regenerate there but also convert themselves into spine or brain neuronal types and integrate functionally into damaged neural circuits. A patent for this transplantation treatment was filed by Rhawn Joseph, Ph.D., of the Brain Research Laboratory in San Jose, California. Human trials have been conducted in Portugal and China, but most efforts in the United States have remained in the basic research phase.

Satiety

The sense of taste is modulated by satiety mechanisms. Scientists know of at least two quite distinct satiety mechanisms. One is a central brain mechanism, and the other occurs in the taste receptors themselves.

In the central mechanism, as you eat and grow full, the item you're eating loses its desirability. This phenomenon is called *alliesthesia* ("changed taste"), and it's a brain mechanism that indicates you're getting full. The mechanism for this can be seen in the reduced firing of orbitofrontal taste/smell neurons to the specific odor of a substance after a substantial amount of it has been consumed.

What's called *sensory-specific satiety* occurs in taste and olfactory receptors themselves. It tends to suppress your appetite specifically for the taste of what you are consuming, acting on a faster time scale than alliesthesia.

Several gastrointestinal hormones regulate appetite via their actions on the brain. Gut-derived hormones, which include ghrelin, insulin, pancreatic polypeptide (PP), and others, stimulate specific areas of the hypothalamus and brainstem. They also modify the sensations that are transmitted by the vagus nerve to the nucleus tractus solitarius (NTS). The hypothalamus receives neural input from brainstem and has receptors for circulating hormones. The hypothalamus and brainstem control appetite via interactions with higher centers such as the amygdala and cortex.

Part III
Moving Right Along: Motor Systems

"Oops."

In this part . . .

Most of the nervous system ultimately exists to allow movement (otherwise we'd be plants), and movement requires muscles, the control of which is the subject of this part. Here, you discover how motor neurons contract muscles and how various coordination centers in the brain and spinal cord organize muscle contraction so that someone like Michael Jordan can, from 30 feet away, take a basketball and throw it through a hoop only slightly larger than the ball itself.

Just as movement itself separates animals from plants, the ability to contemplate and plan complex movements and anticipate its results separates humans from other animals. Not only can we can make tools, but we can think about how to make the tool and solve a problem with it many days later.

In this part, I move from low-level movement control that is not much different from the kind of movement control other animals have to very high-level planning functions that are made possible by our giant frontal lobes.

Chapter 8

Movement Basics

In This Chapter

- Categorizing types of movement
- Looking at how your brain, spinal cord, and muscles work together to control movement
- Examining motor system disorders

Because survival requires finding food, shelter, and a mate and at the same time avoiding predators and dangers, ultimately, the function of the nervous system is to control movement. Neural computation, even abstract thinking, is about planning how to pursue something you need and avoid things that are dangerous or unpleasant.

Nervous systems almost certainly arose in multicellular marine organisms to coordinate the beating of flagella in cells dispersed around the organism's exterior. Translation: You not only need to have all your oars in the water, but they need to move in unison. In advanced animals, movement is accomplished by having muscles that contract, motor neurons that can command the muscles to contract, and nervous systems that coordinate the sequence and amount of contraction.

Muscles and muscle cells are the effectors of the motor system. Muscle cells are similar to neurons in that they have specialized receptors and produce action potentials. The neurons that directly control muscles are called *motor neurons.* These emanate from the spinal cord to control the limbs or from cranial nerves to control muscles in the head and neck. Motor neuron output is coordinated by neural interactions within and between spinal cord segments, and, at the highest level, by the brain. This chapter introduces you to these interactions.

Identifying Types of Movement

Different kinds of movement are controlled by different kinds of neural circuits:

- **Movements that regulate internal body functions,** such as stomach and intestinal contractions. In general, you are not aware of these movements, nor can you voluntarily control them.
- **Reflexive movements,** such as withdrawal from painful stimuli, are generated by local circuits that act faster than your conscious awareness. However, you do become aware of the reflex, and you can suppress it, such as when you hang on to the handle of a hot cup of coffee for a few seconds rather than dropping the cup on the floor and making a mess.
- **Fully conscious, voluntary movements,** such as deciding to get up out of your chair, are the result of "free will," in the sense that the decisions to make such movements are the result of activity that is the product of high-level brain activity.

The following sections explain these types of movements in more detail.

Movements that regulate internal body functions

The autonomic nervous system (refer to Chapter 2) generally controls movements that regulate body functions. Most of these movements, like the contractions that move food and waste through your digestive system or the beating of cilia that remove inhaled debris from your lungs, never reach consciousness and cannot be consciously controlled. As much as you may try and even though you may be intellectually aware that they're working, you can't, for example, physically sense your cilia moving; nor can you get them to beat more slowly or more quickly.

Other functions, such as your heart rate, are functions that you're not generally conscious of. With a little effort, however, you can become conscious of your heart rate, but you can't control it directly. Respiration is another generally uncontrolled movement. You do not generally control respiration and you're usually not aware of it, but you can hold your breath to go underwater, and you control your breath quite precisely in order to speak.

Reflexive movements

Reflexive movements, as the name suggests, generally result as a response to some sort of sensory stimulation. The classic reflex involves jerking away from an input that is painful. Other simple reflexes include the gag reflex, which prevents you from swallowing an inappropriately large object, and the breath-holding reflex, which occurs when your face goes under water.

However, some reflexes are more complicated. For example, unless you're inebriated, you don't have to expend much conscious effort to maintain your balance while standing (even on one leg) or walking in a straight line. Instead, your sense of balance comes from the following:

- **Sensory input from your stretch receptors:** When you walk, your leg muscles are mostly being controlled by spinal cord circuits whose job it is to contract the right muscles the right amount to keep you upright or moving in a straight line. These circuits receive sensory input from stretch receptors in your muscles, joints and tendons about loads, forces, and movement. There are two kinds of sensory feedback signals: proprioception and kinesthesis.
  - *Proprioception* means position sense, while *kinesthesis* means movement sense. Both are mediated by *stretch receptors.* These receptors resemble mechanoreceptors such as Ruffini endings (refer to Chapter 4).
  - The stretch receptors in muscles are called *muscle spindle receptors,* and they signal muscle length (proprioception) and the rate of change of muscle length (kinesthesis). Receptors in the tendons (which connect muscles to bones) that report overall muscle force are called *Golgi tendon organs*, while receptors in connective tissue such as joint capsules indicate the amount of flexion of the joint.
- **Motion information from your visual system and from your vestibular system:** The visual system information is produced by neurons that respond selectively to particular directions of motion. Your vestibular system is based on the three semicircular canals near the cochlea, each of which responds to a particular direction of head rotation which gives you your sense of balance. (When you drink too much alcohol, you disrupt this sensory input about your state of balance.) Refer to Chapters 5 and 6 for more information on your visual and auditory systems.

When you walk, your brain takes advantage of all these balance reflexes. To walk, you transfer your weight mostly to one leg and lean so that you begin to fall forward. Your reflex system causes your other leg to extend and catch your weight so that you don't fall. You then transfer your weight to that leg and "fall" again. For more information on how your body coordinates this complex movement, head to the later section "Stepping up the hierarchy: Locomotion."

Planned and coordinated movements

Humans are capable of making more varied and more complex sequences of movements than any other animal, especially when you consider the fact that most of the complex movement sequences humans execute are ones they have learned.

So why do humans have an enhanced ability for complex movement sequences? Comparing the brains of humans to other animals provides some answers. Almost everyone is aware that humans have big brains, particularly in relation to body size. But brain expansion hasn't been uniform. If you compare primate (apes, monkeys, and humans) brains to the brains of other mammals, you generally see that not only is the primate brain-to-body-size ratio generally higher than other mammals, but the ratio of the frontal lobe to the rest of the brain is also higher. If you compare human brains to other, non-human primate brains, you see the same thing.

What might having an "extra" amount of frontal lobe accomplish? More frontal lobe area may mean more levels in the movement representation hierarchy. Additional levels of abstraction provide a qualitative, and not just a quantitative, advantage, as I discuss in the following sections.

Some researchers have argued that language itself, the ultimate separator of humans from animals, evolved literally hand in hand with manual dexterity. This argument suggests that the left side of the brain skill that enables learning the complexity of language production also makes most of us right handed.

Controlling Movement: Central Planning and Hierarchical Execution

Muscle contractions that produce movement result from many different requirements, as I discuss in earlier sections. The types of movement range from automatic movements that regulate body functions to reflexive movements to planned, conscious executions of complex tasks. In the following sections, I explain the control systems and neural mechanisms underlying each of these types of movement.

Activating non-voluntary muscle movements

Movement functions that are not voluntarily controlled (such as those that regulate body functions) are typically accomplished with smooth, rather than striated muscle. Smooth muscle in the walls of blood vessels, for example, allows the vessels to contract and dilate to assist the heart in moving blood. The ciliary muscle of the eye dilates the pupil. The urinary tract and uterus also have smooth muscles.

In some smooth muscles, the individual muscle cells have gap junctions between them so that the entire smooth muscle unit contracts in a coordinated, simultaneous manner. *Gap junctions* are specialized connections between cells that allow selected molecules to move directly between them.

Gap junctions in muscle cells cause the electrical depolarization in one muscle cell (which is what causes it to contract) to spread directly to neighboring muscle cells, causing them to contract in unison. This allows inputs from a few motor neurons or hormones to cause a large, synchronized contraction in the entire muscle. Smooth muscles that lack or have fewer such gap junctions have more innervating motor neuron terminals that control the contraction of single cells, producing finer, more differentiated contraction patterns. Typical examples of gap junction smooth muscles are those that line the blood vessels, while the smooth muscles that control the iris of the eye (to constrict or dilate the pupil) have fewer gap junctions and finer motor neuron control.

Some smooth muscle cells contract spontaneously because the cells themselves produce their own action potentials through intracellular control of their membrane ion channels. This type of mechanism is usually found where there are rhythmic muscle contractions, such as in the intestines for movement that moves food and waste through your digestive system. You can read more about these systems in Chapter 2.

Activating the withdrawal reflex

The simplest unit of behavior that includes both a sensory and motor component is the withdrawal reflex. You touch something that produces an unpleasant or unexpected sensation, and you flinch away even before your brain engages enough to say "What the. . . ?!" Figure 8-1 shows how the withdrawal reflex works when you prick your finger on a sharp tack (refer to Chapter 2 for a description of another reflex, the stretch reflex).

The withdrawal reflex begins with your finger encountering the sharp point of the tack, setting off this sequence of events:

1. **The point penetrates the skin and activates free nerve-ending pain receptors.**

   Refer to Chapter 4 for more on pain receptors.

2. **The pain receptors send their messages via action potentials along their axons up the arm to the shoulder and then to the dorsal root ganglion that receives touch information from the finger.**

   The dorsal root ganglion that receives touch info from the finger happens to be the second thoracic: T2.

3. **The pain receptor axon releases the neurotransmitter glutamate in the spinal gray area onto spinal cord interneurons.**

   Interneurons are neurons whose inputs and outputs are entirely within one spinal cord segment.

4. **The spinal interneurons contact motor neurons for both the biceps and triceps.**

   Almost all voluntary motion is controlled by so-called *flexor-extensor pairs,* which are muscle pairs like the biceps and triceps that act in opposition. By increasing the firing of its motor neurons, the reflex increases activation of the biceps flexor muscle while, at the same time, decreasing the firing of the extensor triceps motor neurons, allowing that muscle to relax. These coordinated events cause the arm to move.

5. **The motor neuron axons leave the ventral portion of the spinal cord and enter and travel in the same nerve as the sensory input after the dorsal root ganglion.**

   The distal side of the dorsal root ganglion is a mixed nerve that contains both sensory and motor nerve axons. The sensory nerves that go toward the spinal cord are called *afferents* (the name for nerves going toward the central nervous system). The motor nerves are carrying signals away from the spinal cord, toward the muscles, and are called *efferents*.

6. **The end result: The triceps relaxes while the biceps flexes, causing the arm to pivot away from the tack.**

   In most cases, the interneurons in the effected thoracic segment also communicate the pain signal to other spinal segments, which may cause a coordinated withdrawal involving not only your arm but the affected finger and your shoulder, as well. If you were really startled (if you thought the prick was a spider bite, for example), you might actually jump away, requiring flexor-extensor activation of muscles in both your legs. You could also, of course, with sufficient willpower, block the reflex and force your finger onto a sharp tack.

The withdrawal reflex is independent of the brain and acts before the sensory signal is relayed up to the brain by other interneurons. In other words, your arm moves away from the tack before you feel and say, "Ouch!"

Stepping up the hierarchy: Locomotion

While the response to fingertip prick is an example of coordinated spinal cord-mediated activity that occurs mostly within a single spinal cord segment, locomotion is an example of coordinated, four-limbed activity mostly controlled within the spinal cord but distributed across multiple spinal cord segments.

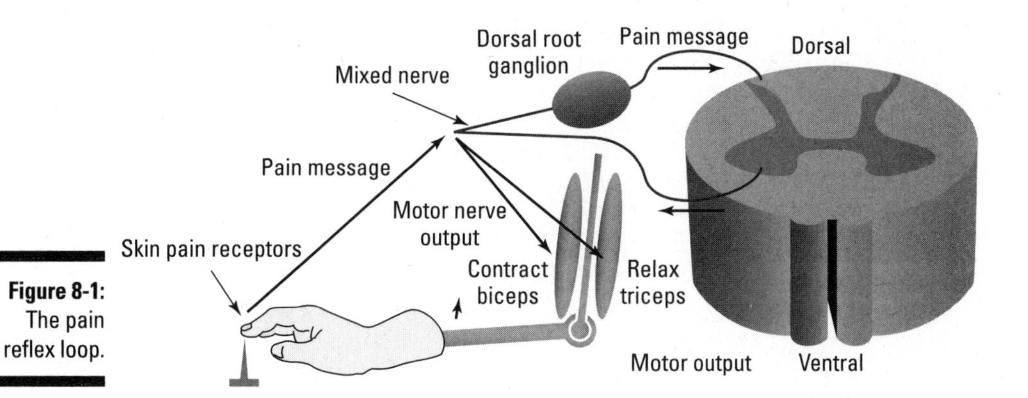

Figure 8-1: The pain reflex loop.

At the end of the 19th century, Sir Charles Sherrington from Cambridge performed treadmill walking experiments on cats whose brains were disconnected from their spinal cords. These experiments showed that sensory feedback from the moving treadmill was sufficient to induce competent walking and that even the gait the cats used switched as a function of the treadmill speed. In other words, the spinal cord is a highly competent peripheral computer or processor for locomotion. Neural circuitry within the spinal cord can produce a set of coordinated oscillations called the *central pattern generator* for alternating control of the four limbs during gait.

In short, you have to decide *to* walk, not *how to* walk. Locomotion is a process involving a multi-level, hierarchical control system in which consciousness controls the overall process but not the minutiae of the steps. There are multiple advantages to hierarchical motor control:

- **Efficiency:** With the lower levels taking care of most of the details, the brain doesn't have to compute and control every single muscle contraction to allow you to move.
- **Speed:** Neural transmission from your joint receptors all the way to your brain and back to your muscles takes time. You can move more rapidly if a shorter, local spinal circuit takes care of most of the movement details, and your brain monitors the overall progress, like, are you about to walk into a tree?
- **Flexibility:** A brain-preprogrammed muscle contraction sequence for walking might be fine, even if slow, on level ground, but what happens when you step into a hole or trip on a root? Your local spinal reflex neural circuits learned to deal rapidly with such contingencies within a few years of learning how to walk.

If the spinal cord central pattern generator can produce coordinated, alternating limb walking, what is the brain needed for? The simple answer is for decision making. We normally don't live on treadmills; therefore, it's the brain that must decide whether to walk (or not), what type of gait to use (walk or run), and where to go. Moreover, we can do things like dance the foxtrot and play hopscotch that involve complex sequences of different steps for which no spinal cord program exists.

The flexibility to hop, skip, and jump, mixed in any order, requires more than the spinal cord is capable of. Moreover, in some animals, like humans and raccoons, the forelimbs are capable of very complex manipulations quite different from what is involved in locomotion. These complex manipulations require observation, thought, error-feedback, and learning. In other words, they require a real brain.

Using your brain for complex motor behavior

Most of the time, large commercial airliners fly on autopilot. A computer gets sensor information about the state of the plane, such as its altitude and course heading, and compares these to the desired altitude and course heading. The autopilot then manipulates the airplane's control surfaces (ailerons, elevators, and rudder) to bring the plane toward the desired state despite wind and other variables that tend to take it off course.

Relying on autopilot works fine when you're several hundred feet off the desired altitude or when taking minutes to execute a correction isn't a problem. Autopilots are much more rarely used during take-off and landing, because split-second decisions have to be made about factors such as crosswinds and other air traffic. During landing, a 30-foot mistake twenty feet off the ground isn't going to be an acceptable margin of error. With the autopilot turned off, the pilot has direct control of the ailerons, elevators, and rudder, and uses the computer between her ears and a completely different set of algorithms than the autopilot to control the aircraft.

Similarly, the brain can take direct control of the muscles from the spinal cord. When it does so, the actions that follow are much more complex and involve planning, knowledge, and the ability to adapt to changing circumstances. At this point, you're beyond the realm of simple reflexive behavior. For details on conscious or goal-generated action, head to Chapter 10.

Pulling the Load: Muscle Cells and Their Action Potentials

Up to this point, I've provided a lot of information about how the brain and spinal cord organize the control of muscles, but I haven't said much about the muscles themselves — a topic I tackle now. Muscles are groups of muscle cells. Muscles contract because the muscle cells contract along their length parallel to the long axis of the muscle.

Your body has two main types of muscle: smooth and striated. I discuss smooth muscles, which are controlled by the autonomic nervous system, in detail in the earlier section "Activating non-voluntary muscle movements." The voluntary muscles are called *striated muscles*. The nervous system controls these muscles.

Muscle cells are excitable cells, something like neurons, meaning that they can produce action potentials. In voluntary muscles, the action potentials are produced when presynaptic motor neuron terminals release the neurotransmitter acetylcholine, which binds to a postsynaptic receptor on the muscle cell. The region of the muscle fiber where motor neuron axon terminals synapse is called the *end-plate*. Virtually all striated muscle neurotransmission in vertebrates is mediated by acetylcholine. The following steps outline how this process works to control movement:

1. **After the motor neuron presynaptic terminal releases acetylcholine, it binds to the postsynaptic receptor on the muscle cell.**

   Acetylcholine binds a receptor called a *nicotinic acetylcholine receptor* on muscle cells. The nicotinic acetylcholine receptor is an excitatory ionotropic receptor (refer to Chapter 3).

   This name doesn't mean the receptor is in the habit of smoking cigarettes. Instead, it means that nicotine is a particularly effective *agonist* for this receptor. That is, nicotine, a substance not normally released onto the muscle receptor, is very effective in producing the same effects as the normal neurotransmitter (ligand), acetylcholine.

   There is typically one motor neuron axon terminal per muscle cell, although a single motor neuron axon can branch and activate many muscle cells.

2. **Upon binding acetylcholine, a channel that is selective for sodium and potassium opens within the receptor protein complex.**

   This produces what's called the *end-plate potential,* which is similar to the post-synaptic potential produced by excitatory connections between neurons.

   Acetylcholine does not stay permanently bound to the muscle cell receptor. If it did, the muscle couldn't relax or contract again. (Some toxins do bind permanently to the receptor, causing either permanent activation or inactivation, both of which are very bad.) The typical binding time of acetylcholine for the receptor is technically referred to by the term *affinity*. After the acetylcholine molecule comes off the receptor, degradation molecules (called *cholinesterases*) in the synaptic cleft destroy the acetylcholine molecule so that it can't bind again.

3. **The depolarizing end-plate potential opens voltage-gated sodium channels in the muscle cell, causing a muscle action potential.**

This action potential has a similar ionic basis as that of the neural action potential, except that it is much longer lasting, on the order of 5-10 milliseconds.

4. **The action potential activates L-type voltage-dependent calcium channels in the muscle cell membrane, causing a structure within the muscle cell to release calcium.**

   This structure is called the sarcoplasmic reticulum.

5. **This increase in internal calcium within the muscle cell causes the actin and myosin myofilaments to slide over each other (refer to Figure 8-2) to shorten their overlap, causing the cell to contract along its length.**

   The process of muscle contraction uses ATP (adenosine triphosphate) as an energy source. *ATP* is the universal energy "currency" within cells for conducting metabolic activities that require energy.

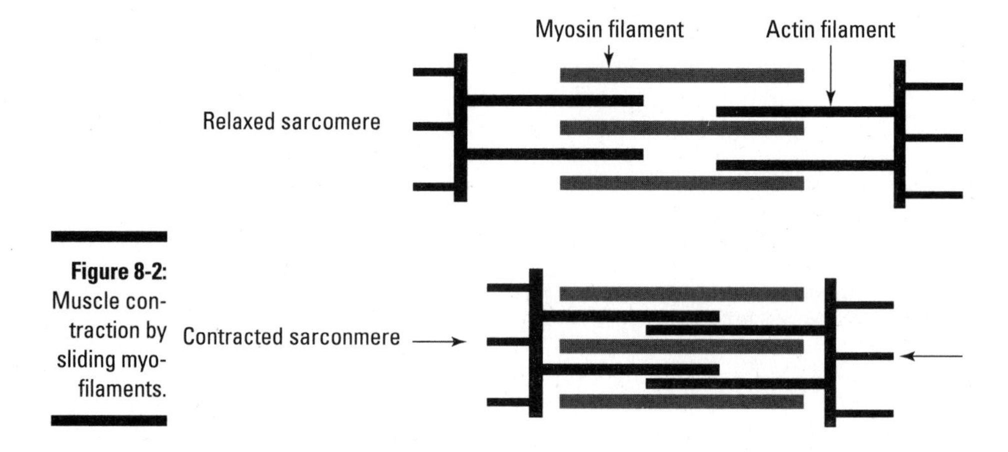

Figure 8-2: Muscle contraction by sliding myofilaments.

Muscle cells are extended, so they are called *muscle fibers,* and a muscle proper is composed of many fibers. The number of contracted muscle cells along a fiber determines its shortening, which is in turn controlled by the number of muscle cells activated (number of active synapses) and their rate of activation (motor neuron firing rate). The force of the contraction is also a function of the number of parallel muscle fibers. Therefore, the number of motor neurons activated and their rate of activation can very precisely control muscle force to produce finely graded contraction forces.

Muscle and Muscle Motor Neuron Disorders

Motor disorders can occur due to damage anywhere in the complex control system. There are also diseases that are specific to motor neurons or the neuromuscular junction.

Myasthenia gravis

Myasthenia gravis is a disease in which the muscle receptors for acetylcholine don't function properly. This disease is characterized by muscle weakness, fatigue, and finally, paralysis. There appear to be congenital and autoimmune forms of myasthenia gravis, with the autoimmune forms being the most common.

Although no cure exists for this disease, anti-cholinesterase agents that inhibit cholinesterase's ability to destroy acetylcholine can strengthen contractions by enhancing the acetylcholine binding to whatever receptors are still viable. These treatments can temporarily mitigate some symptoms of the disease, but the disease is progressive, and eventually anti-cholinesterases don't help very much.

Motor neuron viral diseases: Rabies and polio

When the axon terminals of motor neurons recycle their membrane to make synaptic vesicles, they're vulnerable to viruses which they take up in the recycling process. Two notable examples of viruses that pose a danger are polio and rabies. These viruses enter motor neuron axons terminals and then travel to the central nervous system. Rabies viruses attack other neurons, and the disease is typically fatal unless the infected person (or animal) receives a vaccine that can stimulate the production of antibodies.

Polio tends to cause destruction of the motor neurons themselves, typically resulting in paralysis. Depending on the amount of destruction, rehabilitation may be able to mediate some of the paralysis. One result of motor rehab is that the smaller number of remaining motor neurons sprout more axon terminals and can therefore innervate more muscle cells for enhanced contraction force. A downside of this, however, is that the random death of motor neurons that most elderly tolerate reasonably well are particularly disabling for polio victims because a much small total number of motor neurons are doing all the muscle contraction work.

Spinal cord injury

Damage to any motor control areas in the central nervous system can produce paralysis, *paresis* (weakness), or *apraxia* (lack of motor skill or dexterity). In the brain, such disabilities typically occur from strokes or head injuries. In the spine, such disabilities are commonly the result of falls or other injuries. In cases of spinal cord transection, the result is total paralysis to the body areas innervated below the damaged spinal segment(s).

A number of approaches to dealing with severe spinal cord injuries have been tried. Here are a couple of particularly interesting ones:

- **Regrowing axons:** Many cell biological approaches have operated on the premise that spinal transections typically cut axons, and function could be recovered if the axons that are still alive could be induced to regrow across the transected area to their original motor neuron output targets.

  A particularly striking mystery in this research is that transected peripheral nerves typically *do* grow back after such transections, but central neurons don't, at least not in mammals. In cold-blooded vertebrates such as fish and salamanders, even central neurons regrow axons after transection. Many labs are currently working on this problem and some progress is likely to occur within a very few years of this writing.

- **Stimulating movement mechanically**: This approach to the problem is engineering oriented. One scheme involves implanting arrays of recording electrodes in primary motor cortex to "intercept" the commands to move the muscles, relaying these signals through wires past the spinal cord transection and electronically driving either the muscle itself by direct electrical stimulation or by stimulating the alpha motor neurons (refer to Chapter 2) whose axons leave the spinal cord and synapse on the muscle cells. Considerable research and some progress is now ongoing with this approach.

Chapter 9

Coordinating Things More: The Spinal Cord and Pathways

In This Chapter

- Looking at the different types of reflexive movements
- Understanding the hierarchy of motor coordination
- Examining the role of the cerebellum in correcting motor errors

A central theme in this book is the motor reflex as a model of central nervous system function. In the motor reflex, activation of a receptor causes the appropriate muscle(s) to contract as a response to the receptor signal. The classic stretch reflex your doctor tests when he taps your patellar tendon with a little rubber hammer is one such example. In this reflex, the tap activates a stretch receptor whose activity normally indicates that your knee is buckling. The spinal cord response to this signal is to activate the quadriceps muscle to restore the desired limb position so that you don't fall to the ground.

This reflex is a straightforward way to understand how the brain translates sensory input into motor output. At the lowest level, reflexes are fast and stereotyped: The firing of some *proprioceptor* (a receptor that senses muscle length or joint position) causes some particular muscle to contract. At higher levels, more receptors activate more muscles in a more complex pattern that is extended over space and time.

This chapter explains the different types of reflexive movements, from the lowest level (open-loop systems) to the highest (locomotion — yes, walking involves reflexes) and explains how the cerebellum is such an important part of the motor system.

The Withdrawal Reflex: Open-Loop Reflexes

The withdrawal reflex is what engineers call *open-loop* because it's not controlled to achieve a final position; instead, its purpose is a rapid escape from a particular position with no particular endpoint goal. A classic withdrawal reflex occurs if your fingertip touches a hot surface and your arm snaps back. This reflex is done entirely in the spinal cord loop and, in fact, occurs before your conscious awareness of it.

The withdrawal reflex in the beginning can also be described as being *ballistic*, in the sense that once launched, its trajectory is not controlled (until the end, which I discuss shortly). For example, if you touch a hot pan on the stove, your elbow might knock over the salad bowl on the counter next to it without your meaning to because the whole withdrawal movement is executed as a unit, without ongoing modification (like stopping when your elbow contacts the salad bowl).

Although the initial portion of the withdrawal reflex is ballistic, the end is probably not. We have all learned from a lifetime of knocking over salad bowls and other things not to overreact to such stimuli. You can easily observe that the "violence" of the withdrawal is roughly calibrated with how painful the stimulus is. Keeping in mind the fact that the reflex occurs faster than, and therefore before, conscious awareness, it's clear that there is some low level control of the stop point. This stop control is in the spinal cord primarily but may involve the cerebellum for a long enough movement (a really violent reflexive withdrawal may even involve stepping backwards with the legs). The end of this reflex, then, involves both coordination of multiple limbs and movement patterns learned through previous experience.

Hold Your Position! Closed-Loop Reflexes

A common form of reflex involves maintaining a kind of dynamic homeostasis, such as when you adjust muscle force to maintain a limb in a particular position. Because muscles only exert force when they contract, limb movement usually relies on opposing muscle pairs called *extensor-flexor pairs.*

Opposing forces: Extensor-flexor muscle pairs

A typical example of a flexor-extensor muscle pair is the biceps-triceps system that moves the lower arm at the elbow (see Figure 9-1).

Think about flexing your arm to be approximately at a right angle to your elbow. When your arm is in this position but not moving, the elbow angle is a function of the differences in the forces exerted by the biceps versus the triceps. When the arm is still in this position but moving, additional forces occur from acceleration.

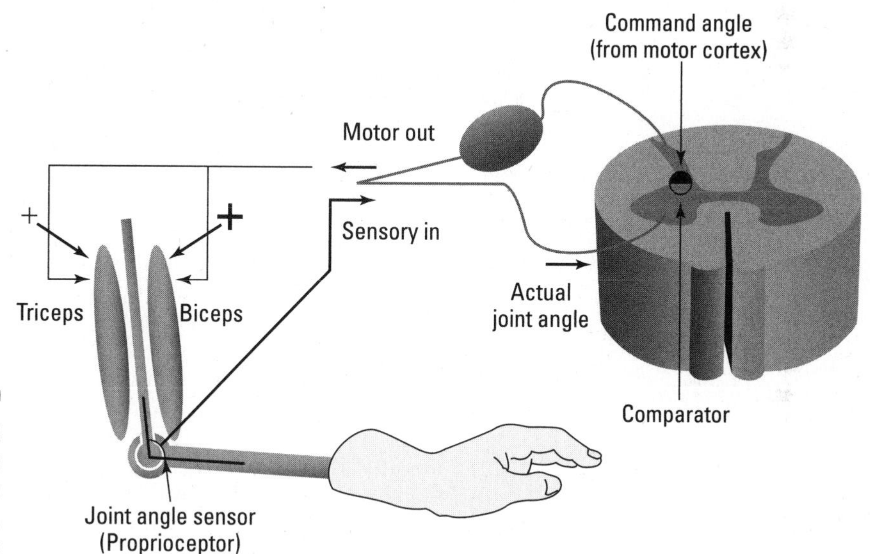

Figure 9-1: The limb position reflex.

Because you can "command" your elbow to be at any angle, you may wonder how holding your arm in a particular position is a reflex. Here's the answer: Unless you're learning to position your body in new ways (learning the fifth position in ballet, for example, or learning to keep your heels down when horseback riding), you don't normally issue conscious commands to your limbs about how to maintain their position.

Instead, the position command comes in the form of the firing of primary motor cortex neurons whose axons descend the spinal cord until they arrive at the appropriate segment. There they synapse on the *alpha motor neurons,* the neurons that actually connect to the muscles. The firing in the motor cortex neurons is thus translated into firing of alpha motor neurons, which is translated into contraction force in the muscle. Feedback from the proprioceptors adjusts the alpha motor neuron firing to keep the actual position of the limb in the command position. This feedback is the topic of the next section.

Determining the correct firing rate with the comparator neural circuit

The problem for the brain is that it has to guess what firing rate in the alpha motor neurons is required to bring your limb to a certain position. Getting an accurate calculation is complicated by the fact that you may be holding something, making your limb heavier than usual, or your muscles may be tired, meaning they don't produce as much contraction per action potential as normal.

This problem is a lot like the problem faced by the furnace that heats your house in the winter. If furnaces were designed to run constantly in order to keep your house at a particular temperature, they would work well only if the outside temperature never varied. If the outside temperature were hotter or colder than this optimum temperature, the additional heat (or cold) added to what the furnace produced would make your house too hot (or too cold).

Instead, furnaces are controlled by thermostats, which are switches that close when the temperature falls below the set temperature but remain open otherwise. If you set the thermostat at 72 degrees, and your house is colder than that, the furnace runs until the temperature of 72 degrees is reached, the thermostat switch opens, and the furnace shuts off.

What does this have to do with brains, arms, and muscles? The motor cortex output, like the furnace, has to be controlled to achieve a set arm angle, which is like the temperature. The thermostat-like device in the motor system is a *comparator neural circuit* in the spinal cord. This neural circuit compares the cortex motor neuron input with a report of the actual position of the limb from the proprioceptors in the limb joint and acts as a switch. If the limb is at an angle less than desired, the comparator gates the alpha motor neurons to put more tension on the triceps (extensor) and less on the bicep (flexor). If the angle is more than desired, the comparator sends more output to the flexor and less to the extensor.

The reason the limb position reflex is *closed-loop* is because the comparator neural circuitry uses sensory feedback from the proprioceptor to keep the limb in a particular position. This mechanism works whether your arm is almost straight out, elbow up or down, or you're holding a cup or a ten pound weight. Moreover, if you want to make a smooth arm movement, say, to throw a baseball, your motor cortex neurons need only to send the sequence of positions desired, and the spinal cord comparator does the rest.

The comparator neural circuit works even better than a furnace thermostat, because thermostats can't generally be manipulated to produce complex temperature variations in a house (it only appears that they can when someone thinks its too cold and someone else thinks it's too hot, and they both have access to the thermostat — which may just be a problem in my house).

In most houses now, the thermostat controls not only the furnace, but also the air conditioner, because the [illegible] can only make the temperature hotter, and the air conditioner [illegible] colder. This is somewhat like the closed-loop system for the [illegible] iceps) pair which involves modifying the drive t[illegible] of the extensor extends the limb, while the fle[illegible] he joint.

In fact, both muscles are alw[illegible] e same time, even though they oppose each ot[illegible] that a real human joint can swivel as well as rotate [illegible] ount of swivel that occurs when you rotate yo[illegible] tivating both opposing muscles, with the lesser activa[illegible] the swivel while the more activated muscle controls the main [illegible] f rotation.

Another, equally important reason for activating both muscles has to do with starting and stopping with both high speed and precision. When you move your arm rapidly, say in extension, the flexor muscle acts as a brake at the end of the movement so that the relative activation goes from all extensor at the beginning, to some combination toward the end of the movement, to all flexor at the end, when you stop. At the stop point, the extensor is re-activated to keep you from overshooting back in the original direction. Patients with tremor, such as Parkinson's, have defective overshoot control systems so their limbs oscillate at what should have been the endpoint of a planned movement.

The Modulating Reflexes: Balance and Locomotion

Another kind of reflex is dynamic. Dynamic reflexes are those that systematically modify limb position set points to accomplish ordered, balanced movement — locomotion. This involves not only changing the set point position command for one flexor-extensor pair controlled by one spinal cord segment but also coordinating numerous flexor-extensor pairs in all four limbs.

Think of reflexes in a hierarchy. At the lowest level is the open-loop escape reflex, where triggering a pain receptor causes a ballistic withdrawal from the pain source with no set end goal position. At the next level is a closed-loop feedback reflex in which the cortex sends the position command to the spinal cord, and the spinal cord neural network modifies the activation of alpha motor neurons that drive muscle pairs to keep the limb in the position ordered by the cortex.

The next, higher level coordinated reflex is the postural reflex. We don't normally think of standing upright as a reflex, but doing so actually involves the closed-loop coordination of multiple reflexes. The goal is still static, however. In order to stand still upright, you must not only keep your legs extended and exerting enough force to counterbalance gravity, but you also need to make sure you don't fall over sideways, forwards, or backwards. Compared to most vertebrates, we humans have complicated this task considerably by insisting on standing on two legs rather than the normal four. In the next sections, I discuss the postural reflex and how a higher control center uses this reflex for locomotion.

Maintaining balance: The vestibulospinal reflex

Imagine you're standing still. Your motor cortex is issuing commands to maintain an upright position. But what actually happens is that you oscillate slightly around this position as this or that muscle tires, or you lean slightly one way or the other as your gaze changes. You not only have to keep particular joints in particular positions; you also have to maintain left-to-right and front-to-back balance, which may require compensation in some joints *away* from their typical position when standing to keep you from falling over. Balance feedback comes from both your visual system, which detects motion such as swaying, and your vestibular system.

The *vestibular system* is a set of three semi-circular canals in the inner ear next to the cochlea. These fluid filled cavities with hair cells are something like those in the Organ of Corti in the cochlea that detect fluid movement (refer to Chapter 6 for information on the inner ear; the semicircular canals and cochlea have a common evolutionary origin). The three semicircular canals detect rotation about the vertical axis, the horizontal axis from front to back, and the horizontal axis from ear to ear, respectively.

The signals from the semicircular canals mediate a balance reflex called the *vestibulospinal reflex* that involves relays through the cortex and cerebellum (discussed later in this chapter). Vestibular signals are also combined with signals from retinal ganglion and cortical cells that detect motion in the same direction that the semicircular canal detects. In this way, the visual and semicircular canal system work together to derive your orientation and motion through space for balance control.

The balance control system controls muscles throughout your trunk. If you start falling to the right, for example, you may, in addition to activating extensor muscles in your right leg, throw your right arm out to generate a counteracting force to restore balance. Animals with tails can use them to restore balance even while in the air, for example.

Balance coordination is accomplished by neural processing that is distributed throughout the spinal cord and includes brain structures such as the vestibular and visual systems and cerebellum. The "simple" cortical command at the top of this hierarchy to "stand still" is accomplished by a complex, multi-leveled, distributed system that takes care of the details and executes some responses in fast local circuits.

Do the locomotion

Of course, most of us want to do something besides stand still all our lives. To actually move, however, another level of the hierarchy is necessary, one that can generate a locomotion cycle, or gait. A *gait* is a repetition of a sequence of leg and arm movements that accomplishes locomotion. Walking, running, hopping and skipping are all different gaits — movement sequences that are repeated in a cycle.

The basics of locomotion

One way to understand locomotion is to think of it as a succession of reflexes, going from lowest to highest level:

- Reflexes from joint receptors that control your position sense (called *proprioception;* refer to Chapter 8) generate enough force in your leg muscles to support your body weight and keep you upright.
- As you lean forward and begin to fall (which is essentially what you do when you initiate a step), another reflex moves the leg that's bearing the least weight forward to catch your fall — a step.
- Neural circuits in the spinal cord control your arms and legs in coordination so that your legs alternate with each other and your arms alternate so that the left arm moves with the right leg (in phase) and your right arm with your left leg.
- At higher levels still you can move in a variety of gaits and you can compensate for rough ground or slippery surfaces. You can also learn new gaits, such as the foxtrot, and, with practice become a proficient dancer.

The fine tuning of movement involves coordination of the entire spinal cord with central brain structures such as the cerebellum and motor cortex.

Alternating gaits: Spinal pattern generators

You can accomplish cyclic movement in a hierarchical system in two ways: You can centrally control the sequence of pre-programmed actions that make up the movement cycle, or you can pre-program the entire cycle. You use both strategies. When you're learning a new movement sequence, such as dancing the foxtrot, you have to consciously plan each step. If you want to step forward with your left foot, you must lean to the right and forward, lift your left foot and advance it forward to catch your fall. Your motor cortical command first momentarily overrides your balance set point to accomplish the lean, and then you restore balance by advancing your left leg and catching yourself. When learning the dance, each step is executed in a similar manner.

Something very different occurs with standard gaits that you've engaged in all your life, like walking, running, and perhaps skipping. In this case, the entire gait pattern has been learned. The location of the learned patterns for the most basic gaits, walking and running, are in the spinal cord itself (with perhaps a little help from the cerebellum), distributed across segments from cervical to thoracic that control the arms and legs.

Locomoting with alternate limb movement

The learned gait pattern in the spinal cord is based on what researchers call a *central pattern generator,* which alternates the two legs and two arms with each other, and the legs and arms in counter-phase (left leg and right arm move forward together, as do the right leg and left arm), which is common for both walking and running. Because the "knowledge" of how to do this is embedded in the spinal cord, the brain only has to command what gait is desired, and the spinal cord takes care of all the details. This is why you can to do things like chew gum and walk at the same time.

Changing the pace: Walking, running, skipping, trotting

Gaits are much more efficient when controlled by a hierarchy in which lower levels take care of lower level details with rapid feedback loops and the higher levels are left to take care of things like selecting the gait and the speed of the gait, without having to program every single muscle contraction.

This division of labor works because most muscle contractions necessary for locomotion are relatively common among gaits and, therefore, can be controlled by lower levels in the hierarchy. This system also allows fast adjustments for errors, such as when you are running and step in a hole. Through spinal cord locomotion circuits, your body automatically compensates for the sudden jolt faster than you could have thought about what to do.

Correcting Errors: The Cerebellum

We've all heard that practice makes perfect. While this isn't strictly true, practice does tend to improve performance. So the question is, how does repetition make us better at doing something? The general process of performance improvement involves two steps:

- Step 1: Recognizing or "tagging" errors
- Step 2: Changing some aspect of the sequence (like timing) in response to the error

The process of repetition-based performance enhancement is typically called *motor learning,* and it depends on a brain structure called the *cerebellum* (which means, "little brain").

The cerebellum is evolutionarily old, existing in non-mammalian vertebrates such as fish and amphibians. It is complex, containing as many neurons as all the rest of the central nervous system (at least 100 billion neurons). Although researchers don't quite understand the details of how its neural circuitry mediates motor-error correction, it is clear that the cerebellum is the center of motor learning and coordination. Cerebellar lesions result in decreased muscle tone, clumsy and abnormal movements, and the loss of balance.

Looking at cerebellar systems

The cerebellum is a sequence-error computer that compares the desired state at any point in a motor sequence with the actual state of the body in the sequence execution. Different parts of the cerebellum receive somatosensory information from all over the body, plus vestibular, visual, and even auditory inputs. External inputs are received almost exclusively in the outer mantle of the cerebellum, called the *cerebellar cortex.* The cerebellar cortex then projects to deep cerebellar nuclei whose projections are organized into three main systems, shown in Figure 9-2:

- **Neocerebellum:** The neocerebellum projects (via the ventro-lateral thalamus) to premotor areas such as the supplementary motor cortex and premotor cortex for error correction pertaining to overall plans; that is, it coordinates the planning of movements that are about to occur in light of previous learning about similar movement sequences. For example, if you are about to try to hit a tennis ball, your cerebellum has stored success/failure information about the entire sequence of needed movements, from racquet preparation setup to follow through.
- **Spinocerebellum:** The spinocerebellum projects to primary motor cortex and down descending motor pathways into the spinal cord for error correction of finer details of movement.
- **Vestibulocerebellum:** The vestibulocerebellum projects into the oculomotor system to control eye movements and into vestibular recipient zones for correction of balance mechanisms.

The anatomical separation between the neocerebellum and spinocerebellum also reinforces the idea of a hierarchy in motor control. The neocerebellum (the evolutionarily newest part) interacts with planning and organizing areas of premotor cortex, while the older spinocerebellum interacts with primary motor cortex and spinal cord coordination areas for error correction at a lower level of control.

Think of it this way: When going to a destination you frequent, you sometimes take the high road and sometimes the low road. Your spinocerebellar pathway makes you run faster and more reliably over the obstacles in both roads, while your neocerebellar pathway prompts you that the low road, in this particular situation, tends to get you there faster.

Interestingly, recent imaging studies show a tight relationship between motor learning in the cerebellum and cognitive activity thought to be quite abstract, specifically, playing chess. When experienced chess players are contemplating particular moves in the game, their cerebellums are activated, suggesting that some of the abstract knowledge of how pieces move is actually represented as movement sequences in the cerebellum.

Predicting limb location during movement

For the cerebellum to perform its error-correction functions, it must maintain neural "models" of how the body's muscles move the limbs in real time. This means that the cerebellum performs a neural computation in which it predicts where the limbs and other body parts will be during a movement sequence. If you're running, for example, and you trip over a root, an error is sent to the cerebellum indicating your leg position is not where it was supposed to be, and multi-level compensatory mechanisms are triggered to maintain your balance and recover your stride.

This prediction function is the key to how the cerebellum allows you to improve with practice. Consider a basketball player who can fling a basketball through a hoop only slightly larger than the ball from over 20 feet away. Practice allows the player's cerebellum to "calibrate" the muscle forces needed. On any particular shot, although the jump won't be from exactly the same place as any shot taken before, the distance measured by the visual system is translated into the cerebellar model, which tunes the output of the motor program coming from the cortex to make the three pointer. Some of us can perform feats like this better than others, of course. The question is, is this possible because of better practice or a better cerebellum? Inquiring neuroscientists want to know!

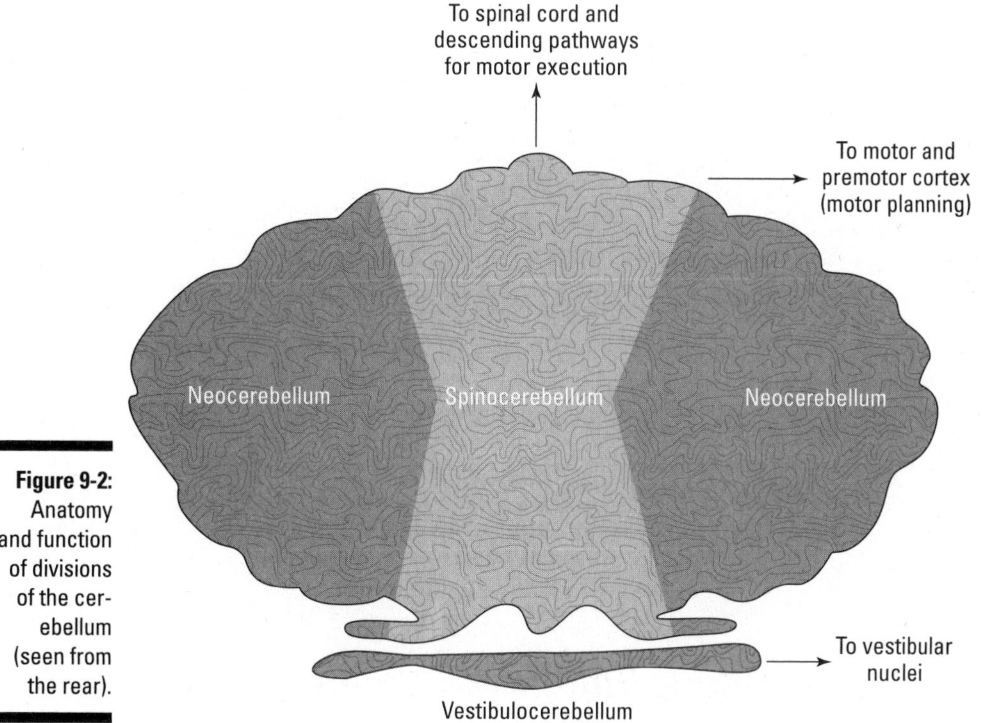

Figure 9-2: Anatomy and function of divisions of the cerebellum (seen from the rear).

Chapter 10

Planning and Executing Actions

In This Chapter

- Understanding the complexity of goal-generated actions
- Looking at the brain areas involved in executive planning
- Examining theories about mirror neurons and von Economo neurons
- Exploring motor disorders related to the basal ganglia

We all do things for various reasons. At one extreme, simple reflexes like jerking your hand away from a hot burner happen so quickly they occur before conscious realization. At the other extreme, an act — such as saying "I do" — may be planned with great deliberation for years. In between, we all sometimes do things we didn't want to do and sometimes later wished we hadn't.

Neuroscience research findings can tell us a lot about reflexes, because virtually all animals have them and because we can study reflexes in animals that act essentially the same as in humans. But human societies don't operate solely on reflex. They operate on the basis of the human ability to choose courses of action based on thought and rationality. So what unique brain location, system, or process exists in humans that supports this "free will" and distinguishes us from other animals?

To gain some traction in this sticky area, consider that behavior is controlled by a hierarchical system. Think of each level of the system as a regulator that maintains an attribute, for example, posture, at some set point. The level in the system that controls that set point has some element of choice with respect to the lower level, which does not.

The questions, then, are these: What is at the top level, and is there really choice there? Do humans have a unique neural level that other animals don't have, some special brain area or set of "consciousness neurons," the hypothesized (often satirically) special neurons unique to humans whose activity creates or supports consciousness? Does consciousness emerge from just having a large enough brain? Or is consciousness not real, not a cause of anything, but a result, an illusion that goes along for the ride with complex brain activity, what some philosophers call an *epiphenomenon?* This chapter considers some of the more relevant data.

Making the Move from Reflexes to Conscious or Goal-Generated Action

As Chapter 8 explains, different kinds of movement are controlled by different kinds of neural circuits. The analogy used there was one of commercial airliners flying on autopilot, which represents a reflexive control system. Reflexive actions include movements that cause you to withdraw from painful stimuli.

Even relatively complex movements like walking involve reflexive action, in which the muscles are controlled by the spinal cord. You may decide you want to walk across the kitchen to get a cup of coffee, but beyond that, how much time do you spend thinking about each step you take, or, for that matter, how to hold a coffee cup? In reflexive movement, a motor program is unfolding in which the current state of the body calls forth the next state.

However, not all movement is reflexive. The brain can take direct control of the muscles from the spinal cord. It does this through projections from the *primary motor cortex,* the strip of brain area just anterior to the central sulcus, the most posterior portion of the frontal lobe (see Figure 10-1). Neurons in the primary motor cortex travel down the spinal cord and synapse on the same motor neurons that mediate the reflexes discussed in Chapter 8. In theory, this direct control allows far more flexibility and adaptability. Going back to the autopilot analogy, having the brain take over is like turning autopilot off, but after you do so, you had better consciously know how to fly the airplane. Where is that knowledge, and how did it get there?

How the frontal lobes function

The primary motor cortex, at the most posterior area of the frontal lobe, is the output of the cortical motor control system. Motor neurons in the primary motor cortex send their axons down the spinal cord and innervate alpha motor neurons, which drive the muscles in the trunk. (For control of muscles in the head, primary motor neuron axons leave the brain via various cranial nerves).

Sending messages from the primary motor cortex to the muscles

Researchers know from stimulation experiments such as those by the world-famous neurosurgeon Wilder Penfield that electrical stimulation of discrete sites in the primary motor cortex causes contraction of discrete muscles in the body according to an orderly map, referred to as the *motor homunculus* (this is very similar to the sensory homunculus in the primary somatosensory area; refer to Chapter 4 for information on this structure).

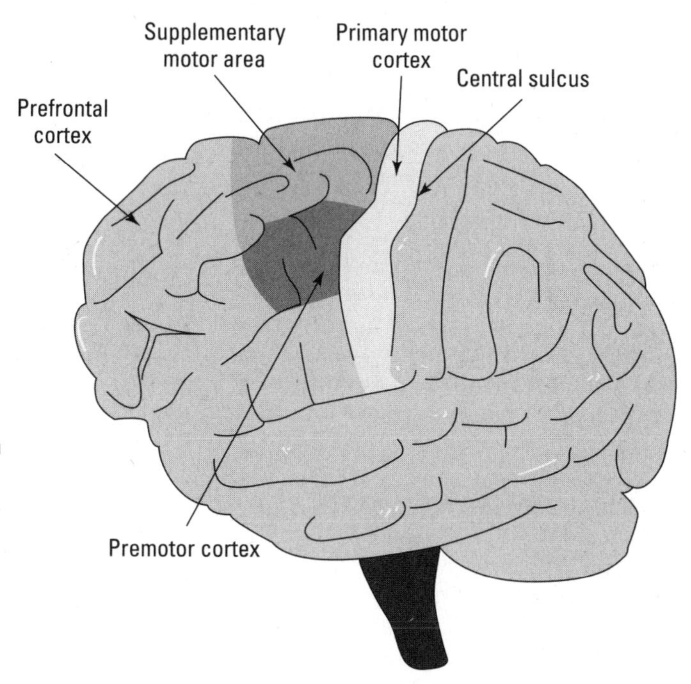

Figure 10-1: The motor control areas of the prefrontal cortex.

The motor and somatosensory maps are similar because where there is dense neural innervation of muscle in the body, such as in the fingertips and lips, there is usually also fine-touch discrimination that allows feedback for learning and performance monitoring.

The cortical motor control system allows flexibility and adaptability in motor behavior that would be impossible with only the lower level spinal cord autopilot processors. Within the frontal lobe are multiple levels of hierarchical control. The hierarchy proceeds more or less from the most anterior part of the frontal lobe at the top of the hierarchy to the primary motor cortex at the posterior end of the frontal lobe at the bottom level, which forms the frontal lobe output.

Setting the goals of motor activity

Something has to tell the primary motor cortex neurons when they should fire and in what order. This involves setting the goals of motor activity (where do you want to go?), determining the strategy (how do you want to get where you're going?), and, finally, executing the plan (flying the airplane by manipulating the ailerons, elevators, and rudder).

The most anterior part of the neocortex is the prefrontal cortex (a name that strikes most students as a rather odd because the prefrontal cortex comprises about half of the entire frontal cortex). How does the frontal cortex motor hierarchy work? Suppose you're at a party and decide to ask someone across the room whether that person would meet you on the patio to chat about something. This goal is abstract, and you can accomplish it in lots of ways. You could yell the request across the room. You could walk over to the person and announce your request in a normal voice. You could send a text message. You could use hand gestures to make your request clear.

Now, most of these approaches are incompatible or, at the minimum, have incompatible components: You can't make large hand gestures and text message. You probably don't want to walk over and then yell at the person. Given the goal you've selected, you must then select one way (*motor program*) of reaching that goal. Each method of reaching the goal involves different muscles in a different sequence. After you select a method — say, walking over — you must then program your own muscle sequence to accomplish that.

But what if the person moves as you pick your way through the crowd on the way over. You may have to change your plan, either altering the program you're pursuing or switching to a different program. This is where having a brain, particularly a large frontal cortex, really comes in handy. The next section goes in the details.

Planning, correcting, learning: Prefrontal cortex and subcortical processors

Where do the goals come from? Complex organisms, like you, have goals that exist in a hierarchy of sub-goals. Say you enjoy listening to Beethoven recordings, and you hear a new version of the Symphony No. 9 that you would like to own. You can reach this goal by going to a music store and buying a CD. Pursuing this goal requires that you know where a music store is that is likely to have the recording you want. The goal of getting to the store could be reached by driving your car, if you have one, or taking the bus. When you get to the store, you will need a means to pay for the CD and a way to get back home.

Anyone could elaborate this scenario for many paragraphs. Although doing so would probably involve tens of steps with numerous alternatives, no normal human would have trouble comprehending how to carry out this plan from the description. No non-human animal could possibly comprehend or carry out such a complex plan, unless explicitly trained in a rigorous, inflexible, step-by-step manner (I take up this kind of sequential learning in the discussion of the cerebellum in Chapter 10).

Complex, flexible goal hierarchies live in the prefrontal cortex. A large prefrontal cortex allows both multi-leveled plans and complexity at each level. Complex plans require long-term memories that can access sub-plans such as how to find a music store, how to drive a car, and how to pay for a Beethoven CD. They also require short-term or working memory (the seven items, more or less, you can keep in mind at one time) to assemble the relevant set of sub-plans into a framework that accomplishes the overall goal.

Working memory

Most of us tend to think of memory as a storage location for data. Working memory in the brain is rather different from this, however. It's more like the temporary maintenance of a dynamic pattern of neural firing. Each goal idea, such as getting Beethoven's No. 9, is a set of related neural representations scattered throughout the brain (Beethoven, music, No. 9, recording, CD player, and so on). Working memory is the creation in the prefrontal lobe of a temporary circuit in which all the related ideas for this recording are activated in a recurrent loop, labeled as a concept. As long as all the related ideas are continuously activated, the idea stays in memory, but if one is no longer activated, the concept is lost.

Think of working memory functioning like a juggler or team of jugglers. As long as the balls (the discrete memory items) remain in the air and get successfully tossed from hand to hand, the show goes on. But stop juggling and the balls fall on the floor (with all the other ideas in the brain). Show's over.

Working memory can hold about seven concepts. In other words, humans have about seven jugglers at our disposal. As the first juggler juggles the Beethoven No. 9 concept, he occasionally throws one of the balls, say, the "get recording" ball, to a second juggler. Juggler number two starts building up the concept of the nearest classical CD store and holds those balls in the air. When these two "juggles" are established, these jugglers may exchange balls with a third juggler who gets the message "drive to 21st St. and Elm, the CD store."

Interestingly, humans don't seem to have a vastly larger supply of working memory "jugglers" than a number of other animals. Crows and other birds, for example, can differentiate between sets of 5 versus 6 "items." However, the complexity of the items that humans can hold in memory (which memory researchers call *chunks*) is vastly greater, so that one item for each human juggler might contain multiple components, each of which would exhaust all the jugglers in a non-human animal.

The neural juggling process that maintains items or chunks in working memory involves interactions between prefrontal cortical neurons and subcortical structures such as the hippocampus and amygdala. In particular, the lateral prefrontal cortex and hippocampus interaction maintains most generic associations (Beethoven, No. 9, CD, recording, and so on) that encode and define items relevant to the present situation. The orbitofrontal cortex (the anterior part of the frontal lobes lying above the eyes) interacts with the amygdala to code the risks and benefits of the sub-plan chunks for the currently considered goal, such as if the CD store were in a bad neighborhood and going there after dark might not be wise.

Initiating actions: Basal ganglia

Given that risks and rewards exist for every action, including which means to use to take the action, where is the choice made? The prefrontal cortex also interacts with another set of subcortical structures called the *basal ganglia.* This interaction is crucial for actually initiating actions appropriate to goals.

The basal ganglia-prefrontal lobe interaction involves two different aspects of goal pursuit: selection and switching:

- **Selection:** How the basal ganglia function as a circuit is one of the least understood areas of system neuroscience, despite their importance in diseases such as Parkinson's and Huntington's (explained in the later section "When the Wheels Come Off: Motor Disorders"). Most models suggest that they execute a "winner-take-all" function among competing cortical alternative goals and sub-goals that are simultaneously activated in any given situation. You may want the Beethoven CD, but you might be tired or hungry or afraid of going to 21st and Elm at night. At each level of the sub-goal hierarchy, THE interaction between the prefrontal activity and basal ganglia causes one goal to be selected and the others inhibited.
- **Switching:** If you happen to be on the way to the store to buy the Beethoven CD and you encounter the pack of wild dogs, your immediate goals are likely to change considerably. The change in your immediate goals requires changes in all your sub-goals, plus, in this case, activation of autonomic fight-or-flight responses. The basal ganglia also control this kind of goal switching. At least some of this control is accomplished through the connection of the basal ganglia to the thalamus.

The basal ganglia are phylogenetically old structures that exist in non-mammalian vertebrates. Animals like frogs and lizards also choose among goals, but their choices are much more constrained by the immediate opportunities and dangers. Mammals in general, and primates and humans in particular, can engage in complex, long-term goals with many steps and decisions and occurring over days, months, and even years. This capability is enabled by the prefrontal cortex, which adds subtlety, complexity and adaptability, and perseverance (over time periods longer than days) in goal pursuit to the phylogenetically older, more instinctive basal ganglia system.

The basal ganglia are a system of nuclei embedded within a larger system of cortical and subcortical areas subserving motor control (see Figure 10-2). Inputs from many areas of the neocortex and the thalamus enter the basal ganglia by relaying through the basal ganglia structures called the *caudate* and *putamen*, which together are referred to as the *striatum*. The *striatum* projects to three interconnected nuclei within the core of the basal ganglia, called the *globus pallidus*, *subthalamic nucleus,* and *substantia nigra*. The output of these three nuclei, particularly the globus pallidus, inhibits motor areas through the thalamus.

The main function of the basal ganglia complex is to monitor the status of many cortical motor planning systems and to inhibit all those except the main one to be executed. For example, you normally won't scratch an itch on your nose while hitting a forehand in tennis. In general, most of us can only do one thing at a time that requires continuous conscious supervision. In cases where those who can "chew gum and walk at the same time" do dual tasks, it is because one or both of the tasks are so automatic or practiced that they can be done with little conscious effort.

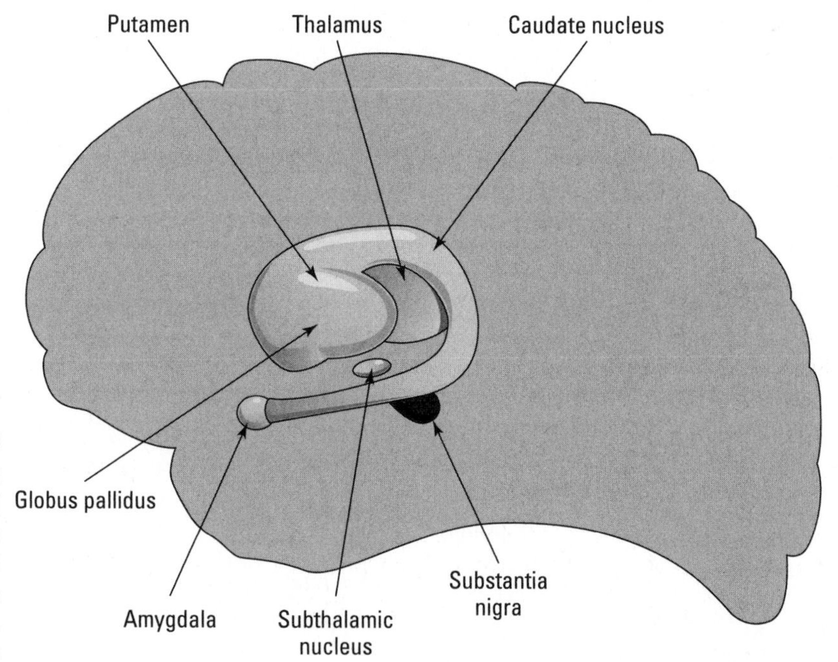

Figure 10-2: The basal ganglia and related brain structures.

In many cases, our apparent multitasking is really a case of cyclic time sharing. Take driving and carrying on a complex conversation. You listen or talk for a few seconds, then pay attention to the road, and then go back to the conversation. You can do this either by automatically shifting from one to the other at set time intervals of a few seconds, or you can do it by letting one task dominate and the other interrupt for brief periods of time. The basal ganglia are one of the main controllers of this type of activity.

In the middle of things: Supplementary and pre-motor areas

Between the primary motor cortex and prefrontal cortex are two cortical areas that mediate movement sequencing. The most medial is the *supplementary motor area* (SMA); lateral to the SMA is the *premotor cortex* (PMC); refer to Figure 10-1 to see these areas.

The premotor cortex: Learning how to get it right

The job of the premotor cortex is to consciously monitor movement sequences, using sensory feedback. After the basal ganglia and prefrontal cortex select the goal, the premotor cortex coordinates the steps to reach that goal. Activity in the premotor cortex helps you learn what to pay attention to while you perform a complicated motor sequence and what to do when you get stuck at some particular point.

Remember back to when you first learned to write your name as a young child. Each letter was difficult, so you worked hard to write each letter, checking its shape as you went along. You also probably had to check where you were in the name, what letter was next, and how much space to leave between letters. All of this monitoring required considerable sensory feedback from your visual system and even from your hand and fingers. Although your cerebellum helped with some of this as you practiced, in the beginning, the writing required considerable conscious concentration and monitoring — all of which occurred in the premotor cortex.

The supplementary motor cortex: Breezing right through

Now contrast signing a credit card slip today with that childhood experience of writing your name. Once you decide to write your signature, you accomplish the task almost unconsciously in a flurry of strokes. In fact, if you think about the individual letters you're writing as you sign, you're likely to stop yourself or make a mistake. Years of practice doing the same signature have caused the whole sequence to be stored as a single executable chunk of movements. This chunk is largely stored in the supplementary motor area.

The cerebellum: Where you coordinate and learn movements

One more subcortical structure deserves mention: the cerebellum. The cerebellum receives sensory input from peripheral receptors as well as from prefrontal and other association cortical areas, and projects to the primary motor cortex.

The cerebellum is a high-level motor-coordination-and-learning area. When you first try to do something hard, like ride a bicycle, you make lots of mistakes not only in what to do but also in how fast to do it. By monitoring your errors and comparing them to the goals you have set, the cerebellum allows you to improve with practice. It also allows you to integrate what you see (such as the flight of the tennis ball) with what you do (the mechanics of the particular swing appropriate to send the ball where you want it to go).

Although the cerebellum looks like a single brain "organ," it actually consists of three quite different areas with different connections:

- At the bottom, the *vestibulocerebellum* performs error correction and motor learning for balance, using inputs from the semicircular canals and visual system.
- The medial, intermediate cerebellum, also called the *spinocerebellum,* performs similar functions for locomotion and coordinated trunk activity.
- The lateral *neocerebellum* interacts with prefrontal and other cortical areas for learning overall plan sequences.

Refer to Chapter 9 for more detail on this brain structure.

Putting it all together

If you put someone in an fMRI magnet and ask that person to execute a set of finger movements he has not extensively practiced, you'd see activation of several prefrontal areas, the premotor cortex, and the primary motor cortex. If you ask your subject to move his fingers in a way he's practiced (such as having a sign language speaker spell a common word), the supplementary motor area, rather than the premotor cortex, would be primarily activated.

Figure 10-3 is a diagram of the brain's motor control system. Think of the basal ganglia/intermediate cerebellum/primary motor cortex subsystem as being one hierarchical level above the spine. The supplementary motor area and the premotor cortex are the next level. Finally, the prefrontal cortex and lateral cerebellum circuit forms yet another control level. Remember also that the prefrontal cortex is a large area with many subdivisions, so the reality is much more complicated than this simple picture.

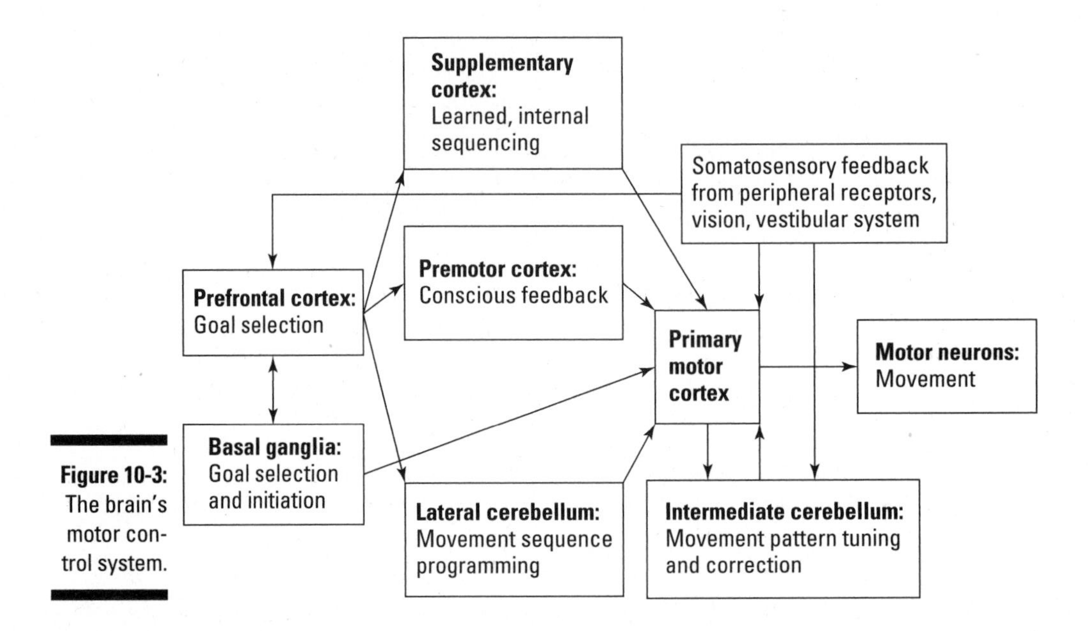

Figure 10-3: The brain's motor control system.

Where Are the Free Will Neurons?

In the beginning of this chapter, I mention the difference between reflexive action that we normally can't control and actions mediated by our free will. Because all actions are ultimately carried out by the firing of a sequence of motor neurons, many people feel that whatever entity is at the highest level of the hierarchy must be the embodiment of the conscious choice selection. The question is, can we identify that entity?

Which comes first: The thought or the action?

Dr. Benjamin Libet, a neuroscientist at the University of California, San Francisco (USCF), conducted a famous series of experiments on this subject in the 1980s. These experiments were controversial then, and they remain controversial now. In these experiments, Libet asked subjects to sit calmly looking at a special clock and, *whenever they themselves decided*, move their hand a bit while noting the position of the clock hand; in other words, the subjects were to record the time when they decided to move their hands. While the subjects were doing this task, Libet made electrical recordings of their brain activity (EEGs).

Libet analyzed the data by superimposing the clock position on the EEG traces so that he could compare the EEGs to the exact times when the participants said they had made their decision to move their hand. Here's what the data very reliably showed: Deflections of the EEG traces occurred typically one-half second *before* the subjects indicated they'd decided move their hands (this deflection is typically now called a *readiness potential*, and it is strongest over the supplementary motor area).

In other words, by looking at the EEG trace, Libet could tell one-half second before his subjects made the decision to move their hands that they would "freely" make a decision to move their hand.

Contemplating the study results

Libet's data showed that brain activity preceded the subjects' conscious awareness of their decision to move. What does this result mean?

What it doesn't mean is that Libet was in any way controlling the minds of the experiment's subjects; he was only recording their brain activity. Some have suggested that the experiment shows that the brain is merely a machine and free will is an illusion or *epiphenomenon* (a side effect, like assuming that the noise a car makes is what actually makes the car move).

What did Libet himself think about this result? Libet's most frequently cited interpretation of the experiments is that the process of deciding to take an action, such as moving your hand, must start unconsciously but that a "veto" point is reached over which you have conscious control. In other words, a hodgepodge of potential actions bubble away in your cortex until, at some point, the basal ganglia inhibit all actions except for one. This one action then receives all the cortical processing and, about a half-second later, rises to conscious awareness.

For the subjects in the experiment, the action potentials bubbling in their cortices included a variety of possible actions, like moving their hands, getting up and leaving, looking out the window, and so forth; then the basal ganglia inhibited all the actions except moving the hand, which received all the cortical processing and rose to consciousness, at which point they made the v to move their hands and marked the time.

Nobody believes that this experiment places the seat of consciousness in the basal ganglia or the supplementary motor cortex (where the readiness potential is strongest). What most neuroscientists believe is that the common perception of a linear hierarchy from conscious decision to motor neuron firing just isn't tenable. We're all a mass of urges and inhibitions that are products of our constitutions and experiences.

You're still accountable!

If no central, conscious controller controls of what we do, how can we be held accountable for our actions?

Accountability stems from the fact that the moment of conscious decision before an action conjures up memories and rational thought processes that involve the entire brain and allow us to make some prediction of the consequences of what we're about to do. Because this predictive rationality derives from our brain as a system, it is robust and involves our moral and cultural values.

But then is the veto process — the point when the basal ganglia inhibit all actions except for the one you're going to take — really conscious? Without trying to define consciousness itself, what researchers can say is that the veto process activates large brain areas and that you're aware that it's occurring. Because this sustained activity reaches your awareness, the "choice" — that is, the non-inhibited action — is informed by your background, which includes your experience, memories, and values. This in itself doesn't mean that whatever choice you make is really "free," but it is an informed rather than a subconscious or reflexive choice in which you can weigh alternatives and their costs and benefits.

You may still impulsively ignore the costs of a certain choice or make a bad choice despite being fully aware of the consequences. But you are at least generally aware of this, and practicing self-discipline and introspection clearly improves your ability to make good choices. Interestingly, abnormalities in the frontal lobe, particularly the orbitofrontal cortex, seem to reduce a person's capacity to make wise choices without reducing the capacity to do the mental calculation of the costs and benefits of the choices.

Discovering New (and Strange) Neurons

One of the fundamental principles of neural science is the *neuron doctrine*. This doctrine says the following:

- Neurons are the computing units of the brain.
- Neurons, though specialized in some cases, operate by common biophysical principles.
- The power of the brain to think and control behavior is mostly a circuit property of the interconnections within the nervous system.

An obvious corollary of the neuron doctrine is that you shouldn't expect brains to be primarily distinguished by having unique cells types, but by having different circuits. For the most part, this is true: The basic neuronal types are quite old evolutionarily. In the retina, for example, the same five basic cell classes exist in vertebrate species from frogs to humans.

However, there have been discoveries of neurons which, although not unique to humans, appear only in animals with large brains like humans, apes, and, in some cases, whales. The following sections will discuss two "strange" neuronal types: mirror neurons and von Economo neurons.

Mirror neurons

One of the most exciting neural discoveries in the last two decades is that of *mirror neurons*. Mirror neurons are visual neurons that fire not only when a monkey is performing a task requiring visual feedback, but also when the monkey observes a human (or presumably another monkey) performing the same task.

Researchers have recorded mirror neurons in monkeys in the supplementary motor area (SMA) and premotor cortex (PMC), as well as in the anterior parietal lobe. These neurons are active when the monkeys performed manual manipulation tasks requiring visual feedback (the supplementary motor area and premotor cortex are clearly involved in motor coordination). What shocked the researchers was that these neurons are activate not only when the monkeys are performing the task on which they were being trained, but that they're also active when the monkeys watch a researcher doing the task in the training session (a discovery that was made accidentally).

The basic properties of mirror neurons have been replicated in many laboratories, and they've even been extended into the audio domain (mirror neurons respond to the sound of particular tasks that make noise, like operating a grinding crank). Brain imaging experiments in humans have also shown similar do-and-view activation patterns in the same areas as the single-cell recordings in monkeys.

The function of mirror neurons is unclear, but here are some hypotheses:

- They may aid in task learning, by providing a substrate with which a monkey can learn by imitation.
- They may participate in a neural circuit by which a monkey (or human) can understand the purpose of another's actions. This suggestion is related to the idea of a *theory of mind* by which humans (and perhaps higher primates) understand the actions of others as being intentional and purposeful, or psychological, rather than as a result of ordinary physics.

  This idea has led to some speculation that some deficiency in this mirror neuron system could be associated with the autistic tendency to lack empathy, to treat others as though they are unfeeling objects rather than entities similar to oneself whose feelings can be hurt and goals frustrated — an idea that is pure conjecture at this point and not based on any direct evidence. Nevertheless, attempts have been made to "treat" people with autism by stimulating brain areas where mirror neurons reside, using techniques such as transcranial magnetic stimulation (TMS: use of a focused, pulsed magnetic field generated outside the head to affect brain activity in particular areas).
- The mirror neuron system may be involved in language learning, during which, over a span of years, infants constantly imitate the speech they hear around them. Some scientists postulate that the mirror neuron system enabled human language by providing a sufficient representation of complex behavior (language) that could be imitated and then mastered.

Because mirror neurons are found in multiple brain areas and therefore are almost never eliminated by any single lesion, confirming or refuting any of these suggested functions has been difficult. (If they existed in only one area, researchers could more easily study whether damage to that area negatively affected any of these capabilities.)

Von Economo neurons

The basic neuronal types found in mammalian brains are evolutionarily old. A section of the mouse motor cortex looks a lot like a section from the human auditory cortex, for example. Neuroanatomists have generally not found differences in mammalian brains at the level of unique cells types.

But then there are *von Economo neurons* (VPNs).

Where these neurons are found

These neurons, named after Romanian neurologist Constantin von Economo (1876-1931), are also called *spindle cells.* They appear to be a rather oversized cortical pyramidal cell that have an unusually sparse dendritic arbor (that is, they have relatively few branches), but they have extensive axonal connections throughout the brain. Here's the interesting part:

- These neurons are found only in humans and great apes (chimps, gorillas, and orangutans), some whales and dolphins, and elephants. They are more common in humans than any of the few other species in which they are found.
- They're found only in two brain regions: the anterior cingulate cortex (believed to be an executive control area with respect to other cortical areas) and the fronto-insular and dorsolateral prefrontal cortex, which are also executive control areas.

Speculation about what these neurons do

Considerable speculation has occurred about whether von Economo neurons are simply an extreme structural variant of normal pyramidal cells (the major cortical cell class involved in long distance connections; refer to Chapter 2) necessitated by large brain sizes, or whether they have some unique function and physiology. The function of von Economo neurons is unknown. Some indirect evidence indicates that these neurons are particularly compromised in Alzheimer's and several other types of dementia, but this may be an effect rather than a cause.

Some researchers speculate that, given their extensive projections, von Economo neurons link multiple cortical areas together. The advantage of doing this with a few specialized cells rather than a large number (as is typical) could be that the action potentials from a few von Enconomo neurons would be synchronized in the different cortical areas because they all arose from a single cell. Synchronized neural firing has been implicated as a mechanism of attention because action potentials arriving at a post-synaptic target are much more effective if they occur at the same time versus when they are out of phase. The basic idea is that amidst all the background firing going on in the brain, those von Enonomo neuron-driven neurons that are firing synchronously form a self-reinforcing ensemble that rises above the noise and reaches attention.

If the function and physiology of von Economo neurons somehow uniquely made self-awareness possible, for example, it would produce a revolution in neuroscience. It could also profoundly change the way people view other species in which this neuron is found. The jury is clearly still out on this one.

When the Wheels Come Off: Motor Disorders

Our ability to move around can be compromised by injury or disease that affects any neurons in the chain of control from prefrontal cortex to the alpha motor neurons that exit the spinal cord ventral root and stimulate muscles.

Injuries to the spinal cord and brain

The most common type of paralysis is undoubtedly associated with injury to the spinal cord which severs the axons of primary motor neurons going from the cortex to synapse on the alpha motor neurons that drive the muscles. If the spinal cord damage is below the cervical spinal segments, the typical result is *paraplegia,* where control of the legs is lost but arm and hand control remains. Damage at higher levels can produce *quadriplegia,* loss of control of all four limbs. When these axons are severed by an injury, they don't grow back; a large segment of all neuroscience research today is trying to understand why.

Central damage can produce more subtle but equally debilitating effects. Damage to frontal lobe areas that control movement can result in partial paralysis or *paresis* (weakness rather than total paralysis). Parietal lobe damage can result in *apraxia*, the inability to execute skilled movements, while some albeit clumsy control remains.

Degeneration of the basal ganglia

Motor impairments can also result from degeneration in the basal ganglia. Two relatively common and well known degenerations are Parkinson's and Huntington's diseases.

Parkinson's disease

Degeneration (specifically, the loss of dopaminergic cells) of the *substantia nigra* causes Parkinson's disease. The substantia nigra is one of the basal ganglia (refer to earlier section "Planning, correcting, learning: Prefrontal cortex and subcortical processors" for more on the role of the basal ganglia in movement). This disease is characterized by a stooped postural rigidity and difficulty initiating movements or changes in gait.

Attempts in the mid-20th century were made to treat the disease by injecting dopamine into those with the condition, but this treatment option was ineffective because dopamine doesn't cross the blood-brain barrier. However, the metabolic precursor in the cellular production of dopamine, L-dopa, *does* cross this barrier and provides considerable but only temporary relief of Parkinson's symptoms.

L-dopa treatments eventually cease working because the cells that convert L-dopa to dopamine cease working or die. Within the last two decades, considerable success has been achieved in alleviating Parkinson's disease symptoms with a technique called *deep brain stimulation* (DBS). DBS appears to be most effective when the stimulating electrode is placed in the subthalamic nucleus rather than the substantia nigra. Researchers aren't quite clear why this location is better, and this uncertainty emphasizes a lack of full understanding of the complex basal ganglia circuit and its relation to its connections to the thalamus and neocortex.

Huntington's disease

Huntington's disease involves degeneration of cells in the striatum (caudate and putamen, the input nuclei to the basal ganglia). Huntington's disease is a late-onset hereditary disease which in some ways has symptoms the opposite of Parkinson's.

Whereas Parkinson's patient have trouble initiating movement or movement sequences, Huntington's patients exhibit uncontrollable movements like writhing and gyrating. The term *chorea* in this disease's original name (Huntington's chorea) is related to the word *dance,* describing the continuous, uncontrollable dancelike movement of Huntington's patients.

Chapter 11

Unconscious Actions with Big Implications

In This Chapter

- Understanding the role of the autonomic nervous system
- Getting into the specifics of the sympathetic and parasympathetic subsystems
- Examining circadian rhythms, sleep cycles, and the different stages of sleep
- Tackling sleep disorders and disorders of the autonomic nervous system

Although we are all familiar with our senses of sight, sound, touch, taste and smell, which tell us about the world around us, we also have senses that tell us about the world inside us. These senses detect and regulate things like our body temperature, heart rate, thirst, hunger, and wakefulness. Many of these internally regulated functions are part of what is called *homeostasis*, the active maintenance of various aspects of the internal state of our bodies.

The homeostatic functions of the nervous system came before all the higher cognitive abilities enabled by the brain. Even invertebrates, like mollusks, worms, and insects have nervous systems that perform some internal regulation. These homeostatic functions are necessary for the cells in the body to live. Even minor dysfunction in these regulatory mechanisms can quickly compromise health, and if any of them fail, death usually follows quickly.

Most of these functions are mediated by the autonomic nervous system, which is distinct from the central or peripheral nervous systems. This chapter examines how the autonomic nervous system works, including its involvement in the regulation of sleep, a highly necessary but poorly understood homeostatic function.

Working behind the Scenes: The Autonomic Nervous System

Although we tend to think of the brain as the organ that produces thought and intelligence, it also controls numerous aspects of body metabolism such as heart-rate, respiration, temperature, and circadian rhythms — things that we're generally not conscious of. In very primitive animals, these regulatory functions are mediated by collections of cooperating neurons called ganglia. In advanced animals such as mammals, these regulatory functions are mediated by lower (caudal) parts of the brain, such as the brainstem, and the spinal cord.

The regulatory components of the central nervous system interact with what is essentially another nervous system, the *autonomic nervous system.* The autonomic nervous system is actually a dual system because it has two components, called the *sympathetic* and *parasympathetic* branches, that act in opposition to each other. The following sections explain what the autonomic nervous is, what it does, and how it works.

Understanding the functions of the autonomic nervous system

Your body does many things that you're not generally aware of. Here are just a few examples; there are many more processes almost too numerous to mention:

- Your heart pumps blood around your vasculature (the vast web of blood vessels running throughout your body and organs), carrying oxygen and nutrients to and waste away from the vicinity of every living cell in your body.
- You perspire on a hot day or when you exercise.
- Your digestive system processes food to extract nutrients and then to eliminate the waste that can't be used. Your kidneys filter out other wastes.

Matching the activity levels of different organs

Throughout the body, various organs manufacture, filter, and circulate essential substances that ultimately are either consumed by cells or released by cells to be removed. To work efficiently, the organs must work cooperatively. For example, when you perspire to get rid of excess heat, your kidneys must keep up with the loss of water to keep your body fluids in balance. The blood flow volume, in turn, should at least match the demand to move those gases to and from where they need to go plus circulate whatever other nutrients and wastes are currently necessary.

Thus, one function of the autonomic nervous system is to match the activity levels of different organs to balance multiple input/output balance sheets for cell consumables and wastes. These demands are complex enough, but they aren't the only thing the autonomic nervous system does.

Letting us switch between quiet and active states

Life consists of more than digestion. We need to move around, sometimes quite rapidly. With rapid movement, as much blood as possible must go to the muscles, and heart rate and respiration must scale up to match *their* metabolite demand, not the metabolic demand of the digestive system.

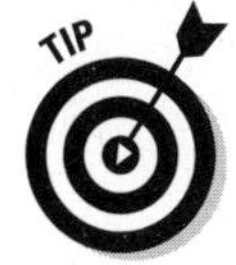

You can think about the duality within the autonomic system as corresponding to our bodies being in one of two major states:

- A quiet homeostatic state in which we are conserving energy resources and efficiently allocating and distributing energy, nutrients, and waste.
- An active, competitive state in which we suppress homeostasis and allocate all the resources we can to energetic activity in order to win or avoid losing some contest, even if we become depleted and have to "pay back" the energy debt and disruption of homeostatic mechanisms later.

Because of the way the autonomic nervous system works (a process explained in the following sections), our bodies are able to shift from one state to another.

Dividing and conquering: Sympathetic and parasympathetic subsystems

The autonomic nervous system is divided into *sympathetic* and *parasympathetic* subsystems (see Figure 11-1). As you can see in Figure 11-1, many of the sympathetic subsystem effects seem to be just the opposite of the parasympathetic subsystem effects. Consider these examples:

Parasympathetic subsystem	***Sympathetic subsystem***
Constricts pupil	Dilates pupil
Stimulates salivation	Inhibits salivation
Stimulates digestion	Inhibits digestion
Contracts bladder	Relaxes bladder
And so on. . . .	And so on. . . .

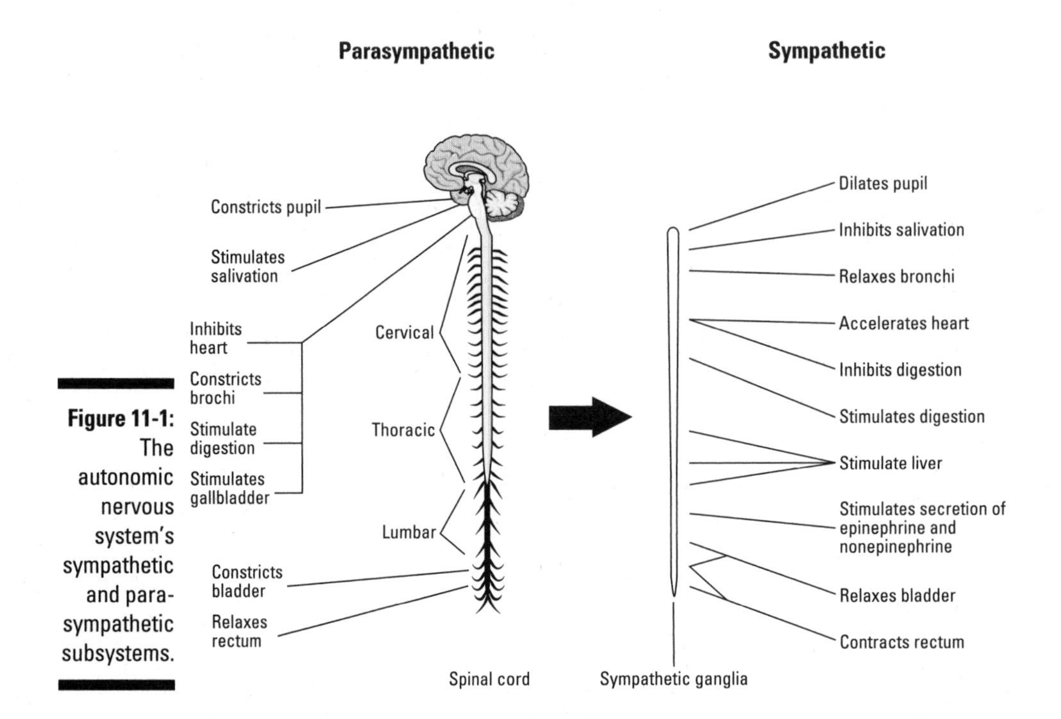

Figure 11-1: The autonomic nervous system's sympathetic and parasympathetic subsystems.

Why does the autonomic nervous system have this dual, oppositional structure? Imagine that, as your ancestors once did, you had to chase down your lunch before you could eat it. The sympathetic "fight or flight" system would increase heart-rate and respiration (among other things) and divert blood flow away from the digestive system to the skeletal muscles to allow you to be very active. After you bagged your meal, it would be time to sit down and eat it. The parasympathetic system would slow down your overall metabolism and direct blood flow to your digestive system, allowing you to extract the nutrients you need from your successful chase. All vertebrate animals have a similar version of this dual system.

Both the sympathetic and the parasympathetic subsystems involve neurons in either particular spinal segments or cranial nerves that use acetylcholine as a neurotransmitter.

The sympathetic subsystem: Fight or flight

Fight-or-flight responses are a shorthand terminology for any activity that requires high metabolic voluntary muscular activity. The perceived need to engage in such activity, such as seeing an approaching bear, activates the sympathetic division of the autonomic nervous system.

This division consists of neurons located in the thoracic and lumbar segments of the spinal cord (refer to Figure 11-1). These *cholinergic neurons* (neurons that use acetylcholine as a neurotransmitter) synapse on neurons in sympathetic ganglia, a system of neurons outside the spinal cord running roughly parallel to it. Neurons in the sympathetic ganglia synapse on target organs such as the heart, lungs, blood vessels and digestive tract. These second-order neurons use noradrenaline (norepinephrine) as a neurotransmitter.

An additional sympathetic target is the adrenal medulla. Post-synaptic neurons there release a small amount of epinephrine as well as nor-epinephrine into the bloodstream. The release of norepinephrine produces very fast effects such that only a small amount of epinephrine needs to be released to balance the system.

The parasympathetic subsystem: All is well!

The parasympathetic division consists of neurons located in the brainstem and sacral spinal cord, which also use acetylcholine as a neurotransmitter. These neurons synapse on neurons in various ganglia near the target organs, such as the heart. These ganglionic neurons also use acetylcholine as a neurotransmitter. The target organ receptors for these parasympathetic ganglia have muscarinic receptors, a metabotropic receptor type for acetylcholine (refer to Chapter 3).

Controlling the autonomic nervous system

Both divisions of the autonomic nervous system have sensory feedback from the target organs that is used in local spinal cord control circuits. Overall control of the autonomic system, particularly the balance between the sympathetic and parasympathetic divisions, comes from higher levels.

Getting sensory feedback from target organs

Just as we have receptors for things outside the body, such as light detectors (photoreceptors) in the retina, we have sensors inside the body that detect things like blood pressure, temperature, and carbon dioxide levels. This information is used to maintain homeostasis.

The carotid artery in the neck, for example, has receptors (called *baroreceptors,* in the carotid sinus) that sense blood pressure. These receptors send high blood pressure signals via the vagus nerve to the nucleus of the solitary tract (NST). The NST projects to the nucleus ambiguous and vagal nucleus, which release acetylcholine as part of the parasympathetic system to slow down the heart.

Another example is temperature regulation. Thermoreceptors in the hypothalamus respond to low or high body temperature by sending signals to autonomic motor neurons to produce responses such as shivering or sweating.

The hypothalamus: Controlling the sympathetic and parasympathic subsystems

One important controller of the autonomic nervous system is the hypothalamus. The hypothalamus is actually a set of nuclei that lie below the thalamus. The hypothalamus is involved in controlling body functions, such as hunger, thirst, body temperature, fatigue, and circadian rhythms.

Activating the hypothalmus

The hypothalamus is controlled or activated by projections from wide ranging brain areas, including the following:

- **Medulla:** Inputs from the ventrolateral medulla carry information from the stomach and heart.
- **Amygdala and hippocampus:** Inputs from the amygdala are typically associated with memories of fearful events that activate the sympathetic nervous division of the autonomic system via the hippocampus. There is a large input from the hippocampus to the mammillary body of the hypothalamus that allows other memory components to modulate the hypothalamus, including remembered smells.
- **Nucleus of the solitary tract:** The nucleus of the solitary tract communicates visceral and taste inputs that control feeding behavior.
- **Locus coeruleus:** The locus coeruleus is the main producer of norepinephrine in the brain. The hypothalamus projects to the locus coeruleus, a site of norepinephrine release, and receives return projections from the locus coeruleus about the level of norepinephrine secretion (which is also modulated by other inputs).
- **Orbitofrontal cortex:** The orbitofrontal input is associated with physiological reactions to reward and punishment.

Some of these signals are subconscious, but others you're aware of. For example, if you become scared watching a horror movie, you're conscious of the fight-or-flight response being activated: You're tense, your heart starts pumping faster, and you may start to sweat.

The activation probably occurs in several low-level neural circuits and is so rapid that the fight-or-flight response kicks in before you even notice it. As the great American psychologist William James (1842-1910) pointed out, it's not that we see a bear, become afraid, and run; rather, we see a bear, run, and become afraid. However, some current hypotheses postulate both cognitive and autonomic reactions occurring simultaneously.

Releasing neurohormones

As mentioned previously, in order for the hypothalamus, and, in turn, the autonomic nervous system to be activated properly, a complex integration of signals from the central nervous system must occur.

For example, upon seeing the proverbial bear, the sympathetic system must be activated based on things like your fear of the bear, memory of the closest shelter to which you might run, and potential weapons that you might use. The sympathetic system then must increase interacting components associated with blood flow, such as heart rate and vascular dilation, and match lung output to the increased blood flow by increasing respiration and dilating the bronchi. Meanwhile, most of the systems increased by the parasympathetic system must be inhibited, such as digestive, liver and gall bladder functions. During the time you're running, your body temperature increases, and that is compensated for by sweating. If you run long enough, you consume most of the available glucose in your bloodstream and you begin to metabolize lactate.

One way nuclei in the hypothalamus exercise this control is through the secretion of *neurohormones* (hypothalamic releasing hormones) that act on the pituitary gland at the base of the brain to release specific pituitary hormones. Some of the numerous hormones secreted by the hypothalamus include the following:

- **Vasopressin:** Vasopressin, a peptide hormone, controls the permeability of kidney tubules to help reabsorb and conserve needed substances in the blood, particularly water, by concentrating the urine. Vasopressin also constricts the peripheral vasculature, increasing arterial blood pressure.
- **Somatostatin:** There are two somatostatin peptides, both of which act on the anterior lobe of the pituitary. These peptides inhibit the release of thyroid-stimulating and growth hormones when their levels are too high. Somatostatin also reduces smooth muscle contractions in the intestine and suppresses the release of pancreatic hormones.
- **Oxytocin:** Oxytocin was first known for its role in female reproduction, triggering distension of the cervix and uterus and stimulating breast-feeding. Synthetic oxytocin (pitocin) is sometimes administered during labor to accelerate the birth process. Circulating oxytocin levels also affect maternal behaviors such as social recognition, pair bonding, and female orgasm. Giving pitocin to males has been shown to enhance socially cooperative behavior.
- **Growth-regulating hormones:** The hypothalamus also controls the release of several growth-regulating hormones, some of which are also controlled by somatostatin release. Growth hormones such as somatotropin have direct effects of reducing lipid (fat) uptake by cells and indirect effects of promoting bone growth.

Crossing signals: When the autonomic nervous system goes awry

Sympathetic fight-or-flight responses divert the body's resources from necessary homeostatic mechanisms like digestion to enable rapid and prolonged voluntary muscular exertion. If this diversion is constant and excessive, however, a price is paid. Chronic stress is a name typically given to continual over-activation of the sympathetic system. The specific physiological effects from chronic stress's over-activation of the sympathetic nervous system include

- Suppression of the immune system
- Reduced growth (due to suppression of growth hormones)
- Sleep problems
- Memory dysfunction

Many of these symptoms are associated with high circulating levels of cortisol released by the adrenal cortex.

The response to chronic stress, both in form and amount, may differ considerably between individuals. In women, the accumulation of fat around the waist frequently accompanies chronic stress, while in men experience a reduction in sex drive and a risk of erectile dysfunction. Both sexes may experience depression, hair loss, heart disease, weight gain, and ulcers. Even alcoholism and fibromyalgia have been linked to chronic stress. These problems can occur even if no real threat is present in the environment. Excessive light or noise, feelings of entrapment, and social subordination can induce stress. Stress can also add to sleep and other health problems.

Don't jump to the conclusion from the problems caused by chronic stress that all stress is bad. Under-stimulation in a boring, unenriched environment has been shown to have negative effects on learning and development, and some recent studies even suggest that enriched environments slow the growth of cancerous tumors.

In many ways, psychological stress is like exercise. Moderate exercise, helps muscles adapt by growing stronger. When exercise is too severe or inadequate time is allocated for recovery, inflammation and lasting damage to joints, ligaments, and tendons can occur. Stress works the same way. Humans (and all animals) have evolved to cope with, and respond appropriately, to a moderate amount of uncertainty in their environments.

Sweet Dreams: Sleep and Circadian Rhythms

All animals with advanced brains, such as all vertebrates and even chordates such as sharks, sleep. While you may think of sleep as a time to get all comfy with your favorite pillow and snuggly blanket, sleep is actually a brain and body state in which consciousness is absent or severely reduced. It is not the same as hibernation or coma.

Even though all mammals and most fish, reptiles, and amphibians sleep, no one is really sure why. Hypotheses have ranged from a need to conserve energy to some advantage in staying hidden and still at night to avoid predators. And no theory for the *function* of sleep that has been proposed is anywhere near universally accepted.

Despite the lack of a clear explanation for a primary purpose for sleep, it is clear that important things happen during sleep. It is also clear that there are different kinds of sleep (stages) and that sleep deprivation produces harmful effects such as cognitive impairment. The following sections have the details.

Synchronizing the biological clock with light exposure

One thing about sleep that scientists understand relatively well is how it is controlled. The sleep-wake cycle is a circadian rhythm based on synchronization between an intrinsic biological "clock" and morning light exposure.

The clock consists of a set of interconnected cells in a nucleus of the hypothalamus called the *suprachiasmatic nucleus* (SCN), which means "nucleus above the chiasm." Although the activity of these cells controls the overall circadian rhythm in humans and all other vertebrates, these cells also exist in primitive organisms and in single-celled entities such as algae.

One function of such rhythms in single-celled organisms is avoiding DNA replication during high, ultraviolet light exposure during the day. Another may be to gain an advantage over other organisms by becoming more active just before sunrise rather than waiting for the light itself.

Intrinsic versus real world cycle times

The circadian rhythm cycle time in the suprachiasmatic nucleus network is approximately 24 hours, but rarely exactly. Different people have different intrinsic cycle times, which may be several hours off the 24-hour approximation. In fact, most people have cycle times of about 25 hours. Our intrinsic cycles are synchronized with the 24 cycle of the real world by exposure to light, particularly bright sunlight early in the day.

Scientists discovered the relationship between the intrinsic cycle times and the real world cycle times through experiments in which volunteers lived in enclosed rooms with no external cues about actual time. These volunteers began to "free run," according to their own intrinsic cycle time and became more and more out of phase with real time the longer they underwent this isolation. Many blind people also free run and have difficulty in coping with real time schedules because they tend to be out of phase. (Interestingly, recent evidence indicates that the constant dim light that was kept on in these experiments affected the apparent intrinsic cycle time and that keeping the subjects in real darkness gives an answer closer to 24 hours.)

The role of retinal ganglion cells

The mechanism for the synchronization of the intrinsic clock with external light involves the activities of a special class of retinal ganglion cells called *intrinsically photosensitive ganglion cells,* which project to the suprachiasmatic nucleus.

These cells are unique in that they have their own visual pigment which allows them to directly respond to light. In other words, they're not driven by the retina's photoreceptors through the normal retinal light processing network (refer to Chapter 5 for more on the function and structure of the retina). Although the intrinsically photosensitive ganglion cells do have this "normal" photoreceptor input as well, this isn't the input that controls circadian rhythms.

Scientists know this because some retinal photoreceptor degenerative diseases in humans and animals result in death of all photoreceptors, yet these humans and animals can still synchronize to light. Diseases that kill all the retinal ganglion cells eliminate the ability to synchronize.

Some cold-blooded vertebrates, such as frogs and salamanders, have what is sometime called a "third eye" or "parietal eye" that emanates from the pineal gland. This eye doesn't "see" in the traditional sense. Instead, intrinsically photosensitive neurons in this eye receive enough photos passing through the thin skulls of these animals to synchronize their circadian rhythms.

Looking at the different stages of sleep

Little is known about any difference in the function of the four sleep phases. What researchers do know is that sleep, which is induced by the circadian rhythm, cycles through several characteristic stages, each of which has unique properties.

If you observe someone sleeping, one of the most obvious differences is whether their eyes are moving. During some stages of sleep, the sleeper's eyes move rapidly and nearly continuously, whereas in other stages, the eyes don't seem to move very much at all. This distinguishes *rapid eye movement* (REM) sleep from *non-REM* (NREM) sleep. NREM sleep is further divided into distinct phases, labeled N1, N2, N3, and N4 (some sources do not include a fourth NREM stage). The NREM stages usually occur in either ascending or descending order, with occasional transitions from N2 to REM.

NREM sleep

The initial phases of sleep consist of the transition from wakefulness to deep sleep through the NREM phases, labeled N1, N2, N3, and sometimes N4.

Figure 11-2 shows a typical night's sleep phase plot, called a *hypnogram.* (Notice that REM sleep is the closest phase to being awake.) Typically the sleeper descends to N1, then N2, N3, and so on in order (although not always reaching N4 in the first cycle). The sleeper then oscillates back up through the N phases to REM (about 90 minutes after falling asleep), then back down. Later during the evening, the phases are often shallower, sometimes not reaching N3 or N4. Occasionally there are also brief transitions from REM to momentary wakefulness. (***Note:*** This figure illustrates a "typical" averaged set of cycles. Different people have somewhat different patterns, and the same person's pattern could vary considerably from night to night.)

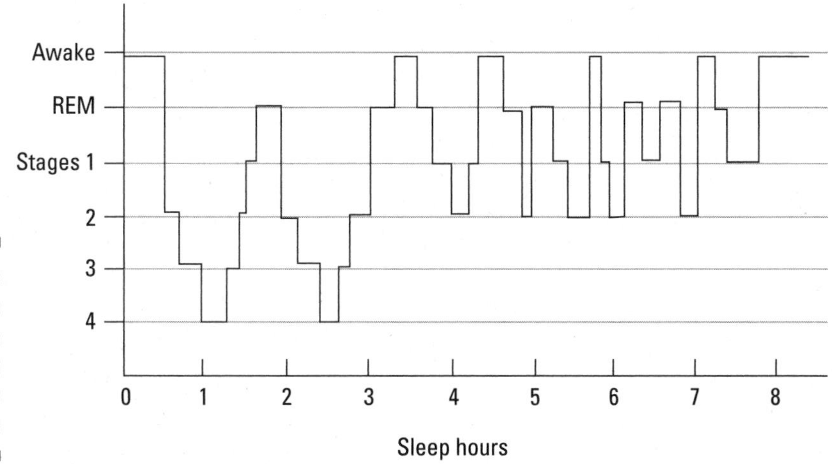

Figure 11-2: Sleep hypnogram showing sleep stages.

The NREM phases are characterized by distinct changes in EEG patterns. Figure 11-3 shows typical EEG waveforms recorded during REM and NREM sleep stages. (*EEG* stands for *electroencephalogram,* electrical recordings of brain activities.)

Normal wakefulness is associated with a predominance of beta rhythms in the EEG, while awake relaxation, such as during meditation, shifts the EEG rhythms from beta to alpha. In the earliest stage of sleep, there are characteristic slow eye movements technically called *slow rolling eye movements* (SREMs). About this time, alpha waves nearly disappear from the EEG and theta waves begin to predominate. Stage N2 of sleep is also dominated by theta waves, but these change to delta waves in N3 and N4. N4 is the deepest stage of sleep, from which it is the most difficult to awaken.

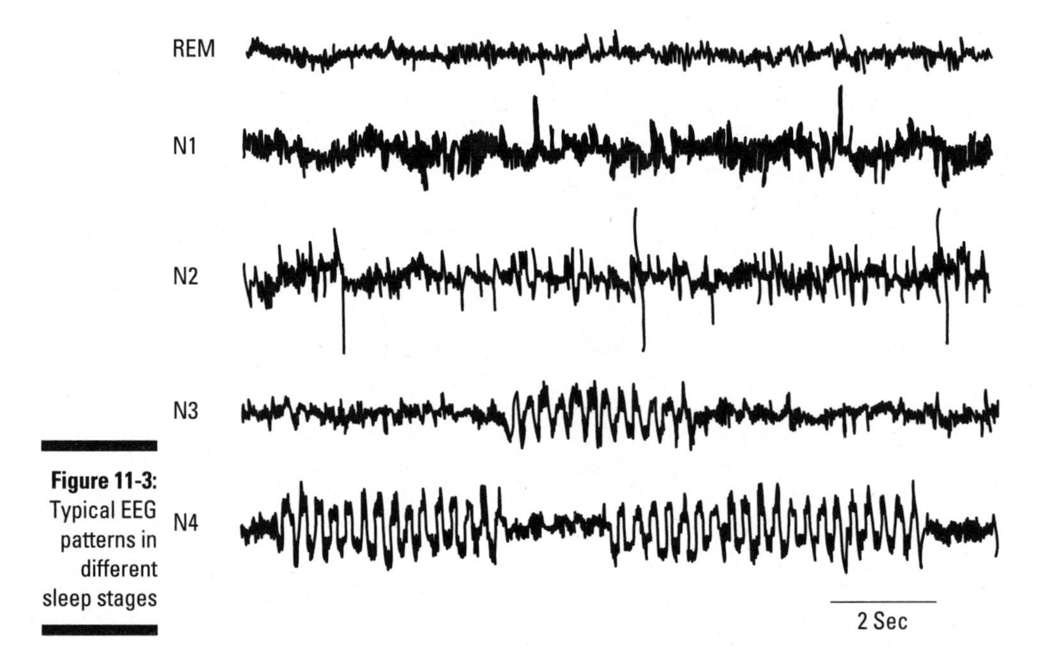

Figure 11-3: Typical EEG patterns in different sleep stages

REM sleep

REM sleep is the phase in which most dreaming occurs. This stage occupies about one quarter of the total time sleeping and is characterized by high frequencies in the EEG. The high frequency EEG activity reflects the fact that the motor activities that occur in dreams, such as running, are associated with frontal lobe activity in motor sequencing areas, but their output is blocked in the spinal cord so that actual movement is inhibited despite the activity in the early movement control brain areas.

REM sleep and learning

Recent research has shown that REM sleep is important for consolidating learning. During REM sleep, what has been learned that day and held in short-term memory is transferred to long-term memory. Researchers believe that short-term memory involves synapses in the hippocampus that have been transiently strengthened by experience via long term potentiation. These are the synapses that received coincident input from various areas of cortex that were part of the sensory representation of what was learned (color and patterns, for example). During sleep these activated hippocampal synapses feed back and activate the areas of cortex that activated them. This causes a kind of "resonant" cortical circuit to exist for that representation. Researchers also postulate that synapses that are potentiated during waking activities are pared back so that the brain is ready for renewed synapse growth during learning.

Here's an example of how the hippocampus transfers memory to long term during sleep. Experiments in which rats were trained to navigate mazes showed that activity sequences in hippocampal place cells, which code for maze location, "replay" the maze navigation sequence during the rat's sleep. Sleep deprivation, specifically of REM sleep, disrupted this playback and the ability of the rats to retain learning that occurred during their daytime practice runs.

REM sleep also appears to be important if for no other reason than sleep deprivation leads to disproportionate compensation for the REM component of sleep. Research in humans shows that when we're sleep deprived, we spend proportionately more time in REM sleep, when we're allowed to sleep, than in the NREM phases. Infants also not only sleep more than adults, but spend a larger proportion of sleep time in REM.

Dream a little dream with me

Dreams typically occur during REM sleep and are best remembered if the dreamer wakes up during or just after the REM stage. Dreams involve images and event sequences such as flying that are literally "fantastic," that is, they disobey normal physical laws.

In many religious traditions, dreams are thought to connect to an alternate reality that portends the future or otherwise informs the dreamer through some extra-normal process. From the point of view of neuroscience, however, dreams are a cognitive phenomenon that accompanies REM sleep, and neuroscience has little to say about the meaning of the cognitive content of dreams. Because the fantastic aspect of dreams resembles psychotic thinking, dreams may represent brain activity that perhaps occurs in all people and is unchecked by rational constraints during the dream process.

A world of dreamers

Researchers believe that all mammals and most, if not all, vertebrates sleep. Although we can't know directly whether non-human animals dream during sleep or what the content of those dreams would be, it is clear that many animals exhibit the EEG characteristics of REM sleep which is accompanied by dreaming in humans. If animals are deprived of sleep they also, when allowed to sleep, enter REM more quickly and spend more sleep time in REM.

You can easily notice REM sleep in the family dog by observing its eye movements through its closed eyelids and the suppressed movements and vocalizations that demark this sleep period from other, deeper non-REM epochs.

From an evolutionary point of view, the near universality of dreaming in humans and evidence that dreaming may also occur in other mammals, suggests that dreams have some function. The counter argument is that REM sleep has a function, but the dreams themselves, which just happen to go with REM sleep, are meaningless. Currently there is no scientific consensus on this issue.

Functional associations of brain rhythms

The EEG is a recording made by one or more (up to hundreds) of electrodes placed on the scalp. These electrodes show constant ongoing brain activity whether the subject is awake or asleep. The frequencies present in the EEG reflect aspects of brain state, such as sleep phase or alertness during waking. The types of waves present in the EEG are often described as *rhythms* and are given Greek alphabetic names like *alpha, beta, gamma, delta,* and *theta.* The strength of each rhythm is the amount of amplitude in the frequency band defined for that rhythm.

Type of Rhythm	***Frequency***
Alpha	8-13 Hz (cycles per second)
Beta	13-30 Hz
Gamma	30-100 Hz
Delta	0-4 Hz
Theta	4-8 Hz

Figure 11-4 shows what a 4-second interval dominated by each of these rhythms would look like. (***Note:*** In real life, several rhythms are typically active simultaneously, and the exact frequency of each rhythm changes over time, so real EEG traces look much noisier.)

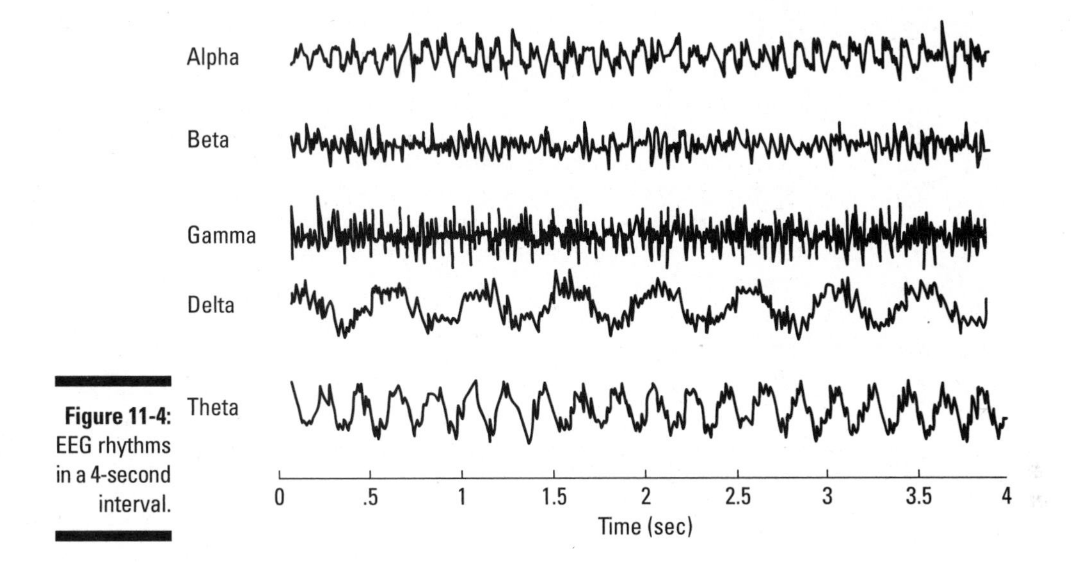

Figure 11-4: EEG rhythms in a 4-second interval.

One of the major functions of EEG recording is to detect what are called *spindle* or *ictal* wavelets that occur in the EEGs of patients with epilepsy. A small number of such spindles are normally produced in N2 sleep, such as the three high-amplitude, biphasic spikes Figure 11-3.

Alpha rhythms

Alpha rhythms are typically associated with brain states like that in meditation, with the person awake but conscious, and are strongest with the eyes closed.

A popular movement in the 1960s and 1970s used home electronic monitoring of alpha waves in biofeedback to increase alpha wave activity as a means of deepening or improving one's meditative state. People, and even animals such as cats, can indeed increase alpha wave production with biofeedback. Investigating whether enhanced alpha wave production leads to cognitive function on a higher plane is somewhat beyond the scope of this discussion, but it is clear that mediation (and prayer) can reduce anxiety.

Beta waves

Beta waves are associated with alertness. They are most prominent in the frontal lobes and occur during concentrated thinking or thinking associated with mental effort.

Gamma waves

Gamma waves have been implicated in information processing in the brain, particularly in "binding" different brain regions together via synchronous firing (see Chapter 14 for a discussion of short-term memory hypothesized as an interaction between theta and embedded gamma EEG waves). Awake, conscious activity is typically associated with beta and gamma wave dominance of the EEG, although there is considerable variation both between subjects and in the same subject depending on mood, task demands, fatigue, and other factors.

Delta waves

Delta waves are almost always characteristic of deep, slow-wave sleep (N3 and N4, discussed above). They appear to originate in a coordinated manner in interactions between the thalamus, cortex and reticular formation under control of hypothalamic sleep mechanisms.

Theta

Theta waves are often associated with inhibition going on in the brain, such as when a usually automatic response to a stimulus is suppressed. Researchers postulate that they are also associated with short-term memory recall.

Controlling the sleep cycles

The control of sleep overall is part of the circadian rhythm mediated by the suprachiasmatic nucleus, which synchronizes the entire body to a single clock. This clock is "set" by exposure to early daylight through the actions of intrinsically photosensitive retinal ganglion cells that project to the suprachiasmatic nucleus via the retinohypothalamic tract.

Control of particular sleep cycles such as REM occurs through activity of neurons in the brainstem, particularly what are called *sleep-on cells* in the pontine tegmentum area of the reticular activating system (a diffuse "structure" that extends from the spinal cord through the brainstem and includes areas in numerous subcortical structures). REM is also associated with reduced release of neurotransmitters serotonin, norepinephrine, and histamine, but higher release of acetylcholine and dopamine.

Among the non-REM phases of sleep, stages 3 and 4 are the deepest sleep. They are sometimes referred to as *slow-wave sleep* due to domination of the EEG by delta waves (refer to Figures 11-3 and 11-4). Slow-wave sleep is known to be prolonged if the preceding duration of being awake is longer than normal. However, paradoxically, some dreams occur during slow-wave sleep, although far less than during REM. This has been ascertained by waking subjects up in this phase (as monitored by EEG recordings).

Although the levels of acetylcholine are elevated in the other sleep stages compared to wakefulness, they are relatively reduced during slow-wave sleep. Scientists hypothesize that high acetylcholine levels in REM sleep gate the flow of information from neocortex to hippocampus, while low acetylcholine levels in slow-wave sleep permit transfer of information from the hippocampus back to neocortex as part of memory consolidation.

Not so sweet dreams: Fighting sleep disorders

About 40 million Americans suffer from sleep disorders serious enough to compromise their quality of life. Beyond that, sleep disorders can be extremely hazardous in professions like truck driving and high-risk factory jobs. Even without any underlying sleep problem in normal circumstances, lack of enough time for sleep, such as during childrearing or rotating shifts, when the time you're supposed to sleep changes frequently, can cause serious sleep disturbances. Many sleep disorders also cause a range of sleep problems even for people given adequate and appropriate opportunity to sleep. Below are some common and/or notable examples of sleep disorders.

- **Primary insomnia:** Insomnia (the inability to sleep) is called *primary* when you have chronic difficulty falling asleep or staying asleep when no other known cause exists (such as pain that keeps you awake). Primary insomnia may result when the stress-induced inability to fall asleep or maintain sleep becomes chronic even if the stress is removed. Primary insomnia may also result from a lesion or damage to the hypothalamic control centers for sleep, such as the suprachiasmatic nucleus or a decrease in melatonin release, as is thought to occur with aging.
- **Narcolepsy:** The tendency to fall asleep during the day at inappropriate times is called *narcolepsy.* Narcolepsy is sometimes accompanied by *cataplexy,* sudden muscular weakness induced by strong emotions. Narcolepsy is known to be caused by genetic mutations. Several strains of Labradors and Dobermans with genetic narcolepsy have been bred for scientific studies.
- **Restless leg syndrome:** Almost all voluntary muscle movement is normally suppressed during sleep. However, people with restless leg syndrome (RLS) experience limb movement during sleep, including jerking, mostly in the legs, but often in the arms and even torso. RLS can range from very mild cases to severe enough to cause significant sleep loss. It can also be progressively worse or stop suddenly. Some RLS is genetic, and many cases are associated with iron and dopamine deficiencies.

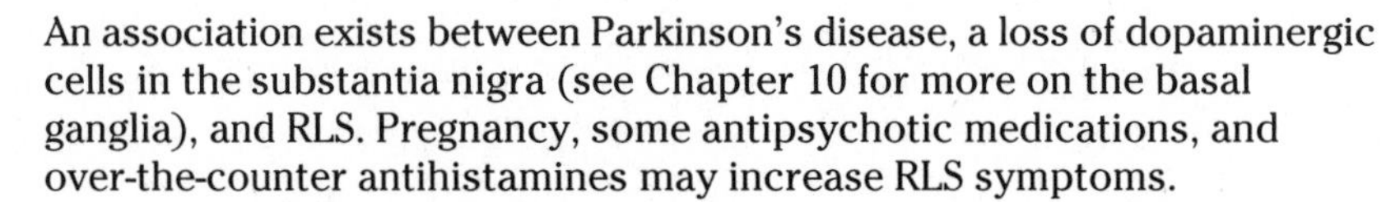

An association exists between Parkinson's disease, a loss of dopaminergic cells in the substantia nigra (see Chapter 10 for more on the basal ganglia), and RLS. Pregnancy, some antipsychotic medications, and over-the-counter antihistamines may increase RLS symptoms.

- **Sleepwalking (somnambulism):** Somnambulism involves engaging in activities normally associated with wakefulness, but occurring during slow-wave sleep. Although its precise cause is not known, it appears to have a genetic component due to clustering in affected families. Somnambulism typically starts in childhood but often decreases after maturation, giving rise to theorizing that it may be associated with some delay in maturation. It is sometimes increased with anti-depressants such as benzodiazepines.

People almost never have normal consciousness during sleepwalking and may not only inadvertently injure themselves but commit violent acts. This phenomenon has presented challenges to the court systems that must determine guilt for such acts, but opportunities for soap opera writers.

- **Jet lag:** Circadian rhythms boil down functionally to light-dark cycles occurring in an ongoing rhythm that is approximately 24 hours long when free running. However, if the light cycle is changed ("daylight" — whether natural or artificial — increased or reduced), shifting the rhythm can take a number of cycles. Our ancestors never experienced this demand in the natural world, of course, but airplane travel and shift changes in artificially lighted work settings can, in one day, change the light cycle humans are exposed to. Such a change causes sleep problems, known as *jet lag* in the case of travel, wherein your body remains in the old circadian cycle despite being physically in a different one. Jet lag type problems are frequently experienced by shift workers who change shifts frequently.

Recently using melatonin to treat jet lag has had some success. Melatonin is a hormone produced by the pituitary gland mostly during darkness and then shut down at first light. Taking melatonin at the time you should fall asleep in the new time zone has been shown in some cases to help induce sleep. Exposure to very bright light in the mornings in the new time zone is also helpful.

- **Sleep apnea:** Sleep apnea is characterized by abnormal pauses in breathing during sleep, often indicated by snoring. In *central sleep apnea,* the breathing is interrupted by insufficient respiratory effort, while in *obstructive sleep apnea,* the interruption is due to physical blockage of the airway, typically accompanied by snoring. Obstructive sleep apnea in males is known to be a significant cause of sleep loss in females.

Part IV

Intelligence: The Thinking Brain and Consciousness

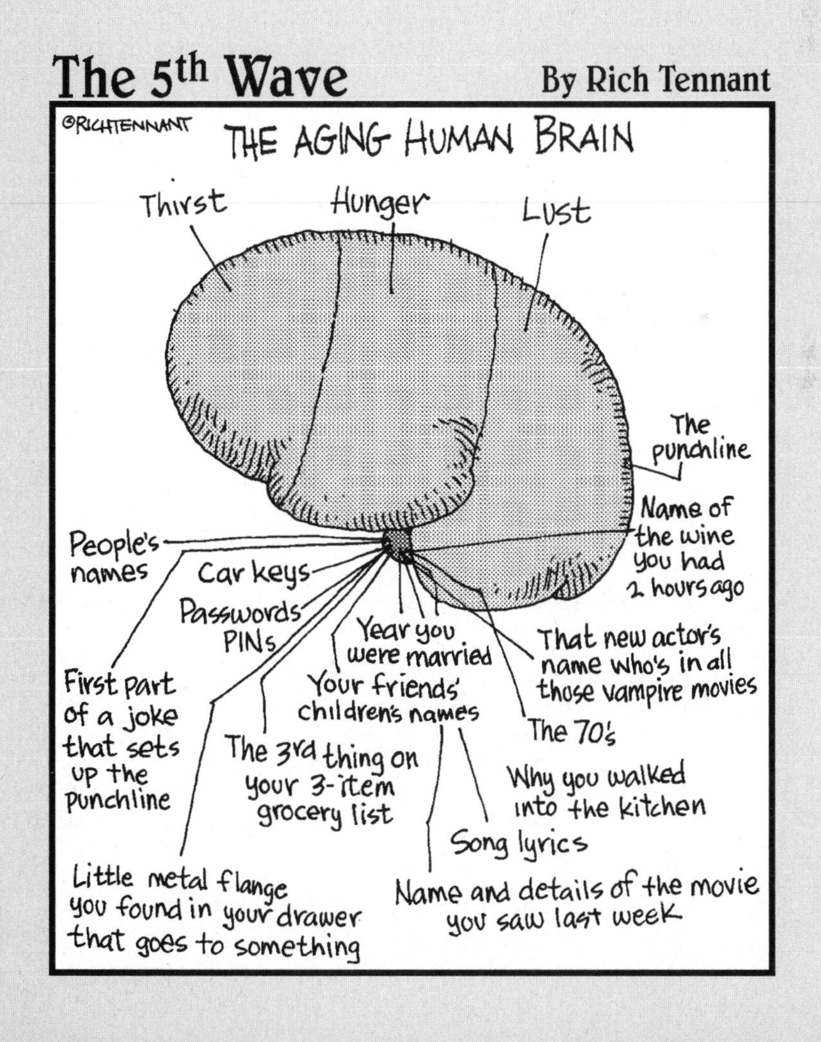

In this part . . .

Humans can do things no animal can do, and some humans have done things no other human previously accomplished. We call much of this intelligence, and consider it the characteristic that makes our species unique. We humans spend decades in schools learning how to think more intelligently. Intelligence exists because of the innate structure of the brain and because that structure can be modified by experience. In other words, intelligence — and our capacity to learn — depends on tweaking the firing of neurons in the brain.

This part covers the important structures in the brain that permit abstract thinking, language, and learning. It also shows how individual neurons participate in this process by making slight changes in the way they communicate with each other. Knowing how the brain learns helps in the effort to improve the rate and amount we can learn and provides insight into how we can deal with problems that occur when some part of the brain is not working properly. There is great hope now for improving normal brain function and correcting dysfunctions.

Chapter 12

Understanding Intelligence, Consciousness, and Emotions

In This Chapter

- Looking at the types and nature of intelligence
- Understanding the role of emotions in cognition
- Processing emotions through the limbic system
- Becoming aware of the details about consciousness

Because humans need to do many different kinds of complex things, there are many different types of intelligence. The intelligence required to make and use tools, for example, is very different from the intelligence needed to understand that your actions are irritating someone who can harm you. Not only do many different kinds of intelligence exist (language versus spatial skills to name just two), but intelligence also exists on many levels. A person can display outstanding intelligence in picking the likely winners in horse races but lack the intelligence to know that betting on horses isn't likely to provide a secure and long-term source of income.

The kind of intelligence we learn in school is mostly rule-based (if these circumstances exist, then this follows. . .) and expressible in language. In contrast to rule-based thinking are emotions. Emotions are feelings, often not expressible in words, that are elicited by circumstances. Although they're thought of as the opposite of reasoning, emotions are a useful and necessary part of cognition. They mediate not only instinctively controlled aspects of behavior, like avoiding social gaffes, but they also permit learning and adaptive behavior that is not rule-based. Damage to particular parts of the brain, such as the orbitofrontal cortex, can produce individuals with normal IQs but deficient emotional processing who are severely dysfunctional in normal life situations.

This chapter delves into intelligence, emotion, and *consciousness,* the explicit awareness of our own thoughts and experiences, which is the ultimate manifestation of intelligence.

Defining Intelligence

Intelligence is something most of us define by reference to examples rather than by any particular rule. Intelligence is exhibited in behavior that is adaptive, that is, appropriately responsive to circumstances, particularly when those circumstances are complex and changing. An important aspect of intelligence is the ability to make predictions. This ability is typically enabled by learning and its result, memory. Upon noticing that clouds are appearing, for example, it would be intelligent to take an umbrella into a movie theater because it might be raining by the time you leave, even though it isn't raining now.

Some power for prediction undoubtedly exists in our genes. Intelligence was selected by evolutionary mechanisms because it produced adaptive behavior, the ability to select more-complex goals and to pursue them in a more complex environment.

Types of intelligence

You can delineate what are called *types of intelligence* in many ways. Standard college entrance exams typically have verbal and math (and sometimes analytical) sections, whereas J.P. Guilford's well-known factor analysis of intelligence had as many as 150 factors in three dimensions (contents, products, and operations). A relatively recent compromise is Howard Gardner's eight types of intelligence, shown here.

Intelligence	***Aptitude for or Competence In. . .***
Linguistic	Comprehension and use of language
Logical-mathematical	Quantitative and logical reasoning
Spatial	Navigation, geometry and patterns
Interpersonal	Social interactions
Intrapersonal	Your own strengths and weaknesses, wisdom
Bodily-kinesthetic	Sports, dancing, athletics and fine motor control
Musical	Singing, musical instruments, composition
Naturalistic	Understanding patterns in nature and surviving in a natural, rather than man-made, environment

While older systems for analyzing intelligence concentrated on factors believed to correlate with success in school and western industrial society in general, Gardner's intelligence types include several distinctly non-academic categories.

Understanding the nature of intelligence: General or specialized?

The environment we humans find ourselves in is complex beyond that of any other species because we have built so much of it ourselves. Humans have modified their physical and social environment to the point of creating numerous microenvironments requiring very different skills: physical manipulation of things like pens and keyboards, social skills for dealing with relationships, language skills for communicating, mathematical and spatial skills for reading maps and navigating, and many others.

One of the longest standing controversies in psychology and cognitive science is whether intelligence is unitary or divisible into distinct sub-abilities.

At one extreme is the behaviorist psychology tradition that regards the brain as a general-purpose learning device whose internal structure is of no significance in understanding learning or behavior. At the other extreme were the *phrenologists,* who believed that development of specific intellectual traits was associated with brain growth that literally made the skull overlying that part of the brain bulge out. By examining people believed to have high or low ability in various traits and then associating these abilities (or lack thereof) with the relative height of various areas on their skull, phrenologists created elaborate maps that identified the location of dozens of traits ranging from mathematical ability to caution (see Figure 12-1).

The fundamental premise of phrenology is now totally discredited. Certain areas of the brain do not map specifically or uniquely to certain types of intelligence. Nor do local areas of the brain (neocortex) grow and bulge out the skull according to the development of particular traits such as caution or secretiveness.

On the other hand, the brain is not a diffuse, non-differentiated association-forming blob, either, as the following sections explain.

Sensory pathways to specific areas of the brain

Sensory inputs and motor outputs follow specific pathways and are processed primarily in particular areas, although most skills involve multiple brain areas. The prefrontal cortex, for example, is crucial for the performance of almost all skills requiring intelligence.

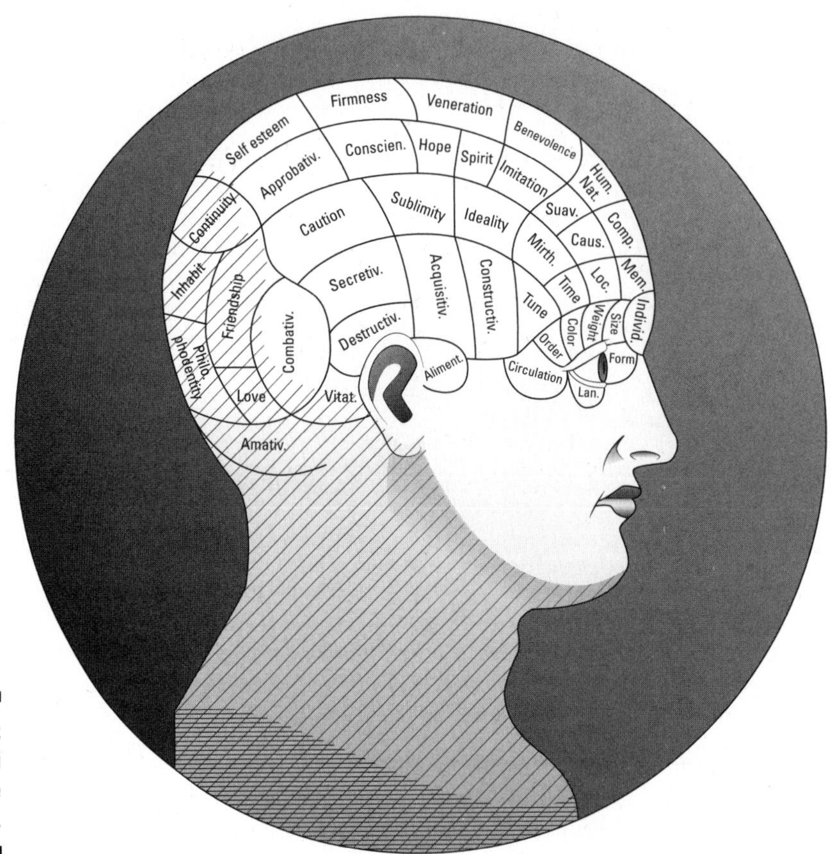

Figure 12-1: Phrenological map of the brain.

Damage to some specific areas of the brain can greatly diminish certain skills, like language comprehension or production, while leaving other skills relatively intact. Besides language, skills such as understanding spatial arrangements and how to manipulate tools and objects also appear to be particularly affected by lesions in specific brain areas. These facts seem to indicate that intelligence is specialized.

Figure 12-2 shows a few cases where specific intelligence traits appear to be particularly compromised by local brain injuries. Scientists know, for example, that spatial skills such as navigation and pattern recognition are highly dependent on the right parietal cortex (top image in Figure 12-2), whereas language depends on Broca's and Wernicke's areas on the left side of the brain (explained in Chapter 13), as shown in the bottom image in Figure 12-2.

One interesting syndrome that appears to be relatively localized is *acquired amusia,* the loss of the ability to recognize musical melodies even when language and other auditory abilities remain intact. Amusia is associated with either unilateral or bilateral damage to high-order auditory cortex in the posterior temporal lobe. It can result from damage on either the right or left side but is more commonly seen with damage on both sides caused by blood flow interruptions in the middle cerebral arteries on both sides.

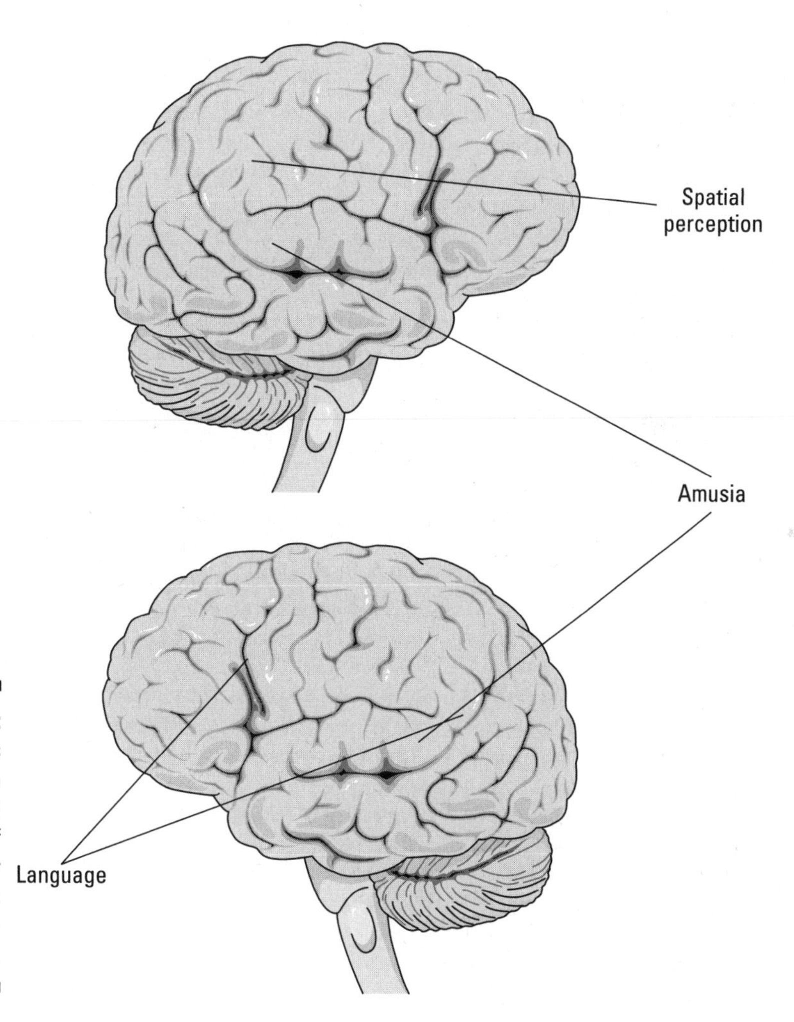

Figure 12-2: Some areas of the brain associated with loss of particular abilities when damaged.

Localization, plasticity, and recovery from brain damage

People may recover from significant brain damage, such as that caused by strokes. This recovery is possible because other brain areas, both locally and globally, take over the function of the compromised brain areas:

- **Local recovery:** The damaged brain area and the area immediately surrounding it reorganize to restore some of the lost function.
- **Global recovery:** This type of recovery is more complicated because it may involve using different brain areas and pathways to accomplish the same task in the same way, or alternate ways of doing things may be learned. People who have suffered motor cortex damage controlling their right hand and arm, for example, may learn to use their hand differently for grasping objects, or they may rely more on their left (spared) hand.

You can read more about plasticity in Chapter 16.

Components of intelligence

Regardless of the outcome of the debate about how many distinct intelligence factors exist, in most real-life tasks, intelligence depends on a number of skills. These include the ability to pay attention, short- and long-term memory, the ability to visualize, and motivation and expectations.

A task that places great demands on intelligence is decision-making, which depends on knowledge, visualization, planning, and abstract thinking. These components typically cross boundaries between linguistic, logical, and spatial types of intelligence. Creativity, specifically, is aided by intelligence but not necessarily a direct product of it. Some very creative people score poorly on intelligence tests, and some very intelligent people appear to be unable to "think outside the box," as the colloquialism goes.

One of the motivations for trying to identify the factors that underlie overall intelligence is the hope that, by enhancing these factors, you could modify intelligence. Another is the hope that identifying the factors that underlie intelligence would help scientists understand it.

Biological variations

When trying to determine the effect biological variations have on intelligence, scientists look at the differences between humans and other animals. In these studies, they've found a link between intelligence and the ratio of brain size to body size, and between intelligence and the relative percentage of prefrontal cortex.

Differences among humans have also been studied, but you can't tell much about intelligence by making comparisons between humans. One of the difficulties is that, although a statistical link does exist between intelligence and brain size, small brain size may be the result of specific genetic pathologies or environmental effects such as poor prenatal nutrition or maternal drug abuse. Because factors such as these also affect intelligence, directly linking brain size and brain power is impossible.

Intelligence does seem to be correlated with the size of working memory and with the complexity of EEG waveforms, and, in males but not females, with the speed of mental processing as measured by specific reaction times. However, like other behavioral correlations, whether these phenomena are the cause of intelligence or the result of it is not clear.

Upbringing

How much does upbringing affect intelligence? One way to approach this question is to look at identical twins raised separately to see whether their genetic makeup caused their intelligence to be nearly the same despite different upbringing. If genetics were everything, the IQs of such twins would be exactly the same. However, such twins who have been studied tend to correlate only about 70 percent statistically, suggesting a considerable environmental influence as well.

This sort of result can and has been misconstrued, however. The twins reared apart are not exactly controlled experiments with random selection. The extent to which the adoptive families, for example, resemble the biological parents makes outcome similarities look genetic. In addition, 50 percent genetic determination of intelligence leaves the other 50 percent environmental. More importantly, even if the environmental component in typical studies turns out to be very small, it does not mean that some non-typical intervention, like a new teaching method, might not potentially make a much larger difference.

Most scientists believe the proper way to view the genetic component of intelligence is as a capacity. There are biological limits to the size of working memory, the rapidity of neural processing, and the information that can be conveyed along specific axonal fiber tracts. These capacities don't seem to vary greatly between individuals, but the ability to use them effectively does vary greatly. Small differences in genetic inheritance can be amplified by experience. For example, if a child reads well, she will tend to read more, and her vocabulary and world knowledge will be greater, as will her powers of concentration and patience. She will do more of the things that result in higher intelligence and do them better during her education than another child with only slightly less raw reading ability.

Savants

Anyone who thinks they really understand intelligence has to be humbled by *savants,* people with inexplicably extraordinary mental capacities, usually in one particular domain: the ability to multiply two 4-digit numbers in their head, instantly calculate the day of the week of a particular historical date, or display near photographic or encyclopedic memories, for example.

Neuroscientists have no ready explanation for how savants can do these things or why the rest of us can't. Some savants have abnormal brain structures (the older brother in the movie 1988 film *Rain Man,* starring Dustin Hoffman and Tom Cruise, for example, had no corpus callosum connecting the left and right brain hemispheres), but others have brains that appear normal in gross architecture.

People with savant skills do not enjoy particularly great success or happiness. Many are burdened by their memory or other skills and are unable to turn them off and concentrate on matters in normal life. Often savant skills seem to be counterbalanced by incompetence in many other areas necessary for success in life. Many, but not all savants are autistic, for example.

If you think of the human genetic endowment for intelligence as a capacity, then savants seem to indicate that either the majority of us are simply nowhere near our real capacity or there is some mechanism or structure in the brain, whose identity is totally not understood, that yields savant capacities in a very few people.

Intelligence as an adaptive behavior

The core of intelligence has to do with adaptive behavior that is broad and flexible, as well as deep. Although savants, for example, score very highly on some IQ tests, their general lack of success in life, and even in coping with normal life, suggests that IQ tests do not capture a fundamental characteristic of intelligence. Consider the parallels between savants, who, for example, can instantly tell you that Christmas day in the year 1221 was a Wednesday but can't dress themselves properly, and people with frontal lobe damage, who often score higher than average on IQ tests but cannot cope with the demands of normal life, particularly when those demands are continually changing.

What IQ tests seem to measure is some combination of the level of some intrinsic abilities and the results from applying those abilities which manifest as specific skills acquired at any particular age. The tests do not measure things like judgment, wisdom, and self control, which are also necessary for success in life. Skills are important and can accomplish almost miraculous things at times, but they are not everything.

Looking at the different levels of intelligence

In addition to intelligence being embodied in different domains, such as language versus spatial manipulation, it also appears to exist in different levels of competence. The lowest level of competence associated with intelligence is that embodied in the lowest life forms, single-celled organisms; the highest level we know of is human consciousness:

- **Homeostasis in the most primitive life forms, such as prokaryotes:** Single-celled living organisms are the first great divide between life and non-life. Prokaryotes (cells without distinct nuclei) such as bacteria maintain homeostasis by regulating the movement of substances through their cell walls. Internal enzyme systems construct intracellular structures, regulate the consumption of energy and release of waste, and allow for reproduction by division. These regulatory functions could be considered the most primitive form of intelligence.
- **Cellular specialization in eukaryotes:** Eukaryotes (cells with nuclei) have internal organelles such as a nucleus and mitochondria. As such, they are multi-component organisms, even at the cellular level. Multi-celled organisms are comprised of eukaryotic cells that specialize individually to function within a larger organization. This internal and external cellular differentiation enormously increases the adaptive power of the organism.
- **Awareness in vertebrates and possibly cephalopods:** Among multicellular animals, vertebrates (and perhaps octopuses and squids) appear to demonstrate awareness. Awareness involves having internal models that allow multi-step but non-instinctual goal pursuit strategies. In mammals, much of this capability is due to the limbic system and its interaction with the neocortex. Currently, scientists haven't been able to discover what processes enable octopuses to exhibit the complex adaptive behavior they do with brains that are small and organized completely differently from human brains.
- **Consciousness:** The ultimate manifestation of intelligence is in *consciousness,* the explicit awareness of our own thoughts and experiences. Consciousness in humans tends to be associated with language that allows us to label and categorize what is going on in the world. Consciousness, derived from language, depends in part on brain size, in part on brain organization, and in part on special abilities such as language unique to humans. Consciousness exists in rudimentary form in primates, but it exists fully only in humans.

- **Metacognition:** Metacognition reflects the highest, most abstract level of mental competence in consciousness. It's associated with awareness of the effects of one's own actions and the ability to manipulate abstract representations. Metacognition involves monitoring oneself and one's actions as an almost dispassionate, external observer. Certain types of meditation and prayer are examples of attempts to enter a metacognitive state.

Intelligence about Emotions

A commonly misunderstood aspect of intelligence is its relationship with emotion. The misconception is expressed in the dichotomy that intelligence is rational and rule based, while emotion is instinctive, animalistic, and irrational.

However, emotions can be viewed in two ways that belie any perception that they are intrinsically inferior to rational, rule-based reasoning:

- **Emotions are crucial for mediating social interactions, which are one of the most complex environments in which humans exist.** Across animal species, the most intelligent animals tend to be social or descended from social species. Emotion is the currency by which complicated interactions such as role playing in social hierarchies occurs.
- **Emotion is a way that the brain can communicate the output of a neural computation to control behavior that is not expressible as a rule.** For example, most mammals, including humans, have an inborn fear of snakes. When we encounter something snakelike as a child, long before we know what a snake is or the difference between poisonous and non-poisonous types, we are instinctively afraid of and avoid it.

  Higher cognitive systems hijack the approach/withdrawal emotion system for much more complex situations that result from learning. After we learn how to drive, for example, we come to understand that a car weaving toward us is probably being driven by someone who is either not paying attention or is mentally compromised in some way, and, feeling fear, we move out of the way even before we're consciously aware of doing so.

Certain brain centers impact emotions. The emotion processing areas of the brain are phylogenetically ancient, such as the subcortical structures included in the limbic system (such as the amygdala, septum, fornix and anterior thalamic nuclei; see the later section "Emoting about the limbic system" for details on these structures). Mammals, and particularly primates, add additional cortical interactions between the older emotion systems and newer areas such as the orbitofrontal cortex.

The orbitofrontal cortex gives high resolution to the emotional response system. Instead of just being afraid of dangers programmed by instincts, such as spiders and snakes, we can become afraid of, and learn to avoid, complex situations that are dangerous, such as insulting our boss or buying a used car without having someone inspect it first.

Tapping into memories of strong emotional reactions

The orbitofrontal cortex-amygdala system is hard to understand as a kind of intelligence because its output is to create an uneasy feeling about a situation that is not expressible in words. This idea has been formalized by the University of Southern California's Antonio Damasio as the *somatic marker hypothesis*. Suppose, for example, that your car runs out of gas at a particular place on a road you travel frequently. The inconvenience and potential danger of the situation create an emotional experience at the time. Days later, when driving past that same area, you are likely to have an uneasy feeling that you cannot understand rationally and that causes you to check your gas gauge.

Evidence indicates that the memories created in the orbitofrontal cortex-amygdala system are very stable. That is, virtually any situation you have experienced that had unpleasant consequences will produce a memory that is activated in similar situations. The rational part of your brain overrides the uneasy feelings when necessary. For example, if you've been in a car wreck, you may, for the rest of your life, experience uneasiness whenever you're in a car, but you override that feeling because you need to drive. Continuing to drive may partially *extinguish* the strength of the feeling, but the unease may never go away entirely.

Emoting about the limbic system

The limbic system is a set of subcortical brain nuclei and their interconnections that were once thought to be an evolutionary system that mediated primitive instinctive behavior.

Brain structures typically enumerated in the limbic system include the amygdala, fornix, hypothalamus, hippocampus, thalamus, and cingulate gyrus (see Figure 12-3). While the first three structures are involved in processing emotionally salient stimuli, like jumping away from a snake, the hippocampus and thalamus are involved in many cortical circuits, such as memory and sensory processing, that have no obvious relationship to instinctual responses. The hypothalamus is involved in lower "reptilian" functions such as temperature and fluid regulation.

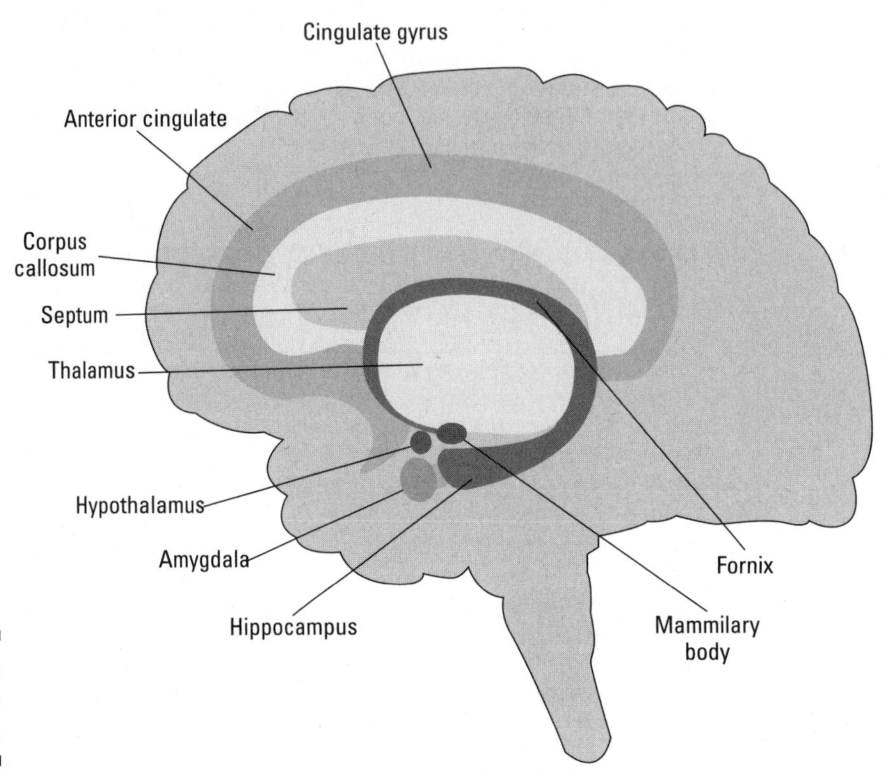

Figure 12-3: The limbic system.

The hippocampus

The *hippocampus* is the most important structure mediating the transfer of information from short-term to long-term memory. It receives connections from virtually the entire neocortex; these connections are mapped onto synapses whose strength can change if multiple inputs are simultaneously active. For example, if some color-detecting area of the cortex signals "gray," another area signals "big," and another signals "has a trunk," the neuron in the hippocampus that all three of those neurons project to could be part of an elephant detector. Neurons in the hippocampus project back to the neocortex, activating exactly the same sensory areas that sent the original signals and doing so in a way that a mental representation of an elephant forms.

The hippocampus is involved in many different aspects of memory, not just those having high emotional salience (which seems to be the function of the amygdala, just in front of the hippocampus). Extensive research on the hippocampus in rats by John O'Keefe and colleagues from the University of Arizona indicates that much of the hippocampus is devoted to navigation in that species. Researchers conjecture that memory for navigation may have been the original evolutionary function of the hippocampus, after which it became involved in other kinds of memory, including particularly episodic memory (see Chapter 13 for information about the difference between episodic and semantic memory).

MacLean's triune brain

One of the most commonly known schemes about brain organization is the *triune brain,* postulated by neuroscientist Paul MacLean in the 1960s and popularized in the late 1970s by Carl Sagan's book *The Dragons of Eden: Speculations on the Evolution of Human Intelligence* (Random House). The basic idea of the triune brain is that the brain has a three-level hierarchical structure in which the top two levels evolved one after the other over the lowest, reptilian level. According the MacLean's theory

- The bottom "reptilian" level (exhibited by lizards and snakes) has the basal ganglia as the highest organizational level (above the brain stem and spinal cord seen in the most primitive vertebrates and even some invertebrates). The basal ganglia mediates instinctive behaviors such as fight-or-flight responses, hunger, defending territory, and sexual pursuit.
- The next level added on top of the reptilian brain is the limbic system, proposed to contain the amygdala, hypothalamus, septum hippocampus, hippocampal complex, and cingulate cortex. These structures arose in early mammals and allowed feelings, which enabled more complex behaviors such as parental nurturing and participation in social dominance hierarchies.
- The final level is the neo-mammalian complex primarily associated with a large neocortex which reached its peak in primates, particularly humans. This level allows complex social behavior, intelligence, varying degrees of awareness, and ultimately, consciousness.

The triune brain hypothesis was influential and contained many worthwhile ideas. However, it has been difficult to see where to draw the line between the limbic and neo-mammalian levels.

The amygdala

The *amygdala* appears to be part of a different memory circuit. It has some structural similarities to the hippocampus, but its inputs and outputs are associated with stimuli that have emotional salience. Many of the inputs to the amygdala reach it by a fast, low resolution route such as directly from the thalamus, olfactory bulb, or other subcortical structures. The amygdala's output goes to numerous subcortical structures such as the hypothalamus, as well as to the orbitofrontal cortex.

The neural circuitry involving the amygdala allows not only the typical inborn avoidance responses such as to snakes but also the ability to learn contexts associated with danger or pain. This is the same circuitry, in fact, that malfunctions in the brains of nearly every horror-movie hero or heroine.

The anterior cingulate cortex

The cingulate gyrus is a mesocortical (older, non-neocortical) area just above the corpus callosum (the large fiber tract that connects the two cerebral hemispheres). The mesocortex is phylogenetically older than the neocortex.

Recent findings about the anterior portion of this mesocortical area suggest that it has an important monitoring role in which it integrates activity in many other brain areas. Specifically, the anterior cingulate cortex is activated in tasks in which error detection and conflict monitoring occur, such as the Stroop task.

In the Stroop task, the participant must report the color of letters that spell color name words whose color is different than the ink of the letters. That is, the letters spelling the word "blue" are some other color than blue, and the test subjects must report that other color. Virtually all subjects performing this task have trouble; they say the color word rather than the color of the letters, despite practice. The implication is that the process of reading words has become so automatic in normal people that it overrides even conscious attempts to report the color of the letters.

The anterior cingulate cortex is also activated by the expectation of receiving pain and appears to occupy a crucial position in neural mechanisms associated with rewards and punishment contingent upon behavior. For example, if you strike a tennis ball improperly so that it doesn't clear the net, your anterior cingulate is activated just from the wrong feeling you had on the swing, as well as when the ball actually hits the net.

Anterior cingulate activation also seems to be dependent on the level of awareness of the quality of performance in various tasks. If you find yourself in a difficult position in a chess game, for example, your anterior cingulate will be activated even though you may be unaware of any particular bad move. In such a situation, it's acting as an executive to summon resources, such as better concentration, to deal with the fact that you are having difficulty with the task at hand.

The orbitofrontal cortex

The orbitofrontal cortex is the anterior and medial part of the prefrontal cortex. The orbitofrontal cortex is essential for risk and reward assessment and for what might be called "moral judgment." Patients with damage to this area may have normal or superior intelligence as assessed by IQ tests but lack even a rudimentary concept of manners or appropriate actions in social contexts; they also lose almost all risk aversion despite clear knowledge of bad consequences.

The role of this area was made famous by an incident in which extensive damage in this area occurred to a railroad worker named Phineas Gage. While using a tamping iron to set an explosive charge, the charge went off and sent the iron completely through Gage's orbitofrontal cortex. Gage lost consciousness only briefly and appeared to recover almost immediately. After several days of treatment to deal with possible infection, he returned to work. After the accident, Gage appeared to be as intelligent as before, but he was not the same man. He became impulsive, abusive, and irresponsible, a total contrast with his previous personality. He also frittered away his income on a series of highly speculative, unwise investments, and was ultimately dismissed from employment.

The orbitofrontal cortex and amygdala form a system for learning about appropriate behavior and avoiding behavior with bad consequences, in part by experiencing fear of those consequences. Many incidents similar to the Gage incident, as well as brain imaging studies, corroborate the importance of the orbital-frontal cortex for learning and being able to display appropriate behavior.

Understanding Consciousness

Neuroscientists used to be loath to discuss consciousness because no one could define it and because there was no hope of determining a neural basis for the lack of consciousness in a dog versus its possession in a human. This has changed considerably for these reasons:

- Brain imaging studies have shown significant differences in brain activation patterns contingent on consciousness.
- Cognitive neuroscience and artificial intelligence have gotten a handle on what awareness is and how it differs from, but supports, consciousness. One of the key ideas is the difference between being aware and being aware that we are aware.

Looking at assumptions about consciousness

Following are a few interesting perceptions people have about consciousness and an assessment of whether these ideas are accurate:

- One of the most universally expressed feelings about the phenomenon of consciousness is its unity. Each of us feels that we are a single, indivisible conscious being who always knows what he knows and doesn't what he doesn't.

  Yet perceptual illusions tell us that our grasp of reality may be distorted. The transient forgetfulness that we all experience indicates that our brain is not a data bank from which we can always reliably access any particular memory. And we sometimes change our minds in the absence of any relevant new piece of information that we know of.
- We also tend to feel that, although impulses and desires may pop into our minds from time to time, an essential core consciousness has final choice or veto power over our actions.

Veto power over impulses is central to legal ideas about guilt and what constitutes sin in many religious traditions. However, impulsive acts that we rationally regret later belie the fact that our consciousness is always paramount. We've all had those "what was I thinking" moments that should convince us that our consciousness is not solid and invariant but dependent on invisible forces within and without, of which we are not aware. Habits that bring you close to danger may lead you to cross the line, despite your rational belief in your own self control.

- Most of us feel that no matter how intelligent a machine may appear to be (think of the computers that have been created to play chess or Jeopardy!), the machine cannot have this central core consciousness, even in principle.

  Virtually no one believes that any current computer is conscious or even close to it. Beyond this, there are two broad camps. One says that no electronic machine will *ever* be conscious because machines are simply not made of the right stuff. The other camp regards the lack of current conscious machines like the difference between conscious humans and aware but unconscious chimpanzees: That is, it's a matter of sufficient brain power that has yet to be achieved artificially. When computers get powerful enough to match human brains in some instructions-executed-per-second sense, they'll become conscious.

  Futurist and entrepreneur Ray Kurzweil has calculated that this computational ability point (the "singularity") should be reached before 2030, so most people alive today will live to see if this outcome is indeed the case.

Types of consciousness

The most commonly accepted parsing of consciousness delineates between pure awareness and consciousness:

- **Awareness:** Awareness entails a level of perception that can determine behavior and be felt as an experience. What awareness lacks, in relation to consciousness, is mostly language, by which the experience is interpreted in a rational scheme. Some meditative traditions, for example, try to achieve in the mind a state of awareness without any overall layer of interpretation or reaction.
- **Consciousness:** Consciousness is mostly associated with language, in which the language mechanism catalogues and relates the experience created by awareness into an overall, rational scheme linked to memory. Gerald Edelman, the Nobel prizewinner who switched his research from molecular biology to neuroscience, has referred to consciousness as "the remembered present." This remembering links current experience with prior concepts that are embodied in language.

Studying consciousness

Although neuroscientists once shunned discussions of consciousness as being in the realm of philosophy, now different types of consciousness are considered to be at least partly associated with differences in neural activity that are measurable, at least in principle. The question is then how to research something that is an intrinsically private experience. Neuroscientists use two main approaches that might be called "divide and conquer" and "sneak up on it":

- **The divide-and-conquer approach:** This approach involves distinguishing between some brain state in which a person is conscious from the closest equivalent in which the person is not. Examples include sleep versus waking, coma versus paralysis, and damage to the brain that does not eliminate consciousness versus that damage that does eliminate consciousness.
- **The "sneak up on it" approach:** In this approach, researchers typically compare slightly different situations that result in large differences in consciousness. Often underlying this approach is the idea that there is some sort of continuum from non-awareness to awareness to full consciousness.

There are profound implications associated with different models for the essential nature of consciousness. If consciousness is a continuum enabled by brain size, then large-brained animals like chimps have some consciousness, as presumably will computers in the near future. If consciousness is the result of language, then what is the status of a human who has lost language function (not just the ability to speak, but the ability to have verbal thoughts)? If a particular neural circuit or even neural cell type is necessary for consciousness, then you'd want to know whether any other animals have that circuit or neuron, whether any humans lack them, or whether they really could be simulated in machines.

So far the best data researchers have in humans come from comparisons between otherwise similar conscious and non-conscious brain states.

Sleep versus waking

Waking versus sleep states are characterized by different EEG rhythms. In waking, the brain cycles through alpha, beta, gamma, and theta rhythms. In non-REM sleep, delta and theta waves predominate (refer to Chapter 11 for more on the stages of sleep).

During REM (rapid eye movement) sleep, which is the sleep in which dreams occur, the EEG rhythms are much like the rhythms that occur during the waking state. So the question is this: Is dreaming more like consciousness than non-REM sleep?

Most neuroscientists would probably agree that it is, suggesting that the EEG rhythms in REM sleep and waking are linked to some aspect of brain activity in which consciousness is or can be supported. The dreaming in REM sleep is, after all, remember-able. It contains images, actions, plots, dialog, and other mental processes that occur during waking consciousness. If awakened during REM sleep, people can say what was happening in their dream, as though they were interrupted in the conduct of a normal conscious activity.

So, in the sleep contrast scheme, REM sleep has experiential characteristics similar to waking consciousness and similar EEG rhythms. Non-REM sleep has neither.

Neuroscientists have also found that REM sleep exhibits a more coherent pattern of synchronous activity that involves interplay between the frontal and other cortical lobes than non-REM sleep does. The conclusion, which is supported by some other data, is that consciousness is particularly dependent on interactive activity between the frontal and other brain lobes.

Coma versus paralysis

Brain injuries can produce a range of effects — from paralysis to total loss of consciousness. Paralysis can occur with or without loss of consciousness. There are many cases of paralysis, for example, in which the only voluntary movement that can be made by the patient is blinking. Although superficially such patients have often been treated as though they are unconscious, attempts to communicate ("Blink once for yes, twice for no," for example) have often been successful. Patients lacking even the ability to blink are almost always assumed to be in a vegetative state, but brain imaging and other techniques have shown that even some of these patients are aware of what is around them.

A longstanding hypothesis about what brain regions support consciousness is sometimes referred to as the *Cotterill triangle hypothesis.* This hypothesis suggests that high-level consciousness requires the involvement of sensory cortex areas, such as the parietal or occipital lobe; the frontal lobes, particularly the premotor areas; and the thalamus.

A recent study involving patients in vegetative or apparently vegetative states seems to support this idea. In the study, familiar music was played to unresponsive patients while researchers imaged their brains. The early sensory areas of the brain in comatose and anesthetized people usually respond to sensory input. This input is typically also relayed via the thalamus to the frontal lobes. However, researchers were able to distinguish between the truly comatose and the "locked in," unresponsive patients because, in the latter, a feedback signal returned from the frontal lobes back to the sensory areas of the cortex; such a signal did not occur in the truly vegetative patients.

Anesthesia and consciousness

Although people commonly describe being anesthetized as being "put to sleep" or "knocked out," being anesthetized is not the same as sleeping. The purpose of anesthesia is to create an artificial state of unconsciousness in which there are not only no responses to painful stimuli, such as surgical cutting, but also no awareness or subsequent memory of the stimuli. Anesthetics do not suppress all brain activity, just the activity associated with consciousness.

One of the great persistent mysteries in neuroscience is that no one really knows in any detail how anesthetics actually work. Some, like isoflurane and nitrous oxide, are relatively inert gases, while others, such as Ketamine and barbiturates, are agonists for particular neurotransmitter receptors (NMDA glutamate receptors, for example), interfering with signals that would otherwise trigger a response.

One thing that many anesthetic agents have in common is lipid (fat) solubility, suggesting that these anesthetics interfere with neural transmission by infusion into the myelin sheath surrounding many axons, or the neural membrane itself. However, the mechanism for their selectivity with respect to consciousness alternation is poorly understood. Ditto for intoxicants such as alcohol.

The debate about levels of consciousness and anesthesia has produced some very practical ramifications for many parents who have had children requiring some sort of surgery (such as circumcision) shortly after birth. Doctors used to routinely withhold anesthesia in these situations because it was argued that (1) anesthetics are dangerous to newborns both in terms of risk of death and potential damage to the nervous system, and (2) infants were not really conscious beings anyway and would have no memory of any surgery with or without the anesthesia. The tide seems to have turned toward using anesthetics within the medical community, although that change in practice seems not to have occurred because of any particular finding from neuroscience.

Brain damage

Damage to higher brain areas such as the cortex can result in loss of some specific aspect of consciousness or loss of consciousness altogether. While serious damage to the brainstem tends to be quickly fatal (this brain area controls basic homeostatic mechanisms such as blood pressure, heart rate, and respiration), damage to the cortex tends to produce more specific losses. Consider these examples:

- Temporal or lateral prefrontal lobe damage can disrupt short-term memory and learning functions without affecting long-term memory.
- Generalized *ischemia* (loss of blood flow and therefore oxygenation) can produce temporary amnesia.

- Damage to Wernicke's area at the junction between the parietal and temporal lobes (refer to Chapter 6) produces an inability to understand language without a loss of general awareness; of course, not understanding language significantly changes the quality of that awareness.

Two camps and a middle ground

The link between language and consciousness is one of the most hotly debated areas of neuroscience. Folks tend to fall in one of two camps: those who think language is consciousness and those who think that non-linguistic forms of consciousness are possible.

Some suggest that, because humans think in words, thought language is consciousness. The fact that few of us remember anything from our lives before the age of two is suggested to result from the profound reorganization of our brains that occurs after we learn language, which itself is thought to be the foundation for consciousness. Moreover, although animals like chimps can learn to use words as signs for objects, they don't use grammar and hence lack language; therefore, by definition, they are not conscious. This would also imply, as I mention earlier in this chapter, that humans who lacked internal language (not just the ability to speak) would not be considered conscious.

One of the most famous accidental test cases of language and consciousness is that of Helen Keller, who was completely normal until about 19 months of age when an unknown infection destroyed both her sight and hearing. She lived an animal-like existence until Ann Sullivan (recommended by Alexander Graham Bell, a researcher on deafness, as well as the inventor of the telephone) came to Alabama to teach Helen sign language. Ms. Keller's own writings about the transition suggest that she led an almost unconscious, animal-like existence until she had access to language and the ability to communicate.

Others point to visuo-spatial thinking and imagery as examples of non-linguistic forms of consciousness. An architect might design a complicated building in great detail using mental imagery, but not talk to herself in words at all during the process. It is really possible to design a building, using knowledge gained during school, without consciousness?

A middle ground I suggest here is the following. Mental imagery without talking to oneself can be conscious, but it is not sustainable longer than a few hours at most. Ultimately, whatever thoughts occur during the non-verbalized imagery state must ultimately be translated back into words for the thoughts that occurred in this state to be remembered in a useable form. It may also be that even during the supposed non-vocal imagery state, sub-vocalization is actually occurring but memory of it is suppressed.

Consciousness seems to depend on the ability to communicate in numerous ways, from conversations between people, to messages being passed from the frontal lobe to other parts of the brain.

Unconscious processing: Blindsight, neglect, and other phenomena

Most people tend to think that our central consciousness is running the show and it picks what aspect of things going on in our brains to pay attention to. But one of the most important realizations of the last several decades in neuroscience has been the understanding that what we are conscious of comprises only a tiny fraction of what is going on in our brains. Most of our neural processing is done unconsciously in the background and only reaches consciousness in very specific circumstances.

For example, I remember very distinctly the day my younger brother asked me how to tie a necktie, an activity I had engaged in once a day, five days a week for several years. I became completely flummoxed trying to show him how to do it, and, in that session, I was unable to tie a half-windsor. Yet several years before, I had consciously gone through the steps to learn how to tie that particular knot. Learning that sequence certainly required conscious effort and would have been difficult to impossible for a chimpanzee to achieve. Since that time, however, the procedure had become *automatized* or *proceduralized* for me; my only conscious attendance to the process of tying a tie was the decision to start the procedure after I picked out the one I wanted to wear.

A common idea about consciousness is that we see, hear, and smell what we pay attention to (admittedly, attention can be drawn from something else to a new or salient stimulus). The common correlate of this idea is that stimuli that we ignore don't reach memory. In other words, to see and recall something, we must be conscious of it; everything else is in the ignored and forgotten background. As the scenarios in the following sections show, this is not necessarily the case.

The "cocktail party" situation

A classic example of unconscious processing occurs in the "cocktail party" situation. You are engaged in a discussion with one person but within earshot are people engaged in other discussions. In general, you block out the other conversations, and, if asked later, you wouldn't be able to recall any details about them. But if someone in one of those other conversations mentions your name, you instantly become aware and could probably recall the entire, previously unattended sentence in which your name just occurred.

Blindsided by blindsight

A notable clinical finding that shows dissociation between the ability of the brain to register an event and the conscious awareness of that event is the phenomenon called *blindsight.* In blindsight, a subject is able to report some details about a stimulus without conscious awareness of the stimulus being presented.

Studies on blindsight

Blindsight was first extensively documented by Lawrence Weiskrantz, a British psychologist, now emeritus Professor of Psychology, Oxford. Weiskrantz was investigating the sight abilities of a patient who had virtually all of the primary visual area V1 destroyed on one side of his brain (refer to Chapters 2 and 5 for brain areas that mediate vision). Because the main pathway for cortical vision processing passes through V1, such patients are blind in the *hemifield*, or sensory field, opposite the lesion (the right side of the brain processes the left hemifield, and vice versa).

Here's the interesting thing about what Weiskrantz did: After presenting objects in the patient's blind hemifield, in which the patient claimed he never saw anything whatsoever, Weiskrantz asked the patient to guess either the approximate location of the object or which of several possible objects was presented.

In subsequent studies, the same kind of question was asked of the subjects, and although they routinely said they saw nothing and that answering such a question was preposterous, when forced to guess, they correctly identified the object at a level far greater than statistical chance. In other words, objects were presented that the patients claimed they did not see, yet they could, with statistically significant accuracy, verbally answer questions about these objects, even though they did not see them.

Hypotheses about how blindsight works

Considerable debate has ensued about neural mechanisms mediating such blindsight. One hypothesis suggests that subcortical structures that receive information from the retina, such as the superior colliculus, must support the visual processing. A contrary hypothesis suggests that some portion of V1 must have been spared (virtually no accidental lesions in humans eliminate 100 percent of any particular brain area without also damaging some part of any neighboring area) and that that these "islands of spared V1" must support the ability to answer questions about the presented objects.

In any case, it is clear that, without some major portion of V1, there is no conscious awareness of items presented to the visual field area that was served by that part of V1, but visual processing still occurs.

Subliminal perception and priming

One reason neuroscientists are interested in blindsight is to determine whether it is related to the phenomenon of subliminal perception. The classic story about subliminal perception involved inserting one frame showing a picture of popcorn every x-number of frames in a film. Viewers could not consciously detect the single frame, but initial reports suggested that more people lined up at the popcorn stand at intermission during movies that had these added frames. Such "subliminal advertising" was banned by law, although subsequent controlled studies afterward suggested the phenomenon of subliminally influencing buying did not actually exist.

More recent and sensitive studies have found a small effect, and subliminal perception is now lumped with the phenomenon of *priming*, in which unconscious exposure to a stimulus affects subsequent choices. For example, if you've been shown pictures of rivers for a while, when asked to define the word *bank,* your first definition is more likely to involve sides of rivers than places to put money. Given the hundreds of billions of dollars spent on advertising, the potential effectiveness of unconsciously perceived imagery continues to be of considerable interest.

Being neglected

Lesions higher in the visual pathway, such as in the visual part of the parietal lobe, tend to produce a phenomenon called *neglect* rather than blindness. Neglect is the tendency not to notice visual stimuli in the area of the visual field served by the lesioned area, although with focused attention and the lack of distractions, objects can still sometimes be seen there.

Visual neglect is more common for right hemisphere lesions affecting the left visual field than vice versa (this is consistent with a general tendency for stronger visual processing in the right hemisphere of the brain than the left).

An interesting relationship between the ability to consciously imagine a visual scene and damage to parietal areas that process normal vision was discovered and reported by Edoardo Bisiach and Claudio Luzzatti in 1978. They asked a patient with right parietal lobe damage to imagine standing at one end of a piazza (a town square in Italy consisting of an open area surrounded by various buildings) that was familiar to the patient and to report what he would see from that vantage point. The patient described all the buildings on the right side of the square and a few in the middle but none on the left. Bisiach and Luzzatti then asked the patient to imagine standing at the opposite end of the square, looking back in the opposite direction. The patient proceeded to describe all the buildings that he did not in the first case, which were now on the right side in his imagination, but none on the left side, which were those he previously described.

Perhaps it is not so surprising that the ability to consciously summon up an image of a familiar area uses areas of the brain that would have been involved in seeing the area. When one area of the brain that would have represented a particular part of the scene is missing, that part of the scene is missing from the consciousness of the imaginer.

Chapter 13

How the Brain Processes Thoughts

In This Chapter

- Understanding the structure of the neocortex
- Traveling sensory pathways between the neocortex, the thalamus, and the hippocampus
- Examining the specialization of the brain's left and right hemispheres
- Looking at theories related to consciousness

Intelligent, adaptive behavior in mammals is associated with the neocortex, which sits at the top of the neural processing hierarchy and consists of major modules that have sub-modules, all the way down to neurons and synapses. In addition to functioning in a hierarchical manner, the brain also operates with a high degree of parallelism, with many functions occurring simultaneously. Out of all that parallel activity, the conscious processing that we're aware of is based on a vast network of unconscious processing that we're generally unaware of.

This chapter explores the structure, function, and processes of the neocortex and takes a look at the link between language and consciousness. These high level functions depend particularly on the neocortex.

To understand the function of the neocortex, you must understand the lower brain structures with which it interacts. These lower brain structures, such as the thalamus and hippocampus, are products of evolution. Animals that preceded humans, such as cold-blooded vertebrates that don't have an extensive neocortex, do quite well for most behaviors. So you can look at the neocortex as a high-level addition to a system that already can process sensory stimuli and execute appropriate behavior.

One of the most significant issues challenging neuroscience with respect to the evolution and function of the neocortex is whether it enables complex functions like language and consciousness simply because it has crossed some size threshold, or whether its structure and organization is somehow unique.

The Brain: Taking Command at Multiple Levels

Many neuroscientists broadly divide the animal kingdom (from a very chauvinistic point of view, for sure) into mammals, non-mammalian vertebrates, and invertebrates (animals without backbones). Here's a brief review of these different categories:

- **Invertebrates:** The evolutionarily oldest multi-celled animals are invertebrates, such as mollusks, worms, and insects. They comprise the vast majority of all animal species. Invertebrates have neurons that share some similarities with the neurons of vertebrates, but their nervous systems are highly dissimilar from vertebrates and, for that matter, from each other. Few invertebrates have large brains or exhibit complex adaptive behavior (compared to mammals, at least), with the exception of mollusks, like octopuses, that continue to surprise researchers with their ability to use various tricks to get to crabs placed in puzzle boxes.
- **Non-mammalian vertebrates:** Much closer to mammals are non-mammalian vertebrates, which include reptiles, such as salamanders and alligators; amphibians, such as frogs; fish; birds; and probably the ancestors of birds, dinosaurs. Non-mammalian vertebrates tend to have relatively centralized brains with many structures similar to those in mammals, such as the brainstem, cerebellum, and inferior and superior colliculi. These structures are quite different from those of invertebrates.
- **Mammals:** Mammals only came into existence during the middle of the dinosaur era and did not become the dominant land animals until the extinction of the dinosaurs about 65 million years ago. Mammals, in neurons and neural organization, are more similar to each other than non-mammalian vertebrates are to other non-mammalian vertebrates. The big brain difference between mammals and non-mammalian vertebrates is the neocortex.

  The cellular and network structure of the neocortex is surprisingly uniform both within a single brain from motor to sensory areas, and between different mammals. Whether you're looking at a mouse's motor cortex or a whale's auditory cortex, the layers, cell types, and neural circuits are very similar. Although this one neocortical circuit may not be the most efficient neural processor for all of mammals' higher functions, it seems to be a uniquely powerful structure that unifies the processing of different types of perception and planning that has lead ultimately, in humans, to consciousness.

All about the Neocortex

As noted previously, the neocortex is the big brain difference between mammals and non-mammalian vertebrates. Prior to mammals, brains had many highly specialized areas with different kinds of neurons and neural circuits that performed different functions. In mammals, a standard neocortical circuit built around columns and mini-columns expanded to take over the top of the processing hierarchy for nearly all central nervous system brain functions. Primates, especially humans, took this neocortical dominance to an extreme.

The four major lobes of the brain and their functions

Virtually the first thing every student of mammalian neuroanatomy learns is that the brain has four major neocortical lobes: frontal, parietal, occipital, and temporal. Figure 13-1 shows the general structure of the neocortex viewed from the left (refer to Chapter 2 for other gross brain anatomy). Virtually everything you can see from this point of view is the neocortex (except for a bit of cerebellum sticking out from underneath the occipital lobe).

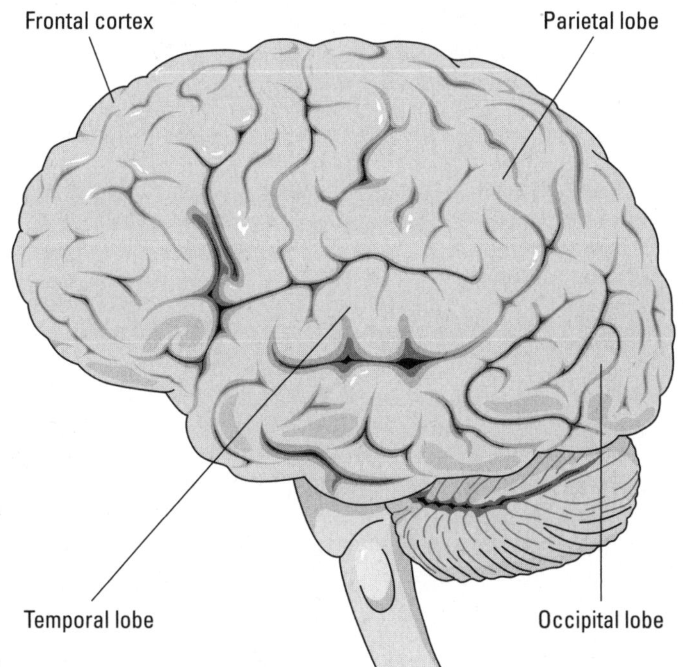

Figure 13-1: The four main lobes of the brain.

Here is a brief outline of the function of the four lobes:

- **Frontal:** All mammals have a frontal lobe with general planning and coordination in the more anterior areas and with the motor output in the most posterior part of the frontal lobe at the primary motor cortex. The fact that the olfactory system, believed to be the oldest neocortical system in mammals, projects directly to part of the frontal lobe without relaying through the thalamus, has suggested to some neuroscientists that this is the phylogenetically oldest part of neocortex, from which other neocortical areas developed.
- **Parietal:** The parietal lobe contains the primary and secondary sensory areas for somatosensation and the secondary sensory areas for visual and auditory processing, particularly the processes associated with spatial location and navigating through space.
- **Occipital:** The occipital lobe contains the primary and some secondary visual areas. It is the only brain lobe devoted to only one sense.
- **Temporal:** The temporal lobe contains the primary auditory and some secondary auditory areas on its superior aspect. Its inferior aspect is devoted to visual processing associated with object perception. Its medial aspect contains some very high-order visual areas such as the fusiform face area, which has cells that respond only to faces, and blends with other structures associated with the hippocampus.

The subdivision of the neocortex into these four lobes is common to all mammals, although the appearance and relative size of the lobes varies considerably across mammalian species. Some small mammals, for example, don't seem to need more neocortical area than you get simply by covering the rest of the subcortical brain areas, so that their cortices are smooth, with no folding. Because increase in neocortical size is primarily increase in area, large complex brains have neocortices that have to be "folded" to fit into the head (think of a wadded up sheet of paper) such that the creases become *sulci* and the areas on the surface become *gyri.*

Gray matter versus white matter

The functional structure of the neocortex is that of a folded sheet. The outermost 3 – 4 millimeters of this sheet forms what is called *gray matter.* This outer layer contains many neural cell bodies that are darker than their axons, hence the gray color. Underneath the gray matter is what is called *white matter* which consists almost exclusively of the axons of neurons going to and from the neocortex to subcortical structures, and going from one area of neocortex to another. Figure 3-2 shows the white and gray matter.

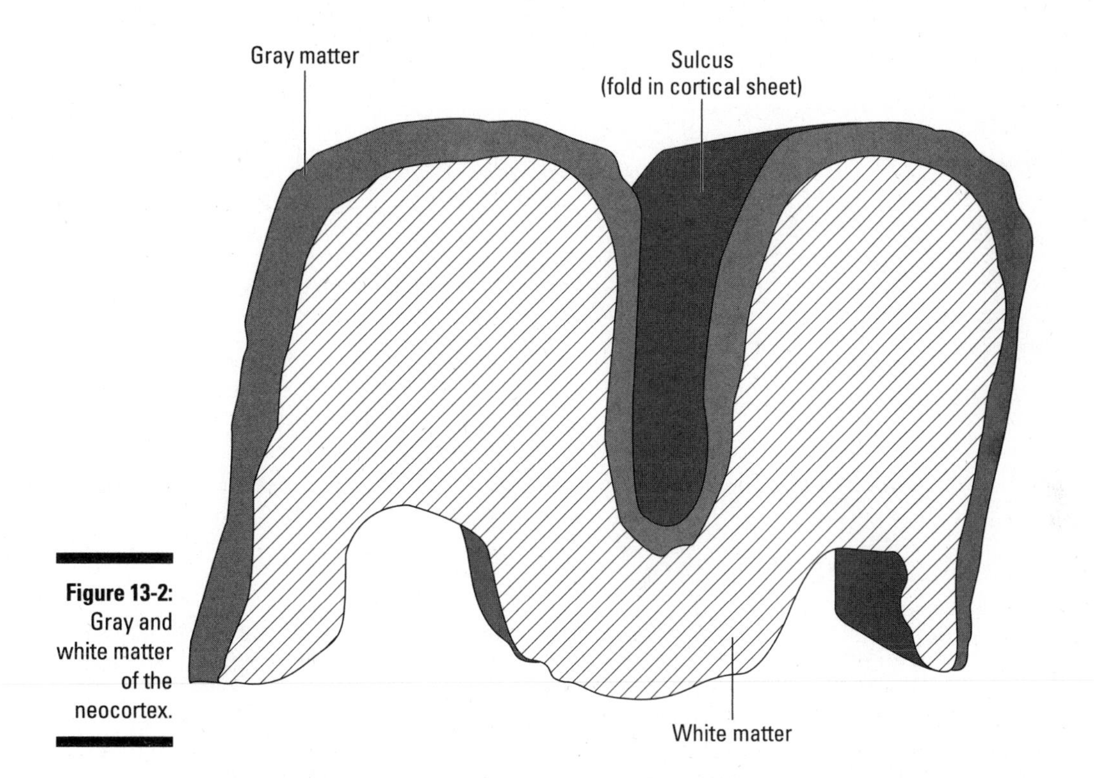

Figure 13-2: Gray and white matter of the neocortex.

Universal versus small-world connectivity

The gray matter thickness in all neocortices, from mouse to elephant, is about 3 – 4 millimeters, but the amount of white matter differs with brain size. The larger the brain, the thicker the white matter.

Imagine that the brain was structured such that every neuron was directly connected to every other neuron (*universal connectivity*). As Figure 13-3 shows, this feat isn't too difficult when you have a small number of neurons, but as the number of neurons grows, the number of axons (links) grows as the *square* of the number of neurons. For a concrete example, say that you add one more neuron to your brain, which already has about 100 billion neurons (more or less). Connecting that new neuron to every other neuron you already have would take an additional 100 billion axons.

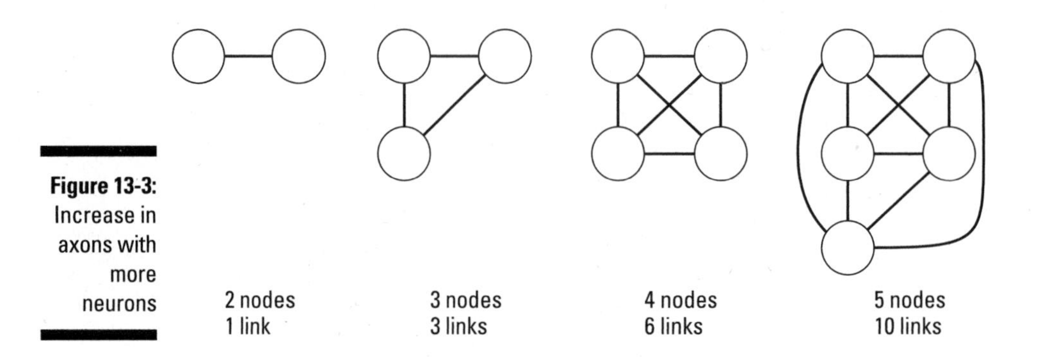

Figure 13-3: Increase in axons with more neurons

For that reason, the brain simply cannot be built on the principle of universal connectivity. If it were — with every neuron connected to every other neuron — the number of axons, or links, would be a function of the number of neurons. Here's the math, with *n* representing neurons:

Number of links = $n\,(n-1)/2$

The upshot? By the time a brain got a few million neurons, the white matter would be the size of a house. A totally interconnected brain with 100 billion neurons would require about 200 billion axon branches.

Still, it's clearly desirable for the activity in any given neuron to be able to affect any other particular neuron; otherwise, the activity in an isolated neuron would be just that: isolated and not very useful.

The brain solves this problem by using what's called a *small-world* interconnection scheme. Any given neuron is connected to a few thousand other neurons (on average), and the densest connections are to nearby neurons, so the axonal lengths are short.

Minicolumns and the six degrees of separation

As mentioned previously, each neuron is connected to only a few thousand others, and the densest connections are to those nearby. These nearby neurons function like a circle of friends and generally follow the six-degrees-of-separation principle, which says that, on average, no more than six steps are required to link you to any person on the earth.

Here's how this principle works: Someone in your circle of friends is linked to someone in a community a little closer to the target person, who is linked to someone even closer, who is linked to. . . well, you get the picture. The key point: Through a series of connections, you can hook up with anyone else on earth — even the Chancellor of Germany, if you're so inclined. Your neocortex works on the same principle.

The small "communities" in the cortex are called *minicolumns*. Neocortical minicolumns consist of about 100 neurons in a vertical column whose surface area is about 40 square micrometers. Roughly about 100 million minicolumns exist throughout the entire neocortex (the 100 billion total count of neurons for the brain includes the cerebellum and subcortical structures as well).

It is likely that activity in any minicolumn can influence activity in any other minicolumn by passing through six synapses or less, the six degrees of separation. This allows your brain with 100 billion neurons to vastly pare down the number of axons so that your head can be smaller than the Great Pyramid of Giza.

Defining the six-layered structure of the cortex

A defining hallmark of the neocortex is its six-layered structure. Figure 13-4 shows a diagram of the six layers, and some of the inputs and outputs. Note that the real cells are much smaller and more numerous than in the diagram. The cells shown are the *pyramidal cells,* named for the roughly triangular shape of their cell bodies. (Note that this figure doesn't show the many other local interneurons that are mostly inhibitory and participate in local circuits.)

The job of the pyramidal cells

Pyramidal cells receive inputs and send outputs outside the local minicolumn, as well as within it. They typically have a roughly horizontal dendritic branching just below the cell body, plus one dendrite that rises from the top of the cell body and extends to a higher layer before branching mostly horizontally there.

The six neocortical layers, numbered I through VI from the cortical surface down, form the gray matter of the cortex. Underneath the gray matter is white matter, which has almost no cell bodies but consists of the axons of neurons going to and from the area of the cortex above to other cortical areas and to subcortical structures. Below the white matter are the subcortical structures.

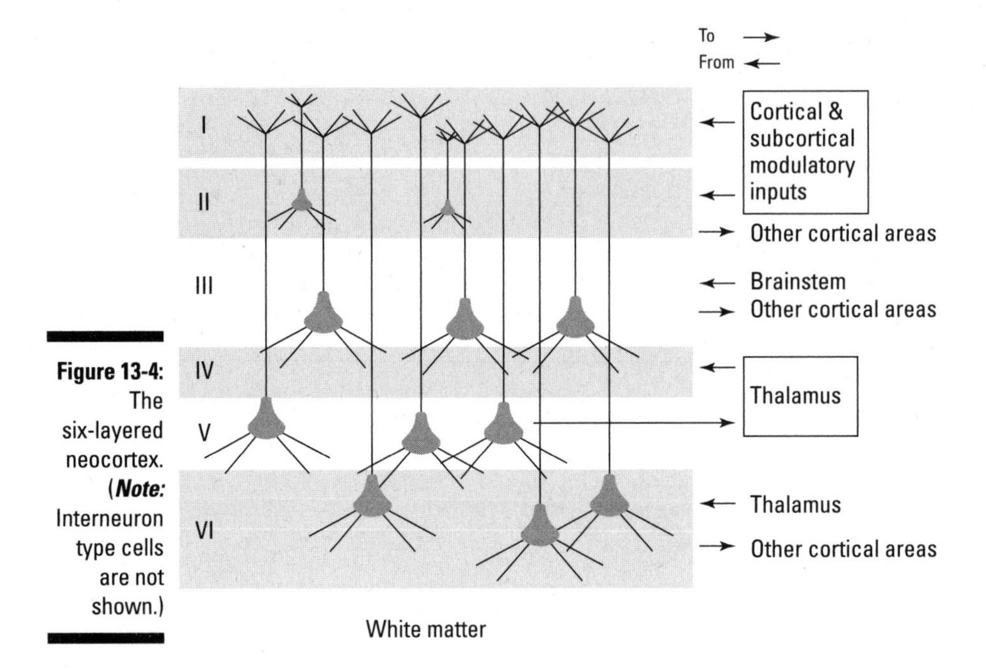

Figure 13-4: The six-layered neocortex. (***Note:*** Interneuron type cells are not shown.)

This basic six-layered neocortical structure, with the same cell types, locations, and interconnections, is relatively constant from mouse motor cortex to elephant somatosensory cortex — a fact that most neuroscientists find astounding. Within these six layers are pyramidal cells, which are the major long-range projection cells in the cortex, plus numerous *interneurons* (neurons only participating in local circuits) that form what's been termed a *canonical circuit* that may be repeated everywhere throughout neocortex.

Here's the general layer scheme:

- **Layer I:** Layer I, closest to the surface of the cortex, tends to receive modulatory inputs that shape the response to the main inputs in layer IV. These modulatory inputs come from diffuse projecting neurotransmitter systems (such as the biogenic amine neurotransmitter systems in the brainstem: dopamine, norepinephrine, histamine, and serotonin), as well as from other cortical areas.
- **Layer II:** This layer has widespread small pyramidal cells that receive inputs from layer VI and project to nearby cortical regions.
- **Layer III:** Pyramidal cells in this layer receive inputs from brainstem modulatory areas (like layer VI) and from other cortical areas. The outputs of pyramidal cells with cell bodies in this layer project to other cortical areas.
- **Layer IV:** Layer IV is the major recipient of thalamic input. This input arrives on the dendrites of pyramidal cells in layer V and is very effective

in driving them. Thalamic inputs also synapse on stellate interneurons (not shown) that then synapse on pyramidal cells in layer V and other layers. Layer IV pyramidal cells project to other layers locally, particularly layer V.

- **Layer V:** Pyramidal cells in layer V receive input from layer IV and make long-distance connections, such as to the spinal cord. They also feed back to anterior integrative areas of the thalamus such as the *pulvinar,* which is involved in the control of attention.
- **Layer VI:** Layer VI pyramidal cells also receive some inputs from the thalamus and project to other cortical layers. The major output of layer VI is to the thalamus.

Details of the input-output connections are not exactly the same in all areas of the neocortex; however, the overall pattern is remarkably consistent.

The canonical circuit

Given the "standard" structure of the input-output connections of the neocortex, neuroscientists such as Kevin Martin and Rodney Douglas have referred to a *canonical* neocortical circuit involving mostly vertical connections among the pyramidal and local interneurons in any small area of the neocortex.

The idea is that this basic circuit performs the same sort of neural computation on whatever input it gets and sends the output of that computation to some appropriate output. Thus, what makes the motor cortex the motor cortex and the sensory cortex sensory is where the inputs come from and the outputs go to, *not any property of the region of the cortex itself.*

For example, when inputs are lined up, the neocortex canonical circuit is good at detecting whether that input is a viewed edge in the visual system or an edge pushing against the skin in the somatosensory system. This circuit is also good at detecting smooth motion, whether it is sequential stimulation across your retina or your skin, or even whether it's a moving sound source coming into your two ears. A local area of the neocortex sends this computation to higher areas that detect the coincidence of several edges, like an "L" or a "T," or the movement of a complex object like a ballet dancer.

David Hubel and Torsten Wiesel at Harvard Medical School established that the visual cortex is dominated by neurons that are orientationally selective and neurons that are directionally selective. That is, some cells, called *simple* cells, respond best to edges of a certain orientation, and other cells, called *complex* cells, respond best to stimuli moving in a certain direction. Later recordings in the somatosensory cortex (where skin receptors project; head to Chapter 4 for more on this) also showed many cells that responded best to an edge of a particular orientation pressing against the skin, while other cells responded best to movement in a particular direction across the skin. Recordings in motor cortex have revealed cells that respond best to movement of a limb in a particular direction.

Hail to the neocortex!

The neocortex seems to be one of the most important evolutionary leaps forward for mammals because it makes mammals so adaptable to changing environments that they have come to dominate non-mammalian land animals.

For example, one obvious advantage of having a common representation for a host of brain functions is that the amount of neocortex devoted to one function can easily be transferred to another by competition.

Suppose, for example, that some mammal is primarily visual but, due to a changing environment, finds itself relying more on auditory cues to survive. The visual and auditory thalamic inputs to the cortex, where they are near to each other, actually compete for cortical synapses on the basis of thalamic activity, attention, and other factors. This competition mechanism enables the previously visual mammal to switch, in one lifetime, some neocortical processing from visual to auditory.

Controlling the Content of Thought: Sensory Pathways and Hierarchies

Thought, the content of consciousness, arises from past and current perceptions. Past stimuli have left their trace in memory not only as recallable objects and events, but also as paths in the brain through which current stimuli are processed. Stimuli received by the senses are transformed into a universal neural currency of action potentials bombarding the thalamus. Thalamic nuclei relay the neural images they receive to different areas of the cortex for each sense. Each thalamic relay cell is like the conductor of an orchestra of hundreds of cortical neurons who play to the baton pattern of relay cell spikes. Second-order cortical neurons listen and respond to sections of multiple orchestras, conducting their own impromptu compositions on the fly. And so it proceeds for billions of brain cells.

When you remember, you remember not only that there were things, but the color and shape of those things and where you were when you encountered them and what those things meant to you. The neural representations of all these aspects are constructed in multiple brain regions. To remember which aspects go with which things, the links, as well as the aspects, have to be stored. These links are akin to the lines in a stellar constellation that indicate which particular shape a group of stars should evoke. Initially, this linkage is represented in the working memory of the lateral prefrontal cortex, but it is processed into a memory by the interaction between the sensory cortical areas and the hippocampus.

This section discusses the organization of sensory processing hierarchies, which reside in all four brain lobes, and their interaction with the hippocampus to form memories. Neocortical areas processing different aspects of each sense and for different senses interact with each other and with phylogenetically older subcortical structures to produce the rich and nuanced representations of the world and ourselves in it that we associate with intelligence and consciousness.

Sensory relays from the thalamus to the cortex

Neuroscientists conjecture that the thalamus, which was once a sensory processing, integration, and attention-control nucleus, started sending projections to the neocortex for added computational processing.

The parietal, temporal, and occipital lobes (discussed in the earlier section "The four major lobes of the brain and their functions") all receive sensory information from nuclei within the thalamus. These neocortical areas perform massive neural computations, the results of which are projected to other cortical areas, back to the thalamus, and to subcortical structures involving memory, motor control and coordination, and goal selection.

One way to think about this is the way a computer program calls a subroutine when a specific, detailed calculation is necessary. The main routine sends the data to the subroutine, the subroutine does some detailed computations and returns the answers, and then control reverts to the main routine which modifies its procedure in light of those answers.

Projecting from the thalamus to each sense's primary cortex

In the typical sensory processing plan, an area of the thalamus specific to each sense, such as audition, vision, or somatosensation, projects to a cortical area called the *primary cortex* for that sense. For vision, the primary cortex is called V1 and is located at the pole of the occipital lobe at the most posterior aspect of the brain (refer to Chapter 5). Similarly, A1, the primary auditory cortex, is located in the mid superior temporal lobe (read more about that in Chapter 6). The outlier is olfaction, where the olfactory bulb projects directly to the frontal lobe which then projects to the thalamus (explained in Chapter 7).

Projecting back to the thalamus and other regions in the cortex

In each of the primary thalamic-cortical systems, there is normally a large projection from the primary cortical area back to the same region of the thalamus that projected up to primary cortex. But primary cortical areas also project to secondary, tertiary, and higher areas within the cortex as well. Each of these areas projects back to the area that projected to it as well.

Neurons in "higher" areas along these streams typically are much more selective in the stimuli they respond to than are the neurons in the "lower" areas. For example, in the primary visual cortex, cells are orientation and direction selective, but the infero-temporal cortex has visual neurons that respond only to faces, hands, or other very complex patterns.

Figure 13-5 gives an example of some of the processing that occurs when you read the phrase "used the ax to." Say you're reading the word "ax." Early in the visual cortex, neurons recognize lines (strokes) going in various directions (vertical, horizontal, leaning left, leaning right). In the infero-temporal cortex, combinations of strokes (highlighted in Figure 13-5) that make up letters are detected. Word detectors at the next level feedback and activate appropriate letters, giving those an advantage over other letter combinations.

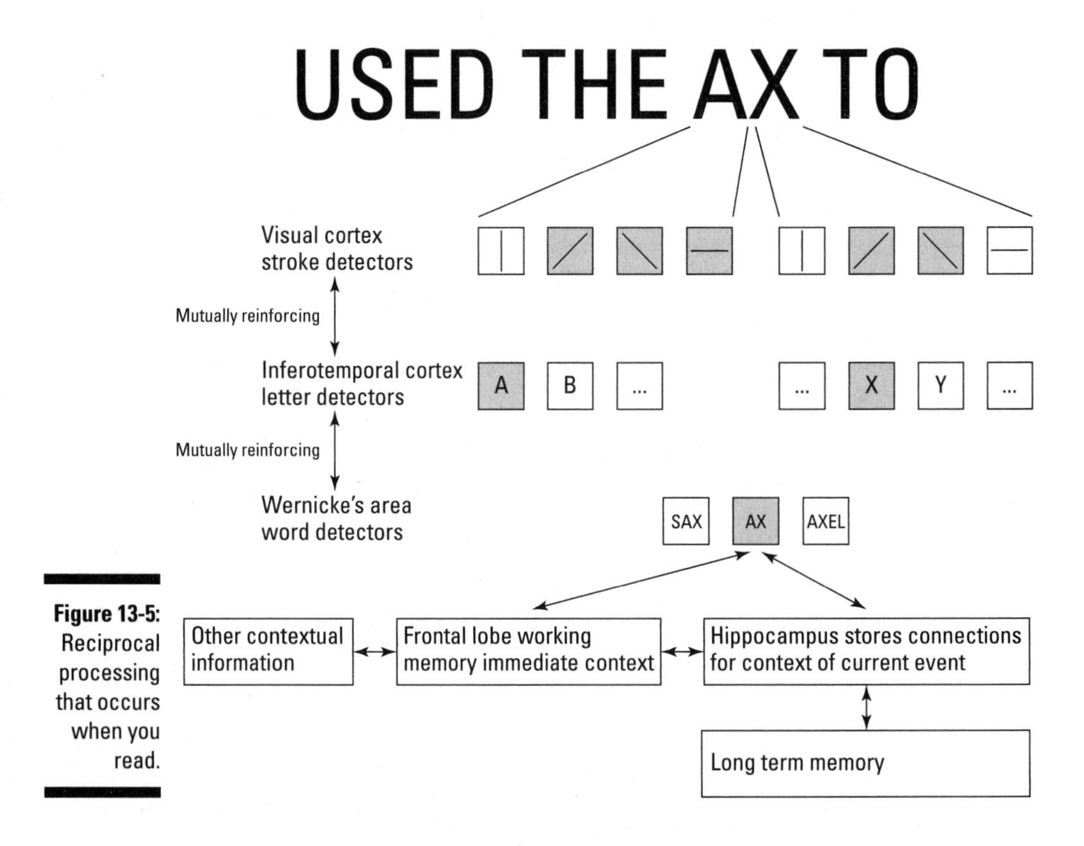

Figure 13-5: Reciprocal processing that occurs when you read.

Although the overall plan is a hierarchy where lower level cells closer to the thalamus detect simple features and higher level cells several synapses later detect more complex ones, the connections are two way. Why? Because neurons are noisy, and the input itself may be partially corrupted. For example, part activation of the "A" detector feedbacks and activates lower level stroke detectors that comprise the parts of the letter "A."

This feedforward and feedback activity occurs between areas just above and below each other in the hierarchy. But there are also connections to working memory in the frontal lobe that hold information about what task it is that you're doing (reading) and what the context is (a story about the forest). Connections to the hippocampus form links of what you're reading and other context so that you can remember that you read this passage today and the details about it.

Thalamic integration and gating functions

One way to think about the thalamus is that it is like the hub of a wagon wheel whose spokes go to and from various cortical areas. In addition to the projections to and from the neocortex, the thalamus has inter-nuclear areas akin to the reticular areas of the brainstem that allow subcortical inputs to modulate the transfer of information from various thalamic nuclei to the cortex. For example, if, while you read a book (in which you're concentrating on the visual input associated with reading), you hear a window break in your house, your focus of attention switches to the auditory domain rather quickly.

In addition to these reticular areas is a central executive processing nucleus within the thalamus itself called the *pulvinar*. Brain imaging studies show that the pulvinar is activated in almost all attention-switching tasks.

One question many students ask when they first learn about cortical hierarchies is where it all ends. Part of the answer to this question is that "it" doesn't end anywhere; the neocortex is a distributed network conducting both serial and parallel processing. But this is only part of the story. Sensory processing hierarchies that have to do with using sensory input to guide actions tend to project ultimately into areas of the frontal lobe that control motor behavior, and sensory processing hierarchies that are involved in object discrimination and identification tend to project to memory structures such as the hippocampus.

The hippocampus: Specializing for memory

Recall something you did recently, like driving home last night. You can remember the sequence of events from getting in the car to arriving home, and many of the sights, sounds, and perhaps even smells encountered on the trip. If asked, you could draw a picture of some of the places you passed or describe the sounds you heard as those of birds, traffic, airplanes or other things you encountered.

To conjure up the memory alluded to in the preceding paragraph, you must form some mental representation of both the things that you encountered and the sequence in time on the trip home last night as opposed to any other night. This kind of memory is called *episodic memory,* which contains not only content (the things remembered) but also context (the particular circumstances of one particular series of events).

One of the most fundamental questions asked by 20th century neuroscience was how people remember. Now, early in the 21st century, neuroscientists have a good idea of the general structure of the answer. Remembering involves activation of many of the same brain structures (visual, auditory, and somatosensory cortices) that processed the *original* sensory input, organized or activated by the hippocampus, and represented in the firing of working memory neurons in the lateral prefrontal cortex.

The hippocampus is located just medial to the medial wall of the temporal lobe and lateral and inferior to the thalamus. It is *not* part of neocortex; it's a phylogenetically older structure that was included in the original limbic system. However, the hippocampus receives inputs from and projects back to virtually the entire neocortex, particularly "higher" areas often referred to generically as the *association cortex.*

These inputs and outputs are partially processed and remapped by several structures on their way into and out of the hippocampus (these structures include the subiculum, parahippocampal gyrus, and entorhinal cortex, but you don't need to know about them for this discussion).

For a detailed look at the role the hippocampus plays in memory formation, head to Chapter 15.

Dividing and Conquering: Language, Vision, and the Brain Hemispheres

The left and right sides of the brain are nearly identical in gross and microstructure. The left side of the brain generally receives inputs from and controls the right side of the body, and the right brain interacts with the left side. Beyond this basic crossed-innervation scheme, the two sides appear to have different styles of processing that are particularly evident in some high-level cognitive tasks. These include a left-side specialization for grammatical aspects of language and a right side specialization for holistic aspects of visuo-spatial processing.

That the right side of the brain deals with the left side of the body and stimuli coming from the left side, while the left side of the brain deals with the right side of the body and stimuli coming from the right is one of the most important principles of overall brain organization. One implication of this principle is that the two sides of the brain should be functionally equivalent mirror images of each other, as our bodies are. However, this turns out not to be the case for some higher cognitive functions, such as language.

Specialized brain systems for language

Language is not merely a more elaborate type of communication that is similar to what animals use when they make verbal calls. While animals clearly communicate with verbal utterances, this communication is not language because it lacks a structured grammar.

Grammatical rules that create nouns, verbs, adjectives, and phrases make the variety of messages that can be transmitted virtually infinite compared to the number of single words in the vocabulary. For example, even in cases where primates such as chimpanzees have learned hundreds of discrete word symbols and have even made up some symbol pairs, their utterances are typically random with respect to word order; thus, they don't obey a grammar. They can only express what they have learned a word for, or, at best a novel word pair.

In virtually all right-handers and about half of left-handers, language depends far more on structures in the left hemisphere than the right (the homologous areas in the right hemisphere have far weaker roles in the same functions). Researchers don't understand the reason for this, although some have conjectured that the speed of neural processing necessary to produce and decode language requires that the processing be done locally, with short distance axon tracts, and therefore, on one side of the brain only.

It isn't clear whether the fact that the left side of the brain processes most language and also controls the dominant right side of the body in right-handers is just a coincidence. If it isn't merely coincidence, a possible explanation could be that a link exists between processing sequences of sounds in language and producing sequences of coordinated movement in manual dexterity. Some have also suggested that left- versus right-brain hemisphere specialization generally follows along the lines of detailed, linear structure versus big-picture structure. Although pop psychology made a real mess of this issue in the 1980s, many studies show results consistent with the different *stylistic* manners of processing by the two hemispheres: The left brain tends to be sequential and rule-based, whereas the right brain tends to do pattern matching.

Humans are unique not only in the use of language, but also in the asymmetry of more than three-fourths of us being right-handed. Chimps show hand preferences, for example, but it appears to be a random 50/50 split.

Two areas of the brain are particularly important in processing and producing language: Wernicke's area and Broca's area. Damage to these areas can result in language-specific dysfunctions, as the following sections explain.

Wernicke's area

Wernicke's area, located at the junction of the temporal and parietal lobes, is necessary for both understanding the meaning of linguistic utterances and for producing utterances that are meaningful.

Damage to Wernicke's area produces patients who have difficulty comprehending language. Depending on the exact location and extent of the damage, the functional loss can range from total inability to understand language to the loss only of the ability to use and understand very specific parts of speech, such as nouns, or even particular types of nouns, such as names of vegetables.

When they speak, Wernicke's aphasics (*aphasic* means having some impairment of language) appear to be fluent, but the actual utterance is filled with random nonwords (clinically referred to as a "word salad"). If you heard Wernicke's aphasics speak in a language you did not know, you might have difficulty discerning that there was anything wrong.

The left side specialization for language in Wernicke's and Broca's areas has led neuroscientists to wonder what those homologous areas on the right side of the brain do. One clinical finding that is rather interesting is that damage to the Wernicke's area homolog on the right side of the brain tends to produce dysfunctions in understanding prosody or tone of voice. Such patients cannot distinguish a statement like "I agree with that" said either in a straightforward or sarcastic tone of voice. They also tend not to get the point of jokes.

Broca's area

Broca's area, in the frontal lobe near the primary motor areas that control the tongue, lips, and other language apparatus, is crucial for producing any fluent or complex language. It also seems to be necessary for understanding some complex utterances.

With Broca's aphasics, their language production consists of short utterances, typically two- or three-word sentences or even single verbs. The original patient described by Dr. Broca could only utter a single word, "tan."

It was once thought that, because Broca's area is in the frontal lobe, damage to it would only produce a motor impairment in the production of language, while comprehension should remain normal. However, in addition to their difficulty producing complete sentences, Broca's aphasics have trouble comprehending complex grammatical constructions such as passive voice.

Seeing the whole and the parts: Visual processing asymmetries

Many spatial manipulation abilities depend more on the right hemisphere than the left. For example, visual hemi-neglect (failing to notice things in one visual hemifield) is much more common after damage to the right rather than the left parietal lobes. The right fusiform face area, a very anterior and medial part of the visual identity processing stream in the infero-temporal cortex, is more important for face recognition than the equivalent area on the left.

The detail- versus big-picture stylistic hypothesis for the difference between the left and right hemisphere argues that the left side processes details, while the right side processes the overall structure. Here's a personal example. I sported a mustache for many years and then shaved it off. In the weeks afterwards, no one who had known me failed to recognize me, but many gave me puzzled looks and uttered things like, "You look different somehow; did you lose weight or get a haircut?" If the recognition of my identity was a right-brain function, it was taking in the overall pattern (they knew it was me) and ignoring the details (the difference was bare upper lip).

Details are, of course, important in many types of visual analysis, but their use typically involves attention. An ordinary onlooker might be able to say whether she thought a building looked attractive or not, but an architect would comment that the windows were such and such a style, the roof treatment was done according to such and such a school, and so forth.

Where Consciousness Resides

One of the deepest questions in philosophy, psychology, and neuroscience concerns how intricately linked are consciousness and language. Although I have touched on this controversy elsewhere in this book, the left-/right-brain data give another bead on this target.

Language sets humans apart from all other animals. Language is more than communication; it is a symbolic system depending on a rule-based grammar. Specialized brain regions, particularly on the left side in most people, mediate the understanding and production of language. Damage to these areas can result in language-specific dysfunctions (refer to the earlier section "Specialized brain systems for language," as well as a reduction in the patient's ability to exercise consciousness.

Language and left- or right-hemisphere damage

The human brain is about twice the size of a chimpanzee brain, which means that each side of the human brain is about equal in size to a chimp's brain. Humans who have had most of their left hemisphere destroyed after childhood generally appear to be profoundly retarded, unable to communicate or respond coherently to sensory stimuli other than basic attraction and avoidance.

Similarly, when Roger Sperry and Michael Gazzaniga (and other colleagues) presented stimuli to the right side of split brain patients who had no corpus callosum to transfer the information to the left, language side, the language capabilities of the right side seemed little better than those of a chimpanzee.

On the other hand, patients with extensive right-brain damage or split-brain patients with stimuli presented to the left side of the brain appear quite normal in cognitive function, other than rather poor performance on visual processing tasks (of course, damage to the right side of the brain will produce paralysis on the left side, just as damage to the left side of the brain will produce paralysis on the right side).

In other words, if you are dealing with only the left side of a human's brain, you are clearly dealing with someone who is just not too good at solving spatial tasks. If you deal with a person in which the right side of the brain is isolated, the impairment is profound.

Understanding the "left side interpreter"

Gazzaniga, among others, espouses a theory, sometimes called the *left side interpreter*. According to this idea, humans are unique in that we have is a system in the sequential, rule-based left side of the brain that constantly tries to make sense of the world by using language. This left side interpreter is constantly making up a verbal story about reality that includes salient events and the role of the person and his or her actions in those events.

In this left side interpreter theory, most of human consciousness is bound up in the story created by the interpreter. Without the interpreter, awareness can exist, but it is like that of animals in that it's semantic rather than episodic. That is, all animals learn by association that some things are good and some aren't, but they have no memory for specific instances in which good or bad things were encountered and this awareness of good versus bad has no context. Their only awareness is the result — reward or punishment — that makes them more or less inclined go toward or to avoid the thing (or person or event) in question.

Simulating intelligence artificially: What exactly is simulated?

Do androids dream of electric sheep? Why don't we have robots that clean our houses, drive our cars, or translate English to Spanish? The problem with the idea of artificial intelligence is that it lumps ideas from very different fields, goals, and methodologies into an overhyped and contradictory mess.

If the term *artificial intelligence* refers to whether a machine can do something that for humans seems to require intelligence, then clearly machines demonstrate intelligence. For example, in the early days of digital computers, it didn't take long for people to program them to play a perfect tic-tac-toe game. At the time, many said, "Sure, computers can play a really simple, rule-based game like tic-tac-toe, but they'll never be able to play a complex game like checkers." A few years later, perfect checkers players were programmed, and people said, "Well, checkers sure, but never chess." Then Deep Blue came along. This is called the *moving target* phenomenon by many in the field.

Another interesting case is language translation. Articles in the 1950s claimed that computer language translators were just around the corner. Yet, just a few years ago, stories of computer-translated technical manuals in which phrases like "the wings were loaded with hydraulic rams" got translated into something like "the wings were loaded with wet male sheep" were common. Why the difficulty? Because good language translation isn't possible unless the translator knows something about the real world. Because computers that have been programmed so far to translate languages lack that knowledge, they have to rely on rules that work sometimes but fail in stupid ways many other times.

Methodologies are also an issue. The IBM chess-playing computer Deep Blue does not play chess like a human. It does not have neurons or even simulations of neurons. We don't know what aspects of human intelligence are due to neural hardware because we cannot simulate the brain with its 100 billion neurons. When a computer does something by using a set of rules that a human brain does with 100 billion neurons, it becomes a matter of definition whether the computer is simulating the human. Does a car simulate a person walking?

The left side interpreter doesn't exactly deny consciousness to the right side, but it asserts that the left side alone sustains consciousness (rather than mere awareness) and the memory of context. This idea suggests that consciousness is not just the "remembered present"; it is the "present remembered in words."

Chapter 14

The Executive Brain

In This Chapter

- Looking at the how the neocortex developed
- Understanding working memory: How it works and its limitations
- Identifying the role of the pre-frontal cortex in decision making
- Monitoring your progress: The anterior cingulate cortex
- Recognizing complex actions that don't require conscious awareness

A key difference between mammalian brains and the brains of non-mammalian vertebrates is the size of the neocortex, in that mammals have a much larger neocortex. Similarly, among mammals, primates have a particularly large neocortex per body weight, and humans have a proportionally larger neocortex than other primates. Much of the cortical increase in humans occurs in the frontal lobes.

Because of our large frontal lobes, we humans can make plans that depend on complex aspects of the current situation and which may require long durations. If a human, for example, wants to move a rock on the ground, he could do so with his left or right foot, with his left or right hand, with a stick in his hand, or by asking someone standing nearby to move the rock for him. He could choose to do it now, ten minutes from now, during the morning when it's generally cooler, or when a large group of people passes by who might help, even though waiting for helpers may take several days. These higher, abstract-level goal pursuits are represented in the prefrontal cortex, the most anterior part of the frontal lobes, and they get translated into specific sequences of muscle contractions in the frontal lobe output, the primary motor cortex that we share with all mammals.

This chapter explains how the neocortex developed and focuses on processing in the two main prefrontal areas, the lateral prefrontal cortex and the orbitofrontal cortex. I also cover the cingulate cortex, a non-neocortical area, that interacts with the neocortex and plays a crucial role in making us aware of what and how we are doing.

Getting the Brain You Have Today: The Neocortex versus Your Reptilian Brain

Prior to the evolution of mammals, cold blooded vertebrates, like lizards, birds, and probably even dinosaurs, had only rudiments of a neocortex. Behavior was controlled in these animals by the basal ganglia (the middle level in the "triune brain," explained in Chapter 12). In mammals, an interaction between this basic basal ganglia system and the neocortex controlled behavior and allowed more complex and learned behavior patterns.

As mammals developed evolutionarily, the neocortex expanded significantly. Many neuroscientists believed the expansion of the neocortex began with an expansion of olfactory processing capabilities in what would become the frontal lobe. The development of olfactory processing capabilities is believed to be due to the fact that early mammals were small, and vision is less useful than smell when you are a very small creature in tall grass.

The other lobes (parietal, temporal, and occipital) then expanded, using the same six-layered design that was so successful in the frontal lobe (refer to Chapter 13 for more on the six-layered structure). Neuroscientists are not sure why this six-layered neocortical neural circuit was so successful that it came to form the majority of brain mass, nor is it clear how the six layers of the neocortex evolved in the first place.

The most popular, though by no means proven, idea about the dominance of the neocortex is that having a similar neural representation scheme for all the senses and for motor control makes it possible to integrate activity across the entire brain more efficiently. In addition, the types of computations that can be done by the neocortex work well in a variety of senses. For example, the visual, auditory, and somatosensory cortex all have neocortical cells that are sensitive to motion in various directions. During development, competition for neocortical synapses occurs not only within each sensory system but also between sensory systems. Exactly how the neocortex enables complex processing in all the senses and in motor control is considered by many neuroscientists to be the most important question in the field.

My neocortex is bigger than yours: Looking at relative sizes

Figure 14-1 shows side views of gross brain structure in the frog, cat, rhesus monkey, and chimpanzee, compared with a similar side view of the human brain (note that the sizes aren't to scale).

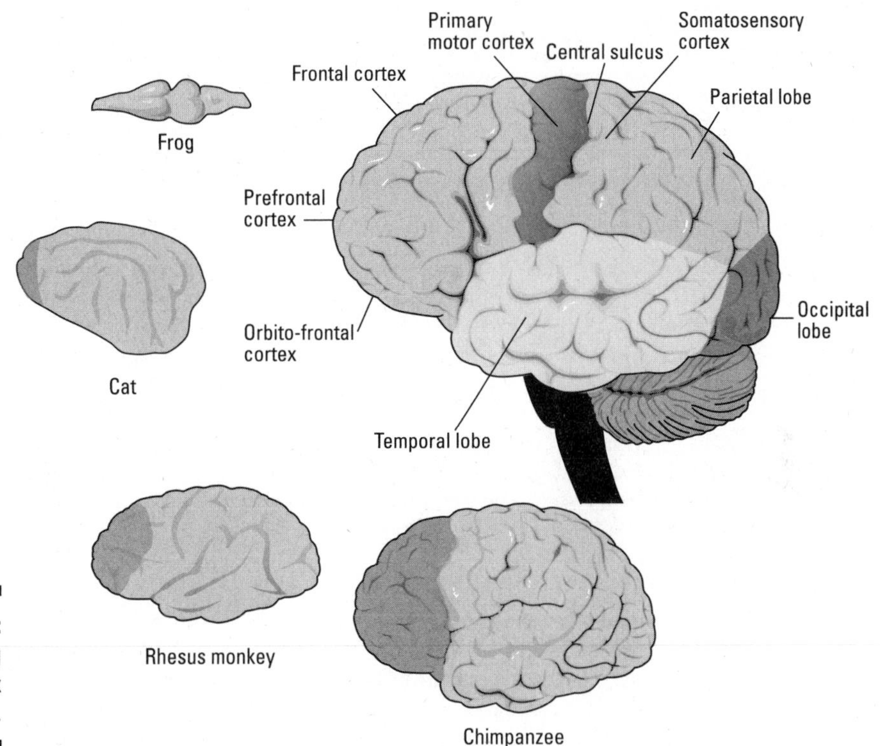

Figure 14-1: Comparing neocortex size.

Here's a quick rundown of the key info about the size of the neocortex in different species:

- **Frog:** In the frog and other cold blooded vertebrates, the brain appears to be little more than an extension of the spinal cord with a few bumps for extra processing, such as the optic tectum for vision. Frogs do have a brain area called the *cerebrum* anterior to the optic tectum, but this cerebrum isn't a true neocortex. A frog's behavior is almost totally stimulus driven. It feeds by snapping its tongue at and capturing flying bugs but will starve to death if surrounded by the very same bugs if they are dead and don't move.
- **Cat:** In the cat, virtually all the subcortical brains structures you can see in the frog brain are covered by neocortex. However, little of the cat neocortex is prefrontal. Cats are fairly intelligent among mammals, but they obviously don't use tools and their communication repertoire is limited to growls, meows, and purrs.
- **Monkey and chimp:** The prefrontal areas expand in monkeys and apes (note the size of the prefrontal areas shown for the rhesus monkey and chimpanzee). Many (but not all) monkeys and apes live in complex

social groups. These groups have extended male and female hierarchies that require hierarchically appropriate behavior whose plots may unfold over days or weeks. Monkeys and apes also demonstrate some rudimentary tool use and variety in vocalization and facial expressions.

- **Human:** In humans, about half of the cortex is frontal, and nearly half of that is prefrontal (areas anterior to the supplementary motor area and premotor cortex).

A sagittal view of the human brain (as though cut right in the middle from front to back), as shown in Figure 14-2, illustrates how the neocortex dominates the subcortical structures. These subcortical structures are similar in anatomy and function to non-mammalian vertebrates such as lizards that are the ancestors of mammals.

Evolution has occurred in the mammalian brain mostly by adding the neocortex to ancestral structures. This is something like the addition of numerous microcomputers and controllers to a modern car engine. The basic cylinders, pistons, and spark plugs are still there, but they are now under more nuanced control.

Figure 14-2 also shows the anterior and posterior cingulate cortices. Note that these areas are not part of the neocortex; they're part of the mesocortex. The anterior cingulate is an important monitoring structure in the brain. For more information, go to the later section "Are We There Yet? The Anterior Cingulate Cortex."

The relationship between prefrontal cortex size and the ability to pursue goals

The enlargement of the prefrontal cortex greatly increased the flexibility with which animals such as primates pursued their goals. Animals with smaller prefrontal cortices than humans and humans with damaged prefrontal lobes exhibit behavior that is much more stimulus driven.

Humans with prefrontal damage, for example, may exhibit what's called *utilization behavior,* in which their response to the presence of an object is to use it, even when doing so isn't appropriate. Upon seeing a hammer and nails lying on a table, a frontal lobe patient might proceed to drive the nails into a nearby wall. The lack of consideration of the socially appropriate context for this act is similar to how animals respond to cues like food or the presence of a potential mate with stereotyped behavior.

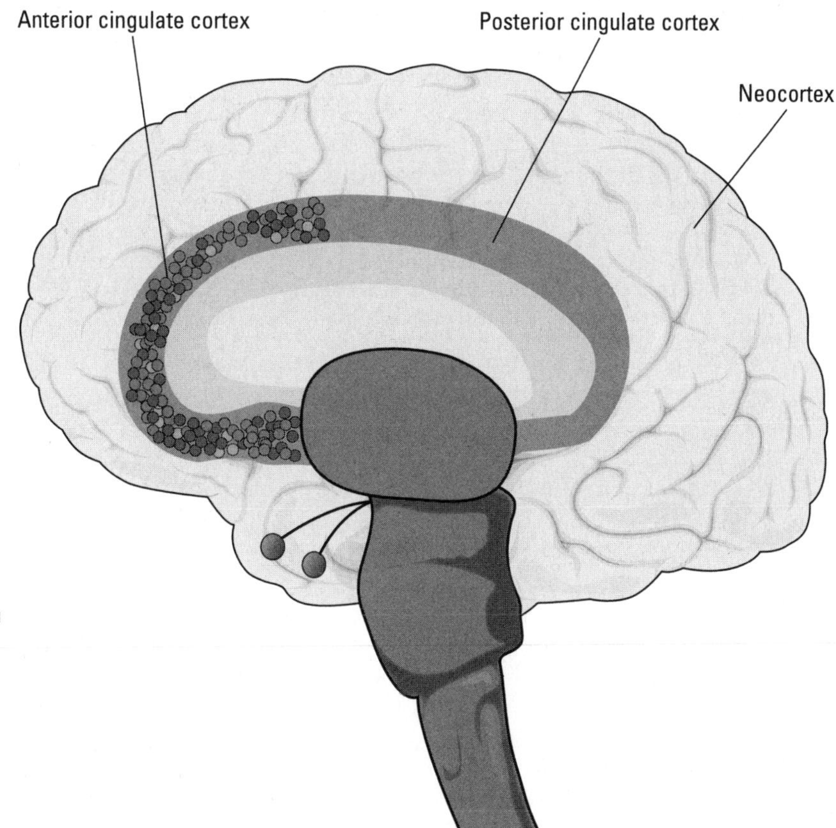

Figure 14-2: The neocortex evolved over the earlier structures.

The "extra" neocortex in the frontal lobes of higher mammals such as primates allows goals to be set with more consideration of the current context. Actions such as pounding nails in a wall may be appropriate in certain circumstances but not in others. Societies and social situations create complex contexts in which small situational differences make enormous differences in whether a behavior is appropriate, like whether you should laugh, at least a little, at an old worn-out joke, if it happens to be told by your boss.

In order to respond to complex and subtle aspects of current circumstances, that information must somehow be represented in the prefrontal lobes of the brain. This means that the brain must form and maintain an image or model of what is going on in the world that is relevant in the current situation, and it must maintain that image even while processing other stimuli. Maintaining the image of relevant items in the brain is referred to as *short-term,* or *working,* memory, and it is the job of the lateral prefrontal cortex, which I discuss in the next section.

Working Memory, Problem-Solving, and the Lateral Prefrontal Cortex

Think about the last time you were involved with a group of people picking a restaurant for the evening. It probably went something like this: Person A said she liked Chinese and Italian. Person B reminded everyone that he is a vegetarian. Person C expressed a preference for steak and Italian but had Italian the night before. Person D didn't care about the food type but didn't want to wait too long for a table if the restaurant was too popular and crowded.

Solving this sort of common problem requires short-term, or working, memory. This type of memory maintains a number of relevant facts or items in consciousness long enough to solve the problem. It's called *short-term,* or *working,* memory because the contents may not be held in memory beyond the solution of the particular problem. Instead, after the problem is resolved, the memory is then "released" to be used for the next problem.

The next sections explain how working memory works and what its limitations are.

Brain processes managing working memory

Neurons in the lateral prefrontal cortex allow the brain to create a temporary active link between the various brain areas that respond to or code for an item to be remembered.

The idea is that any given thought exists as the result of activation of a unique set of neurons in various areas of the neocortex. Projections from virtually the entire neocortex get transmitted into the prefrontal lobe. Even if these projections from other areas are random, some sparse set of neurons in the prefrontal cortex will be specifically activated by this particular combination of inputs. Figure 14-3 shows in a highly schematic way how this is done. (The prefrontal cortex is the shaded area).

The connections between the prefrontal working memory neurons and the rest of the neocortex are reciprocal. For example, a visual stimulus activates the visual cortex. Some areas of the visual cortex project to the prefrontal cortex and activate working memory neurons there. The prefrontal working memory areas project back to the visual cortex and activate visual cortex neurons so that a similar representation of the visual stimulus exists, driven by prefrontal cortex rather than the eye. In this way, activity in the prefrontal neurons can maintain the activity of the other cortical areas that are its inputs.

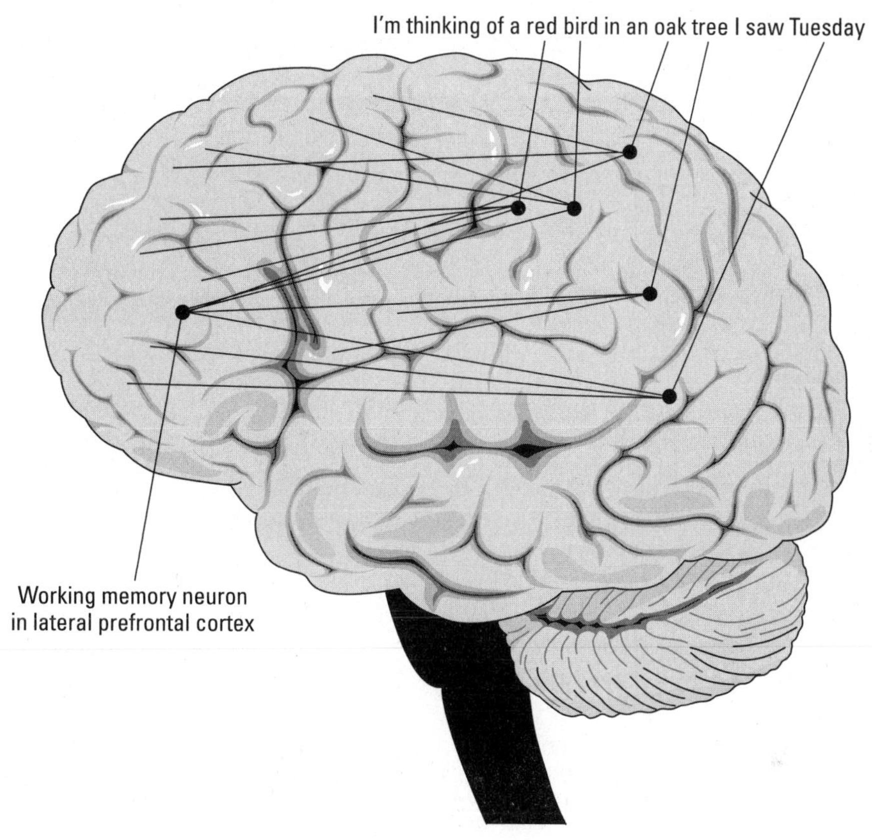
I'm thinking of a red bird in an oak tree I saw Tuesday
Working memory neuron in lateral prefrontal cortex

Figure 14-3: Working memory neurons in lateral prefrontal cortex

The neural activity that's maintained by the prefrontal neurons creates a brain state similar to that evoked by the original stimulus itself, except this activation can be maintained for some time after the stimulus has disappeared. What this means is that a goal can be pursued without the goal being continuously in sight (or in scent!).

RESEARCH

UCLA professor of psychiatry and biobehavioral science Joaquin Fuster's lab conducted an experiment that monitored the firing of a lateral prefrontal cortex neuron in a monkey while the monkey performed a memory task. In this task, the monkey received a stimulus that told him what lever to press for a reward, but he had to wait to press the lever until a go signal was given. While waiting for the go signal, he had to remember the stimulus in order to push the correct lever. Figure 14-4 shows the results of the experiment: The neuron recorded in the prefrontal cortex started firing as soon as the brief stimulus was given and continued firing for the duration of the delay interval after the stimulus was turned off, regardless of the duration of the delay. When the go signal was finally given and the monkey pressed the lever, the neuron stopped firing. Thus, the prefrontal neurons begin firing at the stimulus and continued firing to maintain the memory, and then stopped firing when the memory was no longer needed.

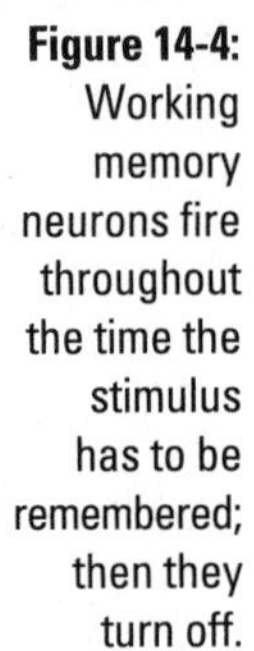
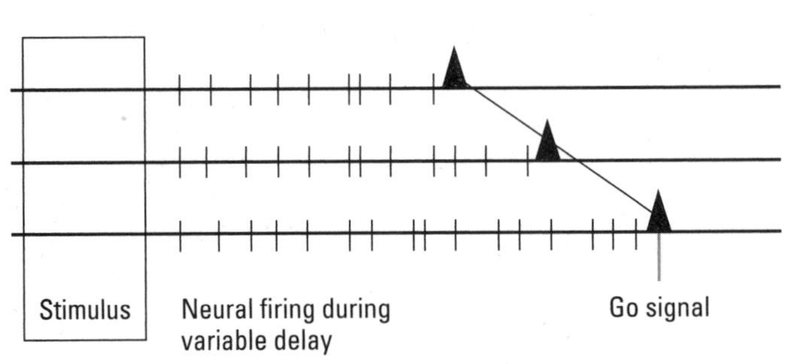

Figure 14-4: Working memory neurons fire throughout the time the stimulus has to be remembered; then they turn off.

To take this back to the making-dinner-plans example, the neurons start firing as soon as A, B, C, and D begin explaining their preferences, continue firing through the discussion, and stop after you're all heading to Applebees. The lateral prefrontal cortex contains millions of neurons coding for short term memory in this manner.

The limits of working memory

Besides the fact that working memory is short in duration, it also is limited in capacity. We can only juggle so many items at once. An average person can store a limit of only seven memory items in working memory.

The neural basis for a limit of seven memory items has been a mystery to neuroscientists for many reasons. The following sections take a closer look at two aspects of this mystery.

How scientists know that the human brain can store seven bits of information in working memory is linked to why local phone numbers in the United States have seven digits. This digit length came from now famous studies by a psychologist named George Miller who was funded by the phone company to determine how many digits an average person could remember long enough after looking them up in the directory or hearing the operator say them, to dial the complete telephone number. The answer he found was about seven.

Despite the fact that humans have much larger brains than many other animals, the capacity of seven short-term memory items doesn't appear to be much greater than that of many other animals, such as crows!

Numerous tests have been made of animals' abilities to distinguish between different numbers of objects. Many animals can distinguish four from five objects, for example, and some animals, such as chimpanzees and even crows appear to be able to count to and remember about six items. It seems odd that humans can only deal with seven items given that human brains have billions more neurons than those of crows.

What constitutes an item of memory is itself rather difficult to specify. In standard memory tests, for example, subjects can remember about seven random letters (on average). On the other hand, subjects can also remember about seven random words, each of which, of course, is composed of multiple letters. Moreover, if short sentences are used, subjects can remember about seven such sentences!

The reason this particular puzzle is considered so important is that scientists don't have a very good idea how memory works in detail, and most neuroscientists feel that these counterintuitive facts (that discrete memory items may themselves contain multiple parts) reveal something important about the mechanism.

One interesting idea about the memory limit was developed during the last decade by John Lisman and colleagues at Brandeis University. Their hypothetical scheme has three major features:

- The transient links that constitute a short-term memory exist during a single cycle of a brain oscillation called a *gamma oscillation.*
- The sequence of gamma oscillations are embedded in another slower oscillation called the *theta rhythm.* There are about seven gamma oscillations within one theta oscillation, and each gamma oscillation controls the representation of one of the set of items that are held in short term memory.
- The gamma cycles constitute what computer scientists call *pointers* that can evoke any memory that has enough mutually reinforcing connection strength. This means that the number of items remembered is relatively independent of the complexity of the items, as long as the items are either familiar enough or learned well enough to be evocable as a chunk. A discrete letter of the alphabet is a chunk, but so is a familiar word, whereas a string of nonsense letters is not a chunk because you have to remember each letter individually.

This scheme is shown schematically in Figure 14-5. Notice that the ongoing gamma oscillations (say, 42 per second) are locked to a master theta oscillation (say, 6 per second) so that there are exactly 7 gamma cycles per theta cycle. Each gamma cycle activates one neuron or neural group, the next gamma activates the next group, until the cycle is complete and the new theta peak restarts the cycle. Each gamma-activated neural group points to a self-reinforcing cluster of neurons that constitutes what is to be remembered.

The theta master oscillation is important because it triggers the cycle that returns to item one after item seven. As long as each of the seven gamma memories does not decay completely during one theta oscillation, they can be maintained indefinitely. In fact, if this process is maintained for several minutes, the hippocampus is activated and the process is initiated by which the memory can eventually become permanent.

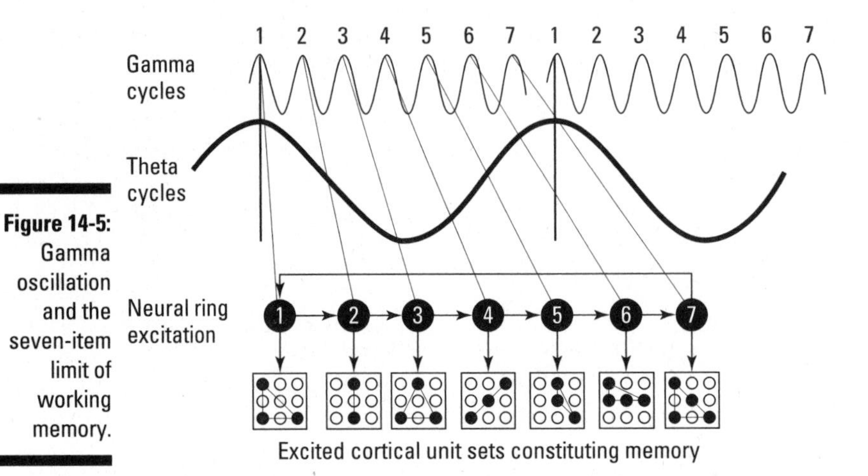

Figure 14-5: Gamma oscillation and the seven-item limit of working memory.

An interesting implication of this idea is that, although the level of abstraction for a memory in a human brain may be much higher than in other animals, the number of "items" that can be remembered may not be so much different. That is a function of some large scale oscillation system in the brain that, for other physical reasons, may not change much with brain size. What does change with brain size is the complexity of the items, or chunks, that can exist as self-reinforcing brain states. This is consistent with the idea that expansion of the neocortex, particularly in the frontal lobe, brings about the addition of higher, more abstract processing levels in a sequence from low-level feature representation to high level-semantic descriptions of abstract concepts.

This is not to say that the human short-term memory capacity of seven is not larger than most mammals. It is. Moreover, intelligence scores among humans are moderately well correlated with working memory capacity, which does show some variation among people.

Perseveration: Sticking with the old, even when it doesn't work anymore

Lateral prefrontal damage is associated with a phenomenon called *perseveration.* The classic test for such damage is the Wisconsin Card Sorting Task, shown in Figure 14-6.

In the Wisconsin Card Sorting task, a test card is matched to the other cards by shape, number of items, or fill color. The patient is given a test card and is asked to put the test card in the correct category according to a "secret" rule, which might be "Match the *number* of items in the test card to the category, and ignore the other features." The subject simply guesses at first and is told whether his guess is correct or not. Frontal lobe patients, like those without frontal lobe damage, have no trouble figuring out the rule after a few tries. The wrinkle, however, is that, after about 10 correct test card placements, the examiner secretly changes the rule to, say, match by color. People without frontal-lobe impairment adjust after a few errors to the new rule, but frontal patients have great difficulty switching to the new rule and persevere with the old one.

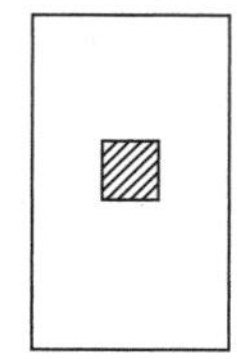

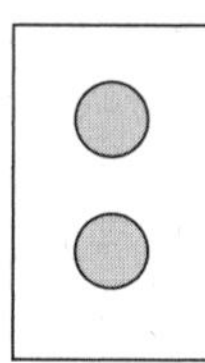

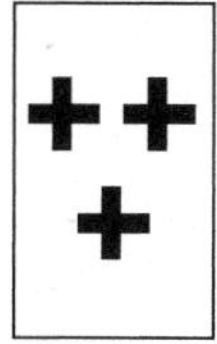

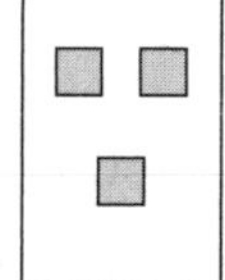

Figure 14-6: The Wisconsin Card Sorting Task.

This kind of frontal lobe function is essential for wisdom and flexibility. You might be driving to an appointment and hurrying because you are late, and, because you are hurrying, you almost hit another car. Your frontal lobe function should kick in and change the driving rule such that being safe overrides being late. Wise people with intact frontal lobes make this change in priorities; teenagers and people with compromised frontal lobe function may not.

Making Up and Changing Your Mind: The Orbitofrontal Cortex

The lateral prefrontal cortex gives you the ability to consider the context of the situation in your choices. But all of life's choices, like who to marry, can't be made by reasoning with only the cold hard facts, even including adequately weighing the present circumstances. Other things — like ideas of right from wrong, gut feelings, or intuition — guide your choices.

Feeling it in your gut: Learned emotional reactions

The medial part of the prefrontal cortex, called the *orbitofrontal cortex,* is involved in decision making that doesn't result from applying rules to facts. Its function is to process life's experience and to report whether your current situation or pending choice might be inappropriate or dangerous.

The orbitofrontal cortex accomplishes this warning function in part through its connections to the amygdala. The amygdala is a medial lobe memory structure just anterior to (and resembling) the hippocampus. The amygdala specializes in learning and producing memories of situations with high emotional content. Suppose, for example, that early in your car driving days, you were run off the road by a red pickup truck. Probably for years after that event, you get nervous when encountering red pickup trucks on the road, *even if you do not explicitly notice the truck or remember the earlier incident.*

Regardless whether red pickup trucks are actually dangerous, you have millions of bits of memory like this from your life's experiences that give you what people call *intuition.* There are good and bad things about intuition. Intuition is good to the extent that certain aspects of situations may generate useful warnings without your needing to consciously remember the circumstance. This is particularly useful when the warning is generated by a constellation of circumstances not easily expressible in words or as rules. You may, for example, get a feeling that it's going to rain without any idea exactly what aspect of barometric pressure or wind or cloud formation you come to associate as a precursor to rain.

On the other hand, intuition can lead to stereotypes and rigidity in behavior. You may then have to use reason to overcome stereotypes or take reasonable risks to accomplish things even if they entail situations having some danger. The amygdala-orbitofrontal cortex system communicates its memory of prior salient experiences with your consciousness through feelings. These feelings signify experienced relationships, but they may confuse cause and effect or be the result of spurious correlations.

Gambling on getting it right: Risk taking, aversion, and pleasure

People with poor orbitofrontal cortex function or damage to this area are prone to behavior such as gambling to excess. Although many people enjoy wagering small amounts of money on sports or horse races, most people limit these wagers, particularly if they lose more than an affordable amount in a short period of time. In this case, the prospect of waging further causes a feeling of uneasiness that approaches being painful. These feeling are weakened in those whose amygdala-orbitofrontal cortex system is compromised.

A sensitive test for orbitofrontal lobe function can be seen in the behavior in a pseudo-gambling task. In this task, the subject is given two decks of cards to play a hypothetical payoff/penalty game. In one deck, both the rewards and penalties are high, with the penalties, on average, accruing more than the rewards. A second deck has lower payoffs and penalties, with a small net reward accruing over time. The subject can choose cards from either deck, and receives the payoff or penalty according to the card randomly chosen. People with normal functioning of the orbitofrontal lobe learn over time to avoid the high payoff/penalty deck, even though they often cannot articulate the reason for doing so. Orbitofrontal patients, on the other hand, do not learn to avoid the high payoff/penalty deck. Instead, the enticement of the occasional high payoff overrides an inappropriate lack of fear for the penalty cards that make the net payout a loser over time.

Case-based reasoning: Thinking about social consequences

Good social intelligence is necessary for survival in complex societies. But this type of intelligence is not rule-based; it's more *case-based,* dependent on precedent and experience. The orbitofrontal cortex informs us when we contemplate socially embarrassing actions by provoking *feelings* of fear and embarrassment through the action of the autonomic nervous system. The unease that causes you to hesitate to tell your boss she is wrong occurs before your conscious awareness of its cause. The data of social intelligence are not easily captured with rules and reasoning, but instead consist of hunches about other people's intentions and vague recollections of their earlier behavior. But intelligent social behavior is crucial for keeping your job, your friends, and even your marriage.

Many things about the way the orbitofrontal cortex and amygdala process and communicate stimuli have the style of right-brain processing, which is case-based and non-linguistic. It is interesting to contemplate the idea that, within this system, as in the right brain generally, a large memory and powerful similarity processor exists that you have no direct access to through the normal manner of thinking in language. Instead these brain areas, despite their potentially powerful processing, communicate via the manipulation of feelings. Within all of us is this intuitive (right) side and another (left) side that wants to know "Why?"

Are We There Yet? The Anterior Cingulate Cortex

"Everybody's got plans... until they get hit." — Mike Tyson

We do many well practiced things, such as riding a bicycle, during which we are not aware of the individual actions, just the overall progress. Well-rehearsed motor sequences are self-reinforcing because each item in the sequence triggers the next. Interactions between the supplementary motor area, premotor cortex, and cerebellum allow learned, practiced sequences to be executed with higher precision and speed than is possible under explicit conscious control. Because such sequences are executed faster than they can be consciously controlled, they are executed with only high level awareness of their progress, without knowledge of the motor details.

But the real world is inherently unpredictable. We all know that even the most carefully considered goals using the most appropriate plans don't guarantee success. The fact that circumstances may change means that, at a minimum, we have to be able to evaluate our progress and change or even abandon plans as needed.

The place in the brain that seems to compare the unfolding of plans with reality is the anterior cingulate cortex (ACC). Brain imaging studies show that this area is activated during the performance of difficult tasks, particularly when errors are made. It is also activated by pain and the anticipation of pain. The ACC is part of the cingulate cortex, a medial region of cortex just above the corpus callosum and below the neocortex (refer to Figure 14-2). It is divided into two regions:

- **The anterior cingulate cortex:** The anterior cingulate cortex has been implicated in the monitoring of progress towards goals, the identification and correction of errors, and the allocation of brain resources in other areas toward meeting goals.
- **The posterior cingulate cortex:** The functions and connections of the posterior cingulate cortex are similar to those of the ACC. Compared to the ACC, activity in the posterior cingulate is more reflective of social context and emotional stimulus monitoring versus internal goals.

Because the functions of the cingulate cortex are associated with consciousness, progress in understanding its activity has been particularly dependent on recent imaging techniques, particularly fMRI.

Logging errors and changing tactics

Modern brain imaging techniques such as fMRI enable researchers to monitor the activity in marble-sized brain regions (called *voxels*) in humans who are performing complex tasks. When these tasks are difficult — and particularly, when errors are being made — the anterior cingulate cortex is activated. This brain area monitors progress toward goals and, as goals are modified or changed, controls the allocation of brain resources, such as working memory.

Say that you're driving along a quiet country road towards a restaurant where you anticipate a nice dinner. You're thinking about your favorite menu choices and the appropriate wine to go with them, which is what occupies your short-term memory. Glancing down at your dashboard, you notice the fuel light is on and you're nearly out of gas. Your anterior cingulate engages, switching the contents of your working memory from wine vintages to calculations of gas mileage and gas station locations.

The anterior cingulate is activated when you do things that are difficult or novel, when you make errors, or when you must overcome habitual behavior patterns. It is at the center of a supervisory control system that mediates goal selection and goal strategies, and allocates brain processing accordingly. It is particularly strongly connected with the lateral prefrontal cortex, allowing it to control the contents of working memory according to current progress and task demands. (Refer to the earlier section "Working Memory, Problem-Solving, and the Lateral Prefrontal Cortex" for more on that brain region.)

Acting without thinking

The fact that you can perform complex motor sequences, such a hitting a tennis ball, without conscious awareness of the action details doesn't mean that you're not processing sensory input. Only a very small percentage of all brain activity reaches consciousness. Consider, for example, the complex activities that Fido can undertake without any consciousness whatsoever. Clearly, supervisory control structures such as the anterior cingulate can operate without consciousness even though, in humans, activity in the anterior cingulate tends to itself produce consciousness about the progress of current activity.

A noteworthy example of how the human brain can process stimuli without a person being consciously aware of those stimuli is the phenomenon of blindsight. People who have extensive damage to the primary visual cortex (area V1; refer to Chapter 5) don't respond to visual stimuli that project to the missing area. When asked what they see, they claim they see nothing. However, when researchers ask them to "guess" where or what is presented, these patients guess accurately at a higher rate than would be expected from mere chance. What this indicates is that, although the patients were unaware of the stimuli, some processing of the stimuli still occurred; otherwise, the accuracy rate of the guesses would have been no better than chance.

Considerable debate exists about whether the unconscious residual processing is being done by subcortical visual structures such as the superior colliculus or by bits of cortical tissue that were leftover in the highly damaged areas. Regardless of the mechanism, the phenomenon points to the fact that unconscious processing of sensory stimuli such as visual inputs occurs without conscious awareness of seeing.

Who's minding the store? Problems in the anterior cingulate cortex

Some patients with depression exhibit low ACC activity. In some of these patients, depressive symptoms have been alleviated by electrically stimulating the ACC. Some studies have shown evidence of lower ACC activity in schizophrenia and OCD (obsessive-compulsive disorder). Like much of the rest of the brain, there are some differences in the functions of the left versus right anterior cingulate cortices. For example, damage to the right ACC has been correlated with the onset of anxiety disorders.

In normal human development, the frontal lobes mature last. For example, myelination of axons is not complete in the frontal lobes until late adolescence. This is the time of the onset of symptoms in many of those who will become schizophrenic or exhibit obsessive-compulsive disorder.

Adolescence is a developmental period notoriously characterized by high raw intelligence but poor judgment typical of inadequate frontal lobe function. One of the oft-stated goals of education is to instill enough good habits and rational thinking capabilities to get adolescents through adolescence without doing harm to themselves or others.

Part of the human maturing process is resolving the conflicts between the innate drive to challenge authority and convention and the wisdom to realize that at least some of society's conventions serve a useful purpose. As Mark Twain said, "When I was a boy of 14, my father was so ignorant I could hardly stand to have the old man around. But when I got to be 21, I was astonished at how much the old man had learned in seven years."

Chapter 15

Learning and Memory

In This Chapter

- Understanding how learning and memory help organisms adapt to the world
- Looking at the brain area and processes that facilitate learning and long-term memory
- Examining amnesia and other memory (and learning) disorders
- Studying ways to improve your learning

Learning is accomplished in the brain primarily by altering the strength (or *weight*) of synapses. Changing the strength of synapses changes the functional architecture of the brain so as to improve the organism's future responses to situations that are similar to the situation that triggered the learning.

Some learning occurs during development. During embryonic growth and the first years of life, the human brain develops by growing and changing its large-scale organization. Starting during early development but continuing throughout later life, nervous system activity itself causes changes in synaptic weights that mediate changes in behavior. Both stimuli from the external environment and internal sources such as sensory feedback in motor control parts of the brain trigger these activity changes.

Through an active process of reconstruction, humans — and perhaps a few other animals — can use their learning processes to store, retrieve, and visualize abstract memory items. Humans have the capacity to use memory to re-activate neural processes that occurred in response to an earlier original stimulus. This retrieval enables abstract thought in which past events, experiences, and images can be manipulated in working memory. In this chapter, I discuss the mechanisms by which learning and memory operate in the brain and the most important brain structures that mediate these processes.

Learning and Memory: One More Way to Adapt to the Environment

Learning and memory allow the nervous system to adapt to the environment to optimize behavior. Looking at the big picture, at least three kinds of adaption to the environment are possible: evolutionary adaptation, developmental adaptation, and classical learning. The next two sections discuss the latter two in detail. For detailed information on how the nervous system evolved, go to Chapter 1 (or if you'd rather get the short-and-sweet version, see the sidebar "Evolutionary adaptations" in this chapter).

Developmental adaptations

All multicellular animals undergo a developmental sequence from a single fertilized cell (zygote) to the multicellular adult form. During this process, the nervous system, as well as all organs and body structures, grows and differentiates into different cell types that form complex structures such as the brain and spinal cord. In higher multicellular life forms such as vertebrates, selection processes within the developing nervous system also affect its final structure.

Evolutionary adaptations

The human nervous system evolved from that of other animals and from earlier organisms that lacked complete nervous systems. Although we think of the nervous system as allowing the ability to sense both the internal and external environment to maintain homeostasis, even single-celled organisms have feedback processes in biochemical pathways that regulate the production of various proteins.

As organisms became multicellular, cells in different parts of the organism communicated by releasing certain substances into the spaces between the cells. Some of these substances became the precursors of hormones and neurotransmitters. Organisms that could act coherently by chemical communication, such as by moving flagella in unison, left more descendants than those that could not.

As multicellular organisms became larger and more sophisticated, neurons took over much of the job of communicating from one part of the organism to another (although circulating hormones that act more slowly still exist, of course). Neural networks were selected by evolutionary mechanisms as fast ways to optimize communication and coordinated multicellular action within an organism. In short, random mutations to these systems that gave the organism a survival advantage ended up getting passed from one generation to the next until those traits became "set" in a species.

Consider that the entire human genome contains about 20,000 *function genes* (genes that are read out to actually form proteins), yet the nervous system has 100 billion cells, each with over 1,000 synapses. The human nervous system is too complex to be specified by any "blueprint" of the genome. Instead, the nervous system develops by using general growth and migration mechanisms, followed by some pruning and stabilizing. The pruning and stabilizing depend on activity within the nervous system itself.

After birth, changes in the size and overall structure of the nervous system continue until adolescence. After adolescence, most changes appear not to take place at the scale of adding cells, but at the synapse level, mostly involving changes in the strength, or weight, of the synapses.

As we age, neurons inevitably die due to random damage as well as from any aging-associated diseases like Alzheimer's. The nervous system compensates for much of this damage by modifying synapses among the remaining neurons at the local level and by changes in problem-solving strategies at higher levels, such as using accumulated knowledge (wisdom) rather than fast processing.

Until very recently, scientists thought that no new neurons were added to adult brains, but in the last decades, numerous findings have shown the birth of new neurons in the adult brain, particularly in areas associated with learning, such as the hippocampus.

Classical learning

Even in the relatively stable adult brain, learning and memory occur at multiple time scales in multiple neural circuits. Short-term versus long-term memory can clearly be distinguished by the differences in their effective time spans (tens of seconds versus a lifetime). Short-term memories are converted to long-term ones via *rehearsal* (the section "Going from short- to long-term memory" has more on that).

A crucial brain area for consolidating learning from short- to long-term memory is the hippocampus in the medial temporal lobe. Specialized neurotransmitter receptors in the hippocampus and neocortex called NMDA receptors are instrumental in mediating synaptic changes that underlie learning and memory. The rest of this chapter is devoted to explaining the brain regions and processes involved in learning and memory.

Sending More or Fewer Signals: Adaptation versus Facilitation

At the most fundamental level, learning is a process by which experience changes an organism's responses. There are two mechanisms by which neural responses change: *adaptation* (or *habituation* in the case of repeated stimuli) and *facilitation* (sensitization in the case of repeated stimuli):

- **Adaptation** is the reduction in firing of a neuron over time despite a constant input or, in the case of habituation, when stimuli are repeated. Adaptation occurs throughout the nervous system. You can easily identify adaptation in sensory systems when something that you hear, see, or feel is quite noticeable initially but becomes unnoticeable over time.
- **Facilitation** is the opposite of adaptation. Facilitation occurs when a neural response increases over time or after repetitions. Facilitation is often associated with stimuli that are slightly noxious (where it is often then called *sensitization).*

Both adaptation and facilitation occur in simple nervous systems like those of invertebrates such as mollusks and insects.

Understanding very primitive mechanisms like adaptation and facilitation is important because it helps researchers understand the neurophysiology of a more complex dynamic process such as learning, which almost always occurs by either weakening or strengthening some neural response, usually at a synapse between neurons. The following sections go into more detail.

Adaptation

Adaptation, which occurs in all multicellular organisms, even invertebrates, is one the most basic forms of experience-dependent change. It often occurs in the sensory receptors themselves, but it can also occur upstream from the receptors. This process is so basic we don't usually even consider it in the context of "learning."

When you first sat in the chair in which you are now reading this passage, you undoubtedly were aware of your rear end contacting the seat and your back resting against the chair back. Now, unless you've fidgeted recently, you are hardly aware of your contact with the chair at all.

You may assume that your lack of awareness is solely due to the fact that you are not paying attention to your contact with the chair. This idea tends to be associated with the folk psychology concept that awareness is located in some deep recess in the brain, and this awareness has decided not to pay attention to your sitting in the chair any longer. But actually, the mechanoreceptors in your skin (see Chapter 4 for more on mechanoreceptors) themselves are no longer firing as much as they did when you first sat down. Much of the reason you pay less attention to your contact with the chair is that fewer signals about that fact are reaching your brain.

Habituation is similar to adaptation in that the neural responses decrease over time, but in the case of habituation the decrease is explicitly due to repetitions of the stimulus. Habituation typically involves changes in the responses of neurons driven by the receptor rather than in the receptor itself. This change is usually mediated by a change in synaptic strength so that equal activation of a receptor after several activations produces a weaker response in the post-synaptic neurons connected to it than before the series of habituating synapses.

Facilitation

The opposite of adaptation is *facilitation.* Facilitation occurs when repeated activation of a receptor causes *more* activity. Facilitation almost always occurs in neurons post-synaptic to the receptor.

Although the neural circuitry that mediates facilitation can be quite complex, the basic experience is universal. Here's an example. You're enjoying standing outside on a warm summer evening, occasionally noticing a slight tingling on your skin. Suddenly one tingling becomes large enough that you realize you are being bitten by mosquitoes. From that point on, every tingling, even from puffs of wind, causes your hand to move to and slap or scratch the area. That's facilitation.

The term *facilitation* is also used in a classical conditioning paradigm, when, for example, a neutral stimulus, such as ringing a bell, is paired with (and there associated with) a noxious stimulus, such as an electric shock so that the avoidance reaction eventually occurs to the bell alone. The other famous kind of conditioning, *operant conditioning*, usually involves pairing a normal behavior with a reward so that a behavior or behavior sequence is learned and triggered in that context.

Judging a study by its name

Some members of Congress tend to pick on research projects with funny titles as examples of government waste. No doubt there is waste in government funded research. On the other hand, deciding this by the titles of the research projects is not a good way of determining what studies deserve funding. One can imagine the early days of Dr. Kandel's research on the gill withdrawal reflex in *Aplysia* being criticized as "fooling around making sea slugs twitch" or some such. But this research provided the first basis for the understanding of synaptic plasticity and was done in this species for important technical reasons (big, easy-to-get-at neurons). Ditto for fooling around with mutations in common fruit flies, which has given researchers the knowledge for the understanding of genetics, genetically caused diseases, and leads for their cures.

Studying habituation and sensitization in sea slugs

Habituation and sensitization were extensively studied at the synaptic level by Eric Kandel and his colleagues at Columbia University (Kandel won the Nobel Prize for this work). Kandel sought out a simple system that he could study in detail at the synaptic level and chose the *Aplysia*, or sea slug. This animal shows habituation of its gill withdrawal reflex when the gill is touched repeatedly, but gently. However, the animal shows sensitization to gill touch if the touch is paired with a shock or other noxious stimulus.

Kandel and his colleagues worked out at the synaptic level exactly how the neural circuits in the *Aplysia* mediate these synaptic strength changes. Mechanisms similar to what Kandel found in *Aplysia* also operate in vertebrates where they mediate synaptic changes underlying learning and memory.

Exploring What Happens during Learning: Changing Synapses

As mentioned earlier, learning boils down to changing how the nervous system works by a mechanism that alters its functional structure as a result of experience (in some way that is appropriate, of course).

Throughout most of the 20th century, one of the most important problems in neuroscience was determining what it was that changed during learning. This dilemma was often referred to as the "search for the *engram*," the memory trace that was the result of the learning process. The idea was that learning and memory leave traces in the brain that could be uncovered and, eventually, understood. Theories for memory traces ranged from changes in protein and DNA synthesis to holographic electric fields in the brain.

Sounds a little science-fictiony, but it turns out that some of these ideas were right. Learning does change the brain. What changes mostly is the strength of synapses that allow neural computations to be done, rather than the numbers and connections of neurons as in development. Changing synaptic weights rather than adding new neurons and connections allows learning to take place rapidly. (New neurons are *sometimes* added in response to learning, but the process is slower than quickly made synaptic changes.)

Neural computation: Neural AND and OR gates

A vital contribution to understanding the mechanism of memory formation came not from traditional neuroscience but from research in the fields of artificial intelligence and what is now called *computational neuroscience,* a field that tries to figure out how nervous systems do computations like recognizing patterns.

To approach the problem of understanding nervous systems as computing devices, computational neuroscientists had to make models of neurons that abstracted their operation so that the model's neural networks could be mathematically simulated and, in many cases, built out of electronic hardware that could perform the kinds of computations that neurons appeared to be capable of performing.

This work was being done at the dawn of the computer age, and the most obvious model for neuronal computation was the computer. Computers work with logic gates, such as AND and OR gates. If neurons could be shown to perform AND and OR computations, then scientists could study the operation of electronic logic circuits to learn how neural circuits processed information.

Here's how the operation of the classical logic gates (OR and AND) work (refer to Figure 15-1):

- **OR gates** can have any number of inputs (A, B, and so on), and the output (O) of the OR gate will be "true" if *any* of the inputs are true. The number 1 represents an input that's true; 0 represents an input that isn't true.
- **AND gates** can also have any number of inputs. Their output is true if and only if *all* (every single one) of their inputs is true (1).

At the bottom of Figure 15-1 is an image of a neuron with three inputs. The question is, what can a neuron compute, and how does it do that computation?

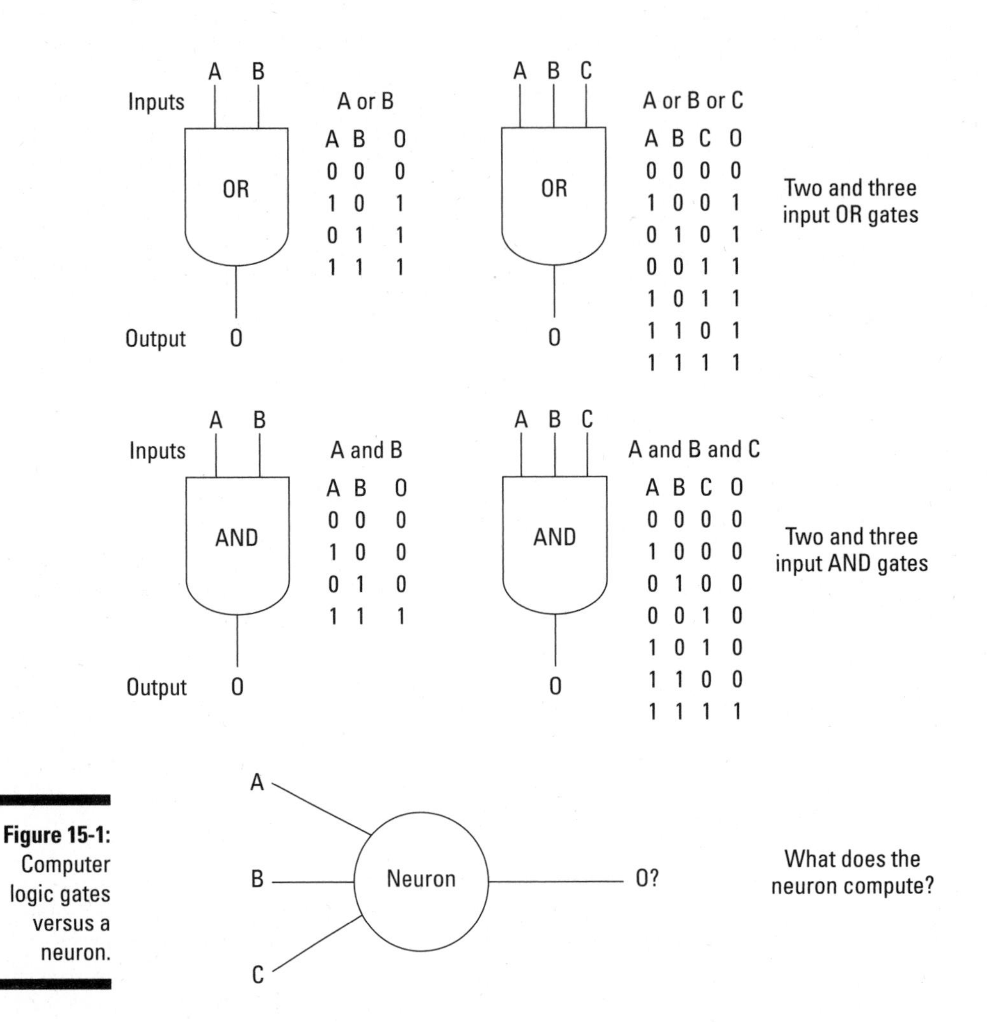

Figure 15-1: Computer logic gates versus a neuron.

Keep in mind the difference between memory in nervous systems and that in computers. In computers, memories are pieces of information represented by a sequence of bits that are stored in a memory location specified by another sequence of bits that act as the pointer to that location. In nervous systems, memories are changes in synaptic weights (usually) that change the way neural circuits operate. Therefore, memories in nervous systems are not necessarily things (or items or tokens) that can be retrieved as such. For example, when you practice and improve your tennis forehand, the memory that permits this consists of small synaptic changes widely distributed from your visual system that is seeing the ball to your motor neurons that are controlling all the muscles in the body to allow you to hit it.

The McCulloch-Pitts neuron

A giant advancement in treating neurons as computational devices was made by Warren McCulloch and Walter Pitts. McCulloch and Pitts proposed a now classical artificial neuron format often now referred to as the McCulloch-Pitts neuron. Figure 15-2 shows one version of this artificial neuron.

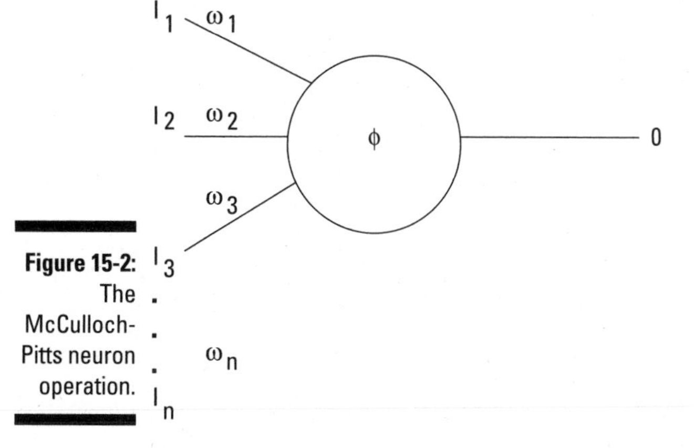

Figure 15-2: The McCulloch-Pitts neuron operation.

What McCulloch and Pitts showed was that the model neurons of this type (described shortly) could emulate standard logic gates such as OR and AND gates *if* the synaptic weights were suitable. They also showed that a system of these neurons, if suitably connected, could perform any function that a digital computer could perform.

Look at two examples, namely, the AND and OR functions. Imagine a McCulloch-Pitts neuron as in Figure 15-2 with three inputs (I_1, I_2 and I_3). Each input (which can be a "1" or "0") is associated with a synaptic weight, indicated by the symbol ω_i. What the McCulloch-Pitts neuron does first is sum (at the cell bodies) each input I_i times its synaptic weight, ω_i. This summation is some number, like 0.4 or 2.1.

The summation number is then passed to what is called a *phi function* (ϕ). A common phi function is a threshold (modeled after the fact that neurons have thresholds by which a minimum excitation level is needed to produce firing). In the neuron of Figure 15-2, say that the threshold is equal to 1. That is, if the sum of the inputs multiplied by each input's synaptic weight is greater than or equal to 1, the output of the neuron equals 1; otherwise, it is zero.

You can make the McCulloch-Pitts neuron an AND gate if you make all the *weights* (w_i) suitably small so that *all* the inputs have to be active for the sum of those inputs times their weights to be greater or equal to the threshold. You can make the McCulloch-Pitts neuron an OR gate if the weights are so high that any one input times its weight exceeds the threshold.

Neural circuits using McCulloch-Pitts neurons can perform the same logical calculations as a digital logic circuit with the addition of one more type of gate, a NOT gate, which in neural terms is simply an inhibitory synapse. Thus, the brain, or, at least some circuits in the brain, can act like (compute like) a digital computer (some neural circuits also appear to operate like analog computers, so the brain is a mixture of both).

Through the McColloch-Pitts neuron model, researchers came to understand that synaptic weights are key in neural computation. Other scientists following the lead of McColloch and Pitts showed how systems of these model neurons could learn to do things like recognize patterns by changing the weights according to certain mathematical rules. The question with respect to learning and memory in organisms then became, how are the synaptic weights set in the brain?

Rewiring your brain: The NMDA receptor

Computers are able to do computations with AND, OR, and NOT gates because engineers very carefully design the circuits specifying the location of every single gate. Most digital computers will fail if a single logic gate fails. But, as has been pointed out, the 20,000 gene genome cannot specify the trillions of synapses in the brain. Moreover, even if the synapses could be specified by some genetic blueprint, the blueprint does not specify how neurons learn (or how the brain adapts when neurons randomly die off).

Setting neural synaptic weights properly means changing their strengths as the fundamental process of learning occurs. Researchers began to think that there had to be *some* synapse *somewhere* in the brain whose synaptic strength could be changed to perform new (learned) neural computations. This realization led to three crucial questions:

- How does a synapse change its strength?
- What and where are these synapses?
- What signal controls this change that is associated with learning?

Starting about in the 1960s, the hunt was on for the memory receptor that would change its synaptic weight to allow learning. Although Kandel's work on facilitation in *Aplysia* (the sea slug, a marine invertebrate with large, easy-to-record neurons) elegantly showed the synaptic mechanisms that mediated the processes of habituation and facilitation, most neuroscientists considered these phenomena too one-dimensional to explain the neural mechanisms of learning generally, particularly in complicated nervous systems such as the human nervous system.

Introducing the NMDA receptor

A cottage industry of laboratories sprang up in the 1970s and 1980s looking for synaptic strength modifications in the brains of mammals like laboratory rats. The typical preparation involved taking a slice from the deceased experimental animal's brain and keeping it alive by immersing it in artificial cerebrospinal fluid (like the normal extracellular fluid in the brain and spinal cord). Electrodes were inserted into single cells and subtle synaptic alterations monitored when cells that had synapses on the monitored cell were stimulated.

The most common neural tissues used were the neocortex and hippocampus. Although many researchers favored the neocortex as the seat of higher cognitive functions, the hippocampus brain slice began to dominate because of its clear role in memory formation. Several learning-related changes in synaptic weights were observed by a number of labs. These often involved an unusual receptor type for glutamate that had properties that had been predicted for modifiable synapses. This was the NMDA receptor. (NMDA stands for *n-methyl D-aspartate,* the name of a chemical agonist that mimics glutamate at this receptor.)

NMDA receptors exist throughout the central nervous system, but they are particularly abundant in the hippocampus, which plays a vital role in memory formation and interacts with the cortex through long-term potentiation and reciprocal synapses.

The NMDA receptor in action

The NMDA receptor, which is abundant in the hippocampus, changes its strength as a function of coincident inputs. This receptor implements a neural AND gate function (see Figure 15-3), and it is also a receptor for the neurotransmitter glutamate. However, unlike most ligand- (neurotransmitter-binding) activated receptors, the NMDA receptor is also blocked by a magnesium ion in the mouth of the pore. This magnesium ion is removed if an adjacent glutamate ion channel is also activated and depolarizes the neural membrane. In other words, activation of the NMDA receptor requires both its presynaptic terminal to release glutamate *and* another presynaptic terminal to release glutamate to an adjacent receptor.

Once open, the NMDA receptor channel has the unusual property (for a typical excitatory ligand-gated channel) of allowing a considerable amount of calcium to pass through the channel as well as sodium. Calcium entry into neurons typically has many secondary effects which are often referred to as *second messenger effects.* One of these effects seems to be to increase, under some circumstances, the potency of the synapse. This synaptic modification depends on both presynaptic and postsynaptic mechanisms, the study of which is the source of many neuroscience journal papers currently.

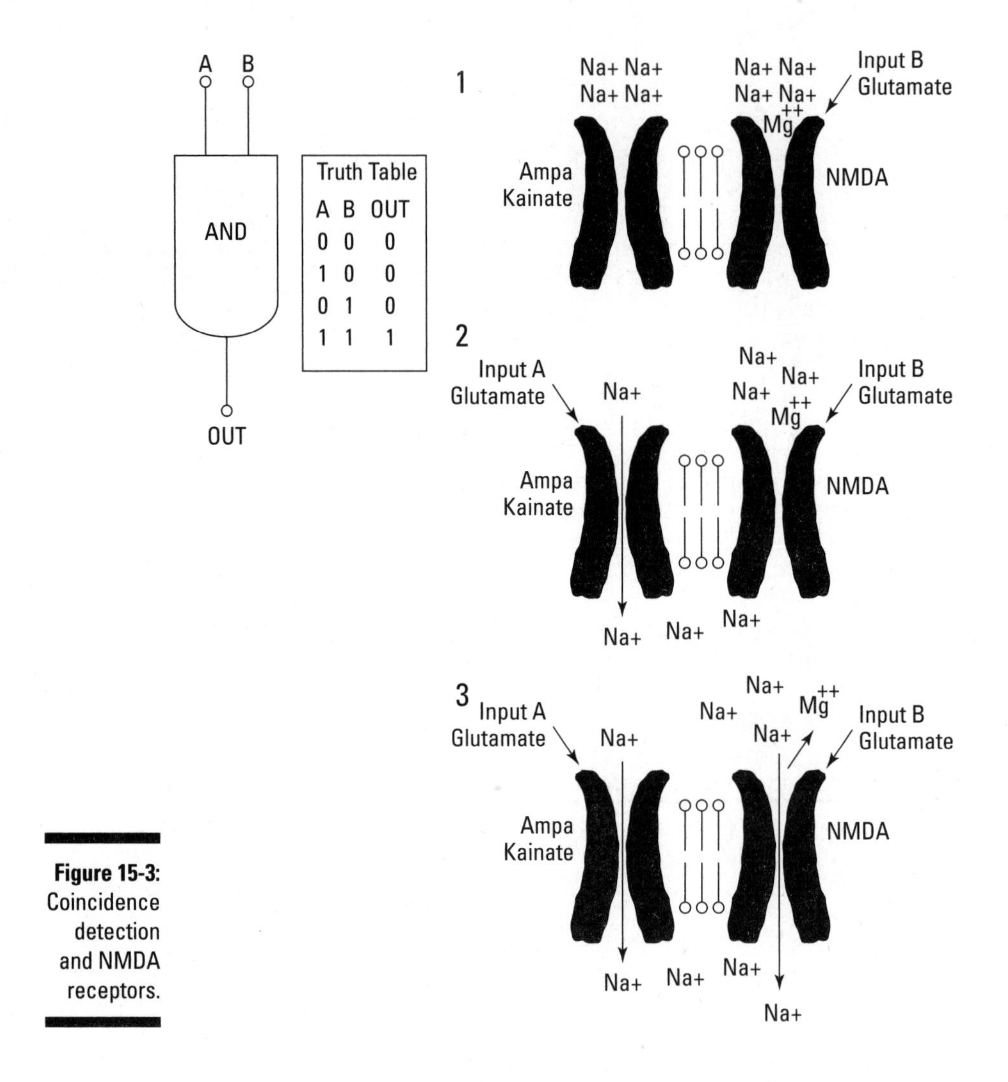

Figure 15-3: Coincidence detection and NMDA receptors.

The left side of Figure 15-3 shows the diagram and truth table for a standard logic gate (as in Figure 15-1), the AND gate. In order for the output of this gate to be true (1), both inputs must be equal to 1.

On the right side of Figure 15-3 is a diagram of two glutamate receptors frequently found together (on the same dendritic branch very near each other). On the left is the common ampa/kainate type of receptor, which is a typical ionotropic receptor for glutamate that has a channel that opens when glutamate binds to the ligand binding site. On the right is an NMDA receptor, which is ionotropic also, but a little different, as I describe shortly.

At the start of the sequence (image 1), a presynaptic terminal signifying whatever it signifies (the color red, for example) has released glutamate that has bound the NMDA receptor on the right. This NMDA receptor not only requires the binding of the ligand glutamate, like many ionotropic receptors, but in its normal state, it alsohas a magnesium ion (Mg^{++}) stuck in the entrance to the receptor which blocks the receptor. The magnesium ion is attracted to negative charges in the amino acids located in that part of the receptor, as well as to the negative net charge inside a neuron in its *resting,* non-depolarized, state. So if only the glutamate input to this receptor is active, the channel is not opened.

However, near the NMDA receptor is another, different type of glutamate receptor, the ampa/kainate type (named for exogenous agents that bind this receptor type). If this receptor receives glutamate from its presynaptic input, signifying, bird, for example, it opens (image 2) and allows positive sodium ions (Na^{+}) to pass through its channel. These positive sodium ions repel the positive magnesium ion in the NMDA receptor, opening that channel if it already has glutamate bound to it (image 3).

Thus, the NMDA channel is a neural AND gate, requiring both its presynaptic input to be active (bound glutamate) and a nearby ampa/kainate channel to be active *at the same time.* In this hypothetical case, this two-receptor complex is detecting a red bird, perhaps a cardinal.

Strengthening the synapse: Long-term potentiation

What happens then is even more interesting and not shown in the diagram. When the NMDA receptors are repeatedly activated by any particular coincidence, the synapses get stronger; that is, it gets easier for the presynaptic input pair to cause the NMDA channel to open. This happens by a mechanism called *long-term potentiation.* (There is also *long-term depression* which balances things out so that all synapses do not become stronger only.)

The Role of the Hippocampus in Learning and Memory

You can remember the phone number of the garage where your car is being serviced not much longer than it takes to dial it, but you do not forget your sister's name for the decades of your lifespan. One of the most intensely studied topics in cognitive psychology is the nature of these two different types of memory, termed short-term and long-term memory. Cognitive neuroscience seeks to locate where in the brain these memories exist and what processes they use. The hippocampus, as it turns out, is crucial for moving memory from the short-term form to the long-term.

Going from short- to long-term memory

When you're driving your car, you're aware of the locations of the cars around you so that you know at all times whether changing lanes is safe or not. But this situation changes constantly. You know (and remember) each situation while you need to deal with it, but you don't remember at the end of the drive every single pattern of cars around you that occurred during the drive.

Most short-term memory of this sort is *disposable,* used while needed and then discarded. Only a fraction of the information that passes through your short-term memory reaches long-term memory, and that usually requires that the information be important enough for you to *rehearse* it (going over it again and again) or so salient that you can't stop thinking about it (another way of rehearsing).

Short-term memory exists in two places, the lateral prefrontal cortex and the hippocampus. Neurons in the lateral prefrontal cortex maintain their activity, representing an input after the input has gone so that you can use the memory to accomplish a relevant task, such as dialing a phone number. (Refer to Chapter 13 for more on the prefrontal cortex.)

The process of transferring information from short- to long-term memory requires the action of the hippocampus.

What the hippocampus does is provide a set of modifiable neural AND gates for neural activity. These hippocampal AND gates receive input from the entire cortex as it represents what is going on in the current situation. These activated NMDA synapses then represent all the cortical activity associated with any particular event. For example, you were in *this* park, *and* sitting on *this* bench, *and this* bird flew by, *and* at *this* time…

A matrix of coincidence detectors

Various areas of the neocortex project to the hippocampus in a matrix fashion to enable coincidence detectors there (see Figure 15-4). The basic idea (highly oversimplified) is that colors of things might be represented in one place in the brain and type of animal in another. If you see a green frog, an intersection will occur in the hippocampus matrix of green and frog at a neuron that will be the coincidence detector for that thing. So seeing a green frog activates the cortical areas for green and frog, which activates the hippocampal green frog cell whose synapses get strengthened.

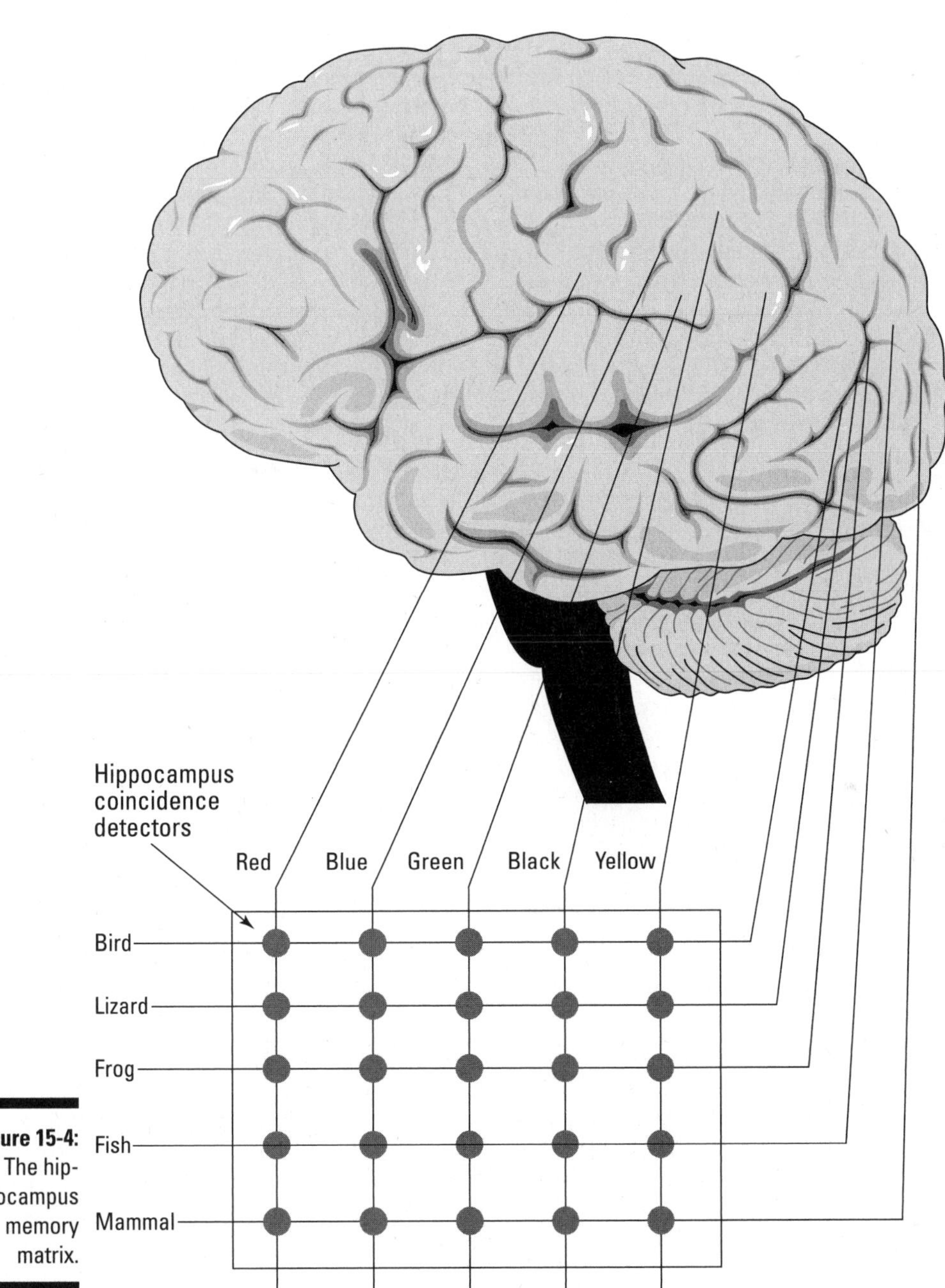

Figure 15-4: The hippocampus memory matrix.

The case of HM

The importance of the hippocampus in learning was dramatically demonstrated by a famous clinical case, that of the patient HM. HM was an epileptic with intractable seizures. The focus (place of origin) of his seizures was determined to be the hippocampus. (This is not unusual; the temporal lobe where the hippocampus is located is commonly implicated as the initiator of seizures for reasons that are poorly understood.)

Although removal of brain tissue is clearly a desperate measure to treat a neurological condition, some cases of epilepsy are so severe that this procedure is warranted. Some epileptics have seizures nearly every hour and are essentially totally disabled, unable to hold a job or engage in any meaningful social relationships. Only a minority of these cases respond to pharmacological treatments. But if the place of origin of the seizures is well localized in the brain, removing this tissue may completely or nearly completely eliminate the seizures. In addition, the diseased brain tissue that's removed often has no normal function and surrounding tissue has taken over its function.

The problem with HM was that the surgeons removed the hippocampus on *both* sides of his brain. The result, studied extensively over many years by Brenda Milner among others, was that, although HM's seizures were greatly reduced, he had no ability to transfer information from short- to long-term memory, although his previous long-term memory was intact.

HM always thought that the date was the date he had had the surgery. He could converse with hospital staff about events in his own past, but if the staff member left for a few minutes and returned, HM had no recollection of the conversation. His life was frozen in the present, and he never recovered any ability to remember anything that happened after the day of the surgery. He remained a patient in the hospital until his death in 2008. Interestingly, he started playing checkers with hospital staff after the surgery and, despite improving as a player, did not remember playing any previous game. The case of HM was a dramatic illustration of the difference in the neural substrates for short- versus long-term memory.

Imagine that some area of visual cortex codes for different colors of things, and those color detectors project to a line of hippocampal coincidence detectors. Some other area of the cortex might have neurons responding to types of sounds or taste of animals (birds, lizards, and so forth). There will be one hippocampal coincidence detector that responds to yellow frogs, one that responds to green mammals, and so on. Of course, the real hippocampus has many similar cells of each preference.

The strengthening of hippocampal synapses works by a process called *long-term potentiation* in which synapses are strengthened when several inputs are simultaneously highly active. A mirror process called *long-term depression* also occurs that reduces synaptic strength when synapses are not active simultaneously.

By replicating this matrix structure, you can get hippocampal coincidence detectors that respond selectively to any list of conditional attributes. If these attributes are presented many times (lots of yellow frogs), then, through long-term potentiation, some group of coincidence detectors will be very activated for the appearance of those attributes that signify a yellow frog — things like small size, found near ponds, eats bugs, and so forth. Now those neurons code for a high order abstract concept after learning, by producing these content cells.

The hippocampus is not so important for the very short-term memory of your surroundings, which seems to be the job of the lateral prefrontal cortex. Rather, the hippocampus is responsible for transferring short-term memory into long-term memory, which is actually stored in the neocortex itself. Neuroscientists know this because of the famous patient HM.

Remembering as knowing: Cortical mechanisms

The crucial aspect of the hippocampal-cortex system when it comes to forming memories is that not only does the cortex project to the hippocampus and cause synaptic strengthening, but the hippocampal neurons whose synapses have been activated also project activity back out to the cortex and can activate it from the hippocampus rather than from the sensory input.

This means you can re-create in your cortex a version of the pattern of activity that occurred when you actually experienced something. When you continue to think about, or *rehearse,* the memory of some experience, the activity reverberates between hippocampus and cortex. If you rehearse enough (which happens during REM sleep, particularly), modifiable synapses in the cortex are changed so that the cortex itself can reproduce the neural activity associated with an experience.

Long-term memory exists in the same cortical areas that represented the initial experience. The hippocampus is a scratchpad for maintaining rehearsal to form the long-term memory. The hippocampus interacts also with the lateral prefrontal cortex which houses working memory, which maintains a memory long enough to allow you to accomplish a task that depends on that memory, such as remembering whether there is someone in your blind spot before you change lanes.

One of the most interesting models for understanding hippocampal function came from multielectrode recordings made in the hippocampus of freely behaving rats running (you guessed it) mazes. Researchers discovered that cells in the rat hippocampus, called *place cells* and *grid cells,* responded when the rats were in a particular place in the maze and that different cells were activated by different places in the maze. After the experimenters figured out that the place and grid cells were linked to particular places in the maze, they could actually tell where the rat was in the maze just by looking at which place and grid cells were active. In the experiments, the rats were learning these mazes to get rewards (probably cheese).

One of the fascinating things about these experiments was that when the rats slept after the training sessions and entered REM sleep, their hippocampi played back the correct maze traversal sequence by activating, in sequence, the place and grid cells and the cortical areas stimulated by the sight, sound, odor, or whatever aspect coded for that maze location. This playback occurred at about seven times the actual speed that the rats ran the mazes. If the rats were prevented from having REM sleep, they did not consolidate the day's training well and did not learn the mazes as well.

Most neuroscientists were surprised that long-term memory resides in the very areas of cortex that process the sensory input during the original experience, and scientists really don't understand this phenomenon very well. NMDA receptors exist throughout neocortex, however, and experiences during development clearly can modify the neocortex. Some mechanisms the mediate this developmental plasticity may also be the basis for learning. Considerable work must be done in the 21st century to get a handle on the specifics of this, however.

Knowing versus knowing that you know: Context and episodic memory

The picture of the inputs to the hippocampus given in Figure 15-4 is missing a crucial component: the role of the frontal lobes, which also project to the hippocampus. This projection provides something other than just the identity of the object of attention as it is represented in the parietal, occipital, and temporal lobes. The frontal lobe input contributes context.

Context is the information about a particular event involving an object, rather than just the information about the object itself. Context is knowing where you learned that the capital of Alabama is Montgomery, rather than just knowing the fact. The name for contextual memory is *episodic memory,* that is, memory associated with an even or episode. General memory about facts is called *semantic memory.*

Rich episodic memory is one of the hallmarks of human consciousness. Animals clearly learn many associations that are equivalent to semantic memories, such as my dog knowing that food arrives in the bowl early in the morning and that rain comes after thunder. However, whether animals have episodic memory at all is unclear, because episodic memory involves awareness of oneself in one's particular surroundings at a particular time. This awareness in turn depends on what's in working memory, which is vastly more complex and deep in humans than any other animal due to our large frontal lobes.

Episodic memory consists of content *and* context. Context includes all the things associated with a particular experience in time. For example, you might have seen a yellow frog yesterday, on a sunny day, at the grade school pond, and with your friend Suzi. Cortical areas that represent all these experiential aspects also project into the hippocampus and can link with each other and the yellow frog sighting. Now you have hippocampal neurons responding to a very particular yellow frog seen yesterday with Suzie, a yellow frog seen years ago at a grade school pond, and so forth. Remember also that the projections from neocortex to hippocampus include both low (realistic, detailed) and high (abstract) meaning associated levels, so that the context for any memory can be quite elaborate and specific.

Once a memory is encoded, how does the activity in the hippocampal cells produce the experience of memory? As I mention elsewhere, not only does virtually the entire neocortex project to the hippocampus, but the hippocampus projects back to the entire neocortex.

Now suppose that you happen to see something yellow. This activates the yellow detectors in your cortex and then the yellow cell network in the hippocampus (per Figure 15-4). Among the hippocampal cells on this line is the yellow-frog-yesterday detector, which had its synapses strengthened. Through a control mechanism that scientists don't yet understand very well, this activation may cause the yellow frog hippocampal cell to be activated, which projects back out and activates all the cortical areas activated in yesterday's experience, which project back to hippocampus, which strengthens the image. If all the brain areas are activated that were activated when you experienced the yellow frog yesterday, you will, of course, experience that event in your mind now.

Losing Your Memory: Forgetting, Amnesia, and Other Disorders

In normal life, our inability to remember things can occur either because the memory itself is too weak (the synapses are not sufficiently strengthened) or because, despite the memory being there, we can't retrieve it. In the latter case, given some clues, we might remember, but if the memory is too weak, we won't be able to recall regardless how many clues or hints we're given.

The inability to retrieve memories can happen temporarily or permanently. We all experience temporary memory loss in the so-called tip-of-the-tongue experience where we're sure we know something but can't retrieve the fact at the moment. This inability can be due to anxiety and distracting factors that make it hard to concentrate. It can also be due to context. Remembering something in the context where it was first learned is easier than remembering it in a very different context. Experiments have been done, for example, with scuba divers who learned a list of words underwater and then were asked to recall the terms later, both on the surface and underwater. The results? The underwater performance was better.

Memory can also be lost from head injuries and epileptic seizures that disrupt brain activity. Because the process of consolidation of memory between the hippocampus and cortex can take weeks, disruption of brain activity tends to affect the most recent memories the most severely.

Electroshock therapy (and recently some forms of transcranial magnetic stimulation, or TMS) is a treatment that deliberatively produces this effect to treat depression. The idea behind this is that, during depression, depressed people interpret and remember everything with negative affect. These negative memories create hopelessness and further negative affect for ongoing experiences. By eliminating the recent negative memories, this therapy resets the patients affect away from the severely negative set point.

People can also suffer the soap opera syndrome of temporary memory loss, known clinically as *transient global amnesia* (TGA). TGA can occur without obvious physiological causes such as strokes or other brain damage, but it sometimes follows extreme physical exertion, particularly in men. Scientists hypothesize that TGA sometimes results from *transient ischemia,* a loss of adequate blood flow to some part of the brain. TGA, if it isn't associated with significant brain damage, may resolve within a few days, with the sufferer then remembering he really is married to Jennifer despite his behavior the last two days.

Repressed and false memories

Repressed memories, by definition, involve some suppression of the retrieval mechanism for a memory that exists and should otherwise be retrievable. A memory might be repressed, for example, if you trained yourself to think of something else at the first recognition that the context for a particular (say, painful) memory was starting to form. This training could become so automatic that the memory becomes effectively irretrievable.

A number of clinical psychologists became well known in the 1980s and later for using hypnosis to "recover" repressed memories of childhood sexual abuse. Several of these cases ended up with court convictions of the accused abusers. This movement lost its credibility, however, when factual evidence showed that some of the convicted abusers could not actually have committed the abuse they were convicted of, because, for example, they weren't in the same town the day the abuse was alleged to occur.

Carefully controlled studies by the psychologist Elizabeth Loftus also showed that the extreme suggestibility that occurs during hypnosis and in certain interrogation techniques can create false memories for implanted events that clearly never occurred. This research doesn't mean that no such thing as repressed memories of childhood sexual abuse exists, only that hypnosis and suggestive interrogation techniques are not reliable methods of uncovering real instances of repressed memories.

Another kind of memory loss involves the inability to experience appropriate feelings from emotionally salient images or experiences. This condition is not often thought of as a memory problem directly; however, compromise of the amygdala-orbitofrontal cortex system (refer to Chapter 14 for more detail on this system) constitutes a kind of memory loss in which patients lack appropriate fear reactions to dangerous situations similar to those they have experienced previously that should evoke fear reactions.

The famous neurological case of the railroad worker Phineas Gage showed that frontal lobe damage could leave normal intelligence relatively intact but severely compromise emotional and social intelligence. After Gage lost much of the function of his orbitofrontal cortex from a work accident, he returned to work but became irresponsible, erratic, and continuously gambled his earnings away on flakey get-rich schemes in which he had had no interest previously. You can read more about his story in Chapter 12.

Getting Brainier: Improving Your Learning

Learning has a lifespan trajectory. It's compulsively easy when you're very young, difficult after adolescence, and very hard for most elderly. We can't do much about the process of aging other than taking care of ourselves, but we can maintain and even increase the ability to learn by engaging in intellectually challenging activities. In this sense, the brain is like a muscle in that you have can use it or lose it.

Learning occurs most rapidly, and we are generally happiest, when faced with moderately difficult challenges that we can overcome. It seems to be particularly important to exercise and to have stimulating, real-life experiences, because learning is not a purely intellectual activity but something embedded in our ability to accomplish tasks such as figuring out how to get to new places and deal with new people.

Important reasons for neuroscientists to study learning mechanisms include improving the ability of normal people to learn and diagnosing and treating so-called learning disabilities. Although no pill has been discovered that will enable you to memorize Wikipedia in an evening, a sound scientific basis now exists for using particular learning practices for better outcomes, as the next sections explain.

The term *learning disorder* is a misnomer for most of the cases in which it is applied. Cognitive dysfunctions such as dyslexia and dysgraphia are not believed by most scientists to be based on any specific dysfunction in learning mechanisms. Rather, they are almost certainly the result of a dysfunction in some aspect of the central representation of the relevant sensory input, which that becomes evident only when the skill is normally learned by children. One reason scientists know this is that children who exhibit these particular deficits learn all sorts of other things perfectly well.

Distributing study time over many shorter sessions

In many real-life situations, people need to learn something but can devote only a set number of hours, total, to learning it. As it turns out, how these hours are distributed makes a big difference. Learning works best when the time devoted to learning is distributed over many shorter sessions rather than done in one single, long session. Sorry, Virginia, cramming may be better than nothing to barely pass an exam, but it produces poor long-term retention. The reason? You're not allowing enough time for the hippocampal-cortex reverberation system to consolidate learning over several evenings of sleep.

Getting enough sleep

Most of the hippocampal-cortex reverberation occurs during REM sleep, and experimental rats deprived of REM sleep do not retain maze learning well (refer to the earlier section "Remembering as knowing: Cortical mechanisms" for details on the rat study). The key role of REM sleep in learning is also bad news for those who stay up all night to cram for an exam, because doing so really just loads up working memory without adequate engagement of the consolidation process required to move the information to long-term memory.

Practicing in your mind

During dreams about movement and when imagining movement, frontal lobe motor activity is very similar to that which occurs when a person is actually moving. The implication of this, which has been experimentally verified, is that, just by mentally practicing an activity such as hitting a tennis ball, you can get better at it. Sports psychologists and trainers now routinely teach athletes to imagine the sequences they will execute in their sport and to incorporate this practice into their training regimens, just as they practice on the field.

Rewarding and punishing

One of the results of decades of artificial neural net simulations of learning is that both success and failure can be instructive. Failure is a signal to change synaptic weights, while success means that some subset should be enhanced. In both cases, the optimal change is very moderate so that failure doesn't induce wholesale destruction of what the network has learned already and success doesn't produce an addiction to pursuit of a single outcome.

One of the functions of the dopamine system in the brain is to provide reward/punishment feedback for adapting the brain to activities in which you are engaged. Unfortunately, drugs of various kinds can directly or indirectly hijack this system and lead to addiction, including addiction to the settings and behavior associated with using the drug as well as the drug itself. Understanding the pharmacology of the brain has permitted both the development of ever more powerful drugs and new ways to combat their side effects. Because drug abuse has existed from long before much was understood about the brain, one can certainly hope that knowledge about the brain and the molecules it uses will have a net positive, enabling effect on humanity rather than contributing to further problems with addiction.

Chapter 16

Developing and Modifying Brain Circuits: Plasticity

In This Chapter

- Understanding the development of the neocortex
- Sorting out the connection details for the trillions of individual synapses
- Examining developmental disorders of the nervous system
- Looking at the aging brain

The most powerful computer we know of, consisting of about a 100 billion cells and a quadrillion synapses, builds itself. This is the brain, of course. The brain builds itself without an explicit design for the final structure. The DNA in the human genome specifies a set of basic rules by which cells divide and migrate to form the nervous system's overall structure. The details arise from interactions among the neurons themselves.

Consider a termite nest or ant hill. These structures may contain hundreds of feet of tunnels and dozens of chambers precisely tuned to circulate air and regulate flood water from rain. But no master termite designer or even construction foreman specifies the design. Individual termites run around and respond to characteristics of the environment they're in and to what nearby termites are doing. There are many false starts. But, by trial and error, these responses suffice to build arches, excavate chambers, and link the entire colony with an efficient tunnel system.

So it is with the brain. The genome codes for rules by which several hundred brain areas will come into existence. A few other rules specify the approximate location of these areas, about how large they are, and about what kinds of connections they will have. This constitutes a general program for *development*. Chemical cues from the local environment and interactions between neurons, like those between termites, determine the final *functional architecture*. The trial-and-error jockeying around of the neurons and their connections until things work in some mutually satisfactory way (to the neurons) is called plasticity, the topic of this chapter.

Developing from Conception

A new organism forms when a sperm and an egg unite to form a *zygote*. Before getting together, the sperm and egg have arisen by a cell-division process called *meiosis* in which each daughter cell gets only one strand of the double-stranded DNA chromosomes the adults have. When the sperm and egg unite, the sperm's chromosomes join the egg's chromosomes to form a new double-stranded set of chromosomes, half from the mother and half from the father. This DNA mix is always a unique being, differing from all other humans, except in the case of identical twins.

From this single cell will develop a many-trillion-cell organism. All the thousands of differently specialized cell types will have the same DNA, but which genes are expressed determines which cells differentiate into blood vessels, kidneys, lungs, or neurons of various types.

Arising from the ectoderm: The embryonic nervous system

After fertilization, the zygote begins dividing and forms a sphere of cells called the *blastula*. At around 10 days, the blastula ceases to be spherical but differentiates into a *gastrula,* which consists of three main tissue layers:

- **The endoderm** forms organs associated with the digestive system, such as the liver and pancreas, the digestive system epithelium, and the respiratory system.
- **The mesoderm** produces muscle, bone, connective tissue, and blood and blood vessels, among other structures.
- **The ectoderm** gives rise to the epidermis of the skin and the brain and nervous system, which is my focus in this section.

The formation of the nervous system begins when the ectoderm germ layer begins to form the neural plate. This neural plate eventually folds inward, forming the neural groove. Finally, the groove closes at the top, forming the neural tube. The nervous system develops from this tubular structure, retaining the central cavity on the interior that gives rise to the ventricles of the brain and the central canal of the spinal cord.

The hindbrain, midbrain, and forebrain: The divisions of the ectoderm

Further cell divisions in the ectoderm partition the developing nervous system into the *rhombencephalon,* which forms the hindbrain; the *mesencephalon,* which becomes the midbrain; and the *prosencephalon* which develops into the forebrain.

All your mitochondria come from your mother

Within the cytoplasm of the unfertilized egg are mitochondria, which all mammalian cells possess to make ATP energy molecules. Mitochondria convert sugar into ATP, the universal energy currency for cell activities. Mitochondria have their own DNA and divide independently from divisions of the cell in which they reside, according to the energy needs of the cell. The sperm's only contribution to the egg to form the zygote, on the other hand, is DNA.

Scientists working on human lineages have used this fact to trace pure maternal lines of descent by analyzing mutations of the mitochondrial DNA. Because mitochondrial DNA is passed unmixed from the mother to children of both sexes, its analysis if one of the most powerful tools available for tracing lineages.

A notable structural change during development is the folding of the embryo from its initial straight line form. These folds are called *flexures,* and they include the cephalic, pontine, and cervical flexures.

The images in Figure 16-1 show the embryonic nervous system at several key stages. Early in development (prior to the third week of gestation), the embryo has a linear, tadpole-like form, as shown in Figure16-1A. Between the end of the third week and the fifth week, the cephalic flexure occurs (Figure 16-1B). Within two more weeks, the other main flexures (pontine and cervical, shown in Figure 16-1C) occur.

As you look as the progression from A to B to C, notice that these three brain regions start off about equal sizes (and, in lower vertebrates such as lizards, they tend to remain the same size). However, later in mammalian and especially human development, the forebrain becomes relatively much larger than the midbrain or hindbrain. The forebrain consists of the *diencephalon,* which will include the thalamus, hypothalamus, and pituitary gland, and the *telencephalon,* which will include the neocortex and basal ganglia.

Ontogeny does not recapitulate phylogeny, but it looks that way

The human embryo in the earliest stages of embryonic development (called *ontogeny*), as in Figure 16-1A, closely resembles similar stages in our evolutionary ancestors such as lizards. After a month or two, the human embryo is similar to that of most other mammals.

Ernst Haeckel (a German biologist of the 19th century) coined the phrase "ontogeny recapitulates phylogeny." Translation: An individual's embryonic development (*ontogeny*) follows the same stages as the evolution of the species as a whole (*phylogeny*). This idea clearly isn't literally true because (1) there are stages and forms in the human (or any other) embryo that aren't present in any ancestor species, and (2) there are ancestor species forms that aren't present in the human embryo.

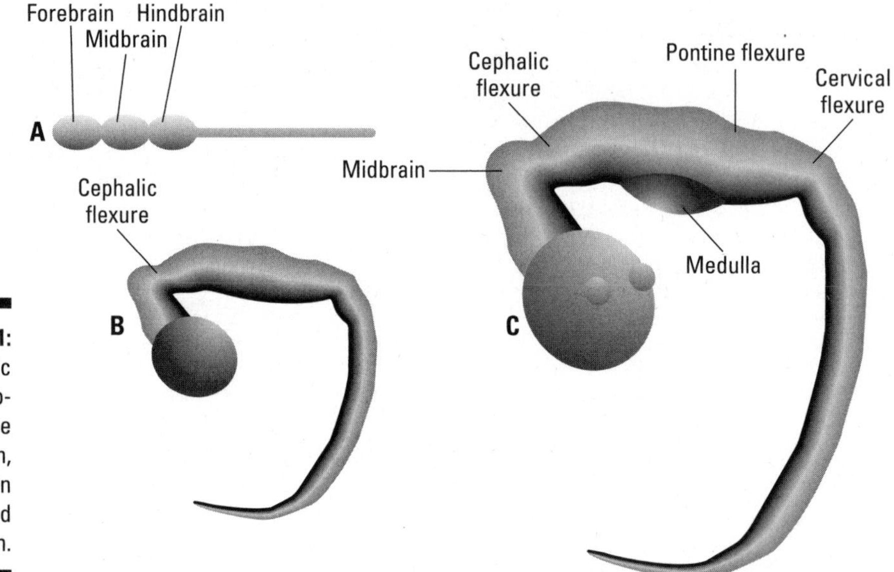

Figure 16-1: Embryonic development of the hindbrain, midbrain and forebrain.

Still, because mammals evolved from cold-blooded vertebrates but have a longer embryonic gestation, much of their early gestational forms are similar to their cold-blooded ancestors because they share the same overall body plan, which was determined early in evolution. For the same reason, primate embryos resemble embryos of other mammals from which they evolved because most of the differentiation between primates and other mammals occurs later in development.

Therefore, although the developing embryo doesn't literally retrace the organism's evolutionary path, you can find some embryonic structures similar to those of evolutionary ancestors in the order in which the ancestors evolved. Here's an example: Whales evolved from land animals. During a whale's embryonic development, limb-bud structures appear and then disappear before birth. Whale front limbs (their flippers) are derived mostly from the forepaws of their ancestors, whereas their rear limb buds appear briefly during embryonic development and then just disappear (except for a few small vestigial bones).

Adding layers: The development of the cerebral cortex

For nearly half a century, neuroscientists have focused a major portion of their time on understanding the development of the neocortex, the largest, most dominant structure in the human brain.

The neocortex has a uniform six-layered cellular circuit organization throughout its extent, suggesting that all areas of the neocortex develop initially, using general layout rules. The reason? The 20,000 coding genes that specify the entire human organism can't possibly comprise a detailed blueprint for the brain's 100 billion neurons and quadrillion interconnecting synapses. For this universal neural circuit structure to process inputs from all the five senses, as well as execute programs for motor control, synaptic connections must be fine tuned during development according to activity patterns and experience imposed on each neocortical region. This developmental fine tuning, which becomes stabilized into permanent neural circuits afterwards, is called *plasticity*.

Neural stem cells and migratory precursor cells

The neocortex develops from a specialized layer of neural stem cells in what is called the *ventricular zone*. This zone is located below what becomes the neocortex white matter. Neural stem cells residing in this zone divide asymmetrically: one daughter of the division remains a neural stem cell, but the other daughter becomes a migratory cell that migrates to a particular neocortical layer and differentiates into a specific cell type, such as a pyramidal cell. This migrating cell is called a *migratory precursor cell*.

All cells in the developing organism have the same DNA. Cells differentiate into different tissues and cell types according to what part of the DNA is actually *expressed*, or read out, by messenger RNA to be converted into proteins. This is controlled, in part, by sections of DNA that do not make protein but instead regulate the expression of other DNA segments. These regulatory DNA sequences are affected by the past and present environment of the cell. Early in development, all cells are *pluripotent stem cells*, meaning that they can differentiate into any cell type. As development proceeds, they become committed to being endoderm, mesoderm, or ectoderm cell types, then to specialized types within those divisions, and so on, until they differentiate into a final cell type and remain so for the life of the organism, never dividing again.

Stem cells

Stem cells show great promise for treatment of many human degenerative conditions because, if inserted into a damaged area, they respond something like they would have during development to repair damage by differentiating into the appropriate cell types and structures. Stem cells are hard to find and identify, however. The first stem cells came from embryos because a high percentage of all embryonic cells are pluripotent stem cells.

In addition to the many ethical issues about transplanting embryonic stem cells, there are also practical issues, such as tissue rejection (because embryonic stem cells come from a donor). One way to solve this and other problems is by using adult stem cells that come from the patient with the degenerative condition. The problem is finding these cells and identifying their developmental pathway and how far along that pathway they have differentiated. Some scientists have made progress recently in converting some cells, such as skin cells, into stem cells by directly modifying DNA expression.

Gluing things together: Glial cells and development

As I mention earlier, the neocortex is a six-layered structure (the gray matter) that sits on top of axons (the white matter). These six layers build from the bottom up, with the help of special glial cells called *radial glia.*

Before neural migration begins, the radial glia extend processes from the ventricular zone (the area just below the white matter) to the surface of the cortex. The neural precursor cells migrate up these processes.

Some neuroscientists believe that each radial glial cell and the precursor neurons that migrate along it form a fundamental unit of cortical organization called the *minicolumn* (see Figure 16-2). A minicolumn consists of about 100 cells dispersed vertically across the six cortical layers.

This group of cells may have arisen from migratory precursor cells that used the same radial glia process to migrate and which are interconnected in a standard way that is reproduced throughout several hundred million mini-columns that constitute the neocortex in humans. In the sensory cortex, for example, all cells in a given minicolumn have the same receptive field location (although they differ in how they respond to stimuli at that location). Some researchers believe that subtle errors in laying out these standard minicolumns properly may underlie some mental disorders such as autism.

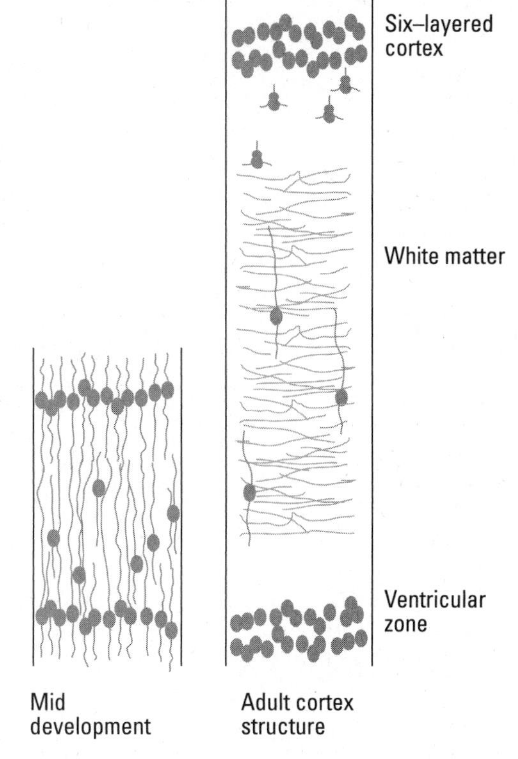

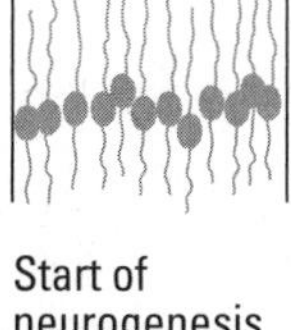

Figure 16-2: Radial glial cells form the scaffolding for development of the neocortex.

Migrating along radial glial cells and differentiating

Here's how the six-layered neocortex is built (Figure 16-3) illustrates this sequence:

1. The first cells leaving the ventricular zone — the neural precursor cells from the asymmetric division described earlier — travel up the radial glial processes to a layer just above where the white matter will end. There, they stop and begin differentiating.
2. Precursor cells born later in the ventricular zone cells migrate up through the already differentiating cells to form another layer on top of the layers already present.
3. This process continues until the last layer is formed. This layer is the most superficial cellular layer of the cortex.

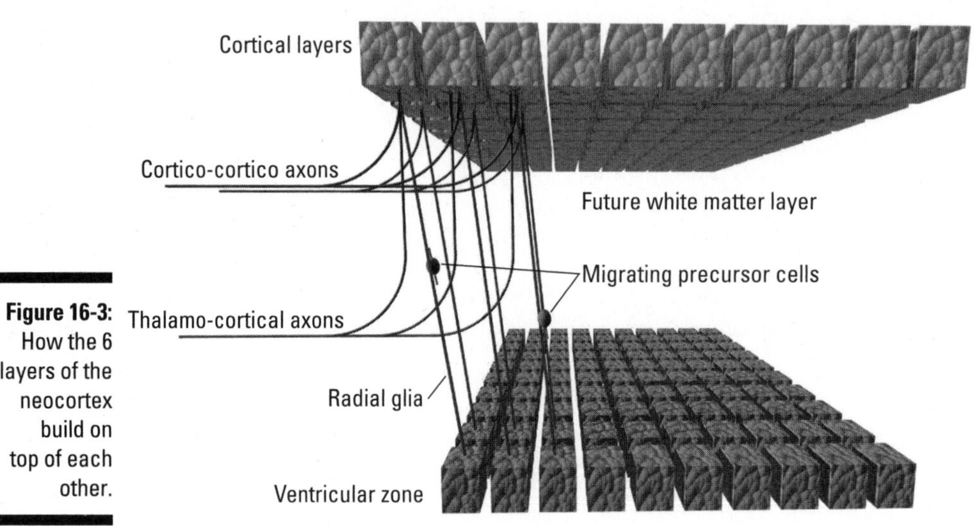

Figure 16-3: How the 6 layers of the neocortex build on top of each other.

Wiring it all together: How axons connect various areas of the brain to each other

After the neocortex's six-layered structure is formed, the axonal wiring takes place. Neurons in all cortical areas send their axons to other cortical areas. Axonal projections also go from the thalamus to cortex and cortex to thalamus. All these (and a few other) axons form the white matter.

A major question in neuroscience concerns how axons know where to go and what to connect to when they reach their target areas.

Chemical affinities and cell surface markers

The frequently mentioned phrase "genetic blueprint" gives the false impression that the genetic code contains a plan of some sort for the structure of the organism. What the genetic code really codes for is a set of cellular responses that comprise rules or procedures that cells follow when responding to their environments, which they do through the manufacture of proteins.

During development, the relevant responses include manufacturing cell surface markers and knowing when (and whether) to move or differentiate upon contact with other cells' surface markers.

During embryonic development, even before the nervous system is formed, cells at the dorsal and ventral poles of the embryo, as well at other key locations, release chemical messengers. Here's what you need to know:

- **The concentrations of the chemical messengers establish gradients across the embryo.** Developing cells sense the multiple chemical gradients and use this information to determine where they should be and what sort of tissue they should become.
- **Cells also express surface markers that act as trophic substances.** These are molecules that are recognized by receptors on migrating cells that attract and guide axonal growth. When the growing axons come in contact with cell surface markers on the paths the axons travel, the surface markers stabilize the axons' growth. The genome encodes the production of a set of chemical gradients that specify particular responses in developing cells, such that cells in nucleus A should send their axons to nucleus B and, when they get there, synapse on dendrites bearing cell surface marker C.

Putting together the basic brain structure

The chemical affinity and cell surface marker systems complete most of the basic brain structure and connectivity in a sequence of stages, which, very broadly, are

1. Target areas develop.
2. These areas release affinity chemicals.
3. Neurons that project to that target send axons toward that affinity target, often using glial cell processes as migration paths.

This complex orchestration by the genome only sets up the basic structure of the brain. Connection details are hashed out by a number of mechanisms. An important one is competition between projecting neurons for target synapses. This process is controlled by activity in the neurons themselves that occurs both during and after embryonic development. The next section has the details.

Learning from Experience: Plasticity and the Development of Cortical Maps

Chemical affinity and cell surface marker systems specified by the genome are enough to get major axon tracts in approximately the right target nucleus. But the connection details at the level of trillions of individual synapses go way beyond what they can orchestrate. The final connections are made after a competitive process in which a superabundance of initial connections is pruned to a final, more functional subset of connections. This process is called *plasticity,* and it works via activity in the developing nervous system itself.

The embryo doesn't develop in a vacuum. The fetus gets skin sensation from its mother's movement and its own. Sound penetrates the womb and activates the auditory system. Where there is no direct stimulation of portions of the developing nervous system (such as for vision), spontaneous firing occurs. This causes muscles to contract and ganglion cells in the retina to fire even before there are photoreceptors. This organized firing helps form correct synaptic connections.

Plasticity works by a kind of handshaking between pre- and post-synaptic neurons. Out of the initial randomness in the superabundant initial neural connectivity, some groups of input neurons will be more effective in driving a post-synaptic neuron than others. These connections will be stabilized and the others removed. Neurons that are so unlucky that they don't end up with enough synapses eventually die.

Plasticity allows the developing organism to compensate for unforeseeable deviations from the general developmental plan due to mutations or injury. Suppose, for example, that either a random mutation or a toxin ingested by the mother destroys the retina. The competition mediated by plasticity during development will remove many visual pathway synapses from areas of the neocortex close to auditory and somatosensory inputs, which will take these pathways over. This may help the particular individual compensate for lack of vision with superior auditory and touch-sense acuity. Plasticity also works within senses to compensate for less severe defects.

The following sections look at how plasticity molds the initial, genetically programmed connectivity to generate organized neocortical processing arrangements such as sensory and motor maps.

Mapping it out: Placing yourself in a visual, auditory, and touching world

One connectivity pattern that occurs throughout sensory areas of the neocortex is the formation of maps. Consider these examples:

- The primary visual cortex contains a *retinotopic map* in which adjacent retinal ganglion cells project (after relaying through the lateral geniculate nucleus of the thalamus) to adjacent cortical cells. Because the optics of the eye guarantee that adjacent retinal locations correspond to adjacent visual field locations, the result is that the visual field is mapped in an orderly way onto the surface of visual cortex. (Chapter 5 has more on the visual system.)

- In the somatosensory system (covered in Chapter 4), the map on the primary somatosensory cortex is an orderly, although distorted, map of the skin.
- In the auditory cortex (Chapter 6), the map from the periphery is frequency coded (or *tonotopic),* corresponding to the primary organization among auditory nerve fibers leaving the cochlea into different frequency bands.

Precise maps are formed in the cortex in two main ways, resulting in a well-ordered topographic relationship between the location across the cortical surface and the location of the sensory receptors in the periphery.

First, through cell adhesion processes, the axons of cells that were near each other in the projecting area tend to stay together when they reach their target cortex area. This automatically produces some topography between periphery and the neocortex. Then this map is further refined by other, activity-dependent mechanisms. These involve an adjustment of connections from input to output cells so that nearby input cells drive nearby output cells. An important aspect of this process is called Hebb's law, discussed in the next section.

Firing and wiring together: Looking at Hebb's Law

Everywhere in the developing nervous system is competition for synaptic space. In the retina, for example, ganglion cell dendritic trees compete for synaptic input from precursor (bipolar and amacrine) cells at the same time that the ganglion cells are sending their axons to the thalamus. Because the thalamus, and the cortex after that, can't know in advance which ganglion cells will win the competition, a mechanism has to exist which allows cortical cells to keep synapses driven by the retinal ganglion cell "winners" and discard those of the losers.

In 1949, Donald O Hebb, a Canadian psychologist, postulated that there had to be a principle by which synapses could be modified according to activity that was local to the synapse, that is, not dependent on any information or process not directly associated with the synapse itself.

Understanding Hebb's law

Hebb stated his idea as follows: "When an axon of cell A is near enough to excite cell B and repeatedly or persistently takes part in firing it, some growth process or metabolic change takes place in one or both cells such that A's efficiency, as one of the cells firing B, is increased." Of course, as the quotation itself indicates, neither Hebb nor anyone else had any idea at the time exactly how this local synaptic modification mechanism would work.

Despite not knowing exactly what mechanisms might mediate this synaptic modification principle, the idea had tremendous appeal because the suggested mechanism was entirely local, and many schemes could be imagined that might do the job. For example, when both pre- and post-synaptic sides of the synapse were frequently activated together, each side could release a complementary metabolic substance that together would strengthen the synapse.

Hebb's law, as it became known, spawned numerous research programs in computation neuroscience and artificial intelligence that assumed two things:

- Plasticity and learning take place by modifying the strength of synapses (synaptic weight modification is discussed in Chapter 15).
- This modification occurs according to correlated activity in the pre-and post-synaptic elements of synapses.

Applying Hebb's law to cortical maps

So what does this have to do with cortical maps? Consider two ganglion cells that are adjacent to each other in the retina. Each of these cells has a receptive field that is the area of the visual world projected onto that part of the retina that causes the cell to fire. Adjacent retinal ganglion cells will have receptive fields that either overlap to some degree or abut against each other. As a result, textures and objects in the world will tend to affect both nearby ganglion cells and produce similar light distributions in their receptive fields.

In other words, the firing of adjacent retinal ganglion cells tends to be correlated. When the axons of these ganglion cells reach the lateral geniculate nucleus (LGN) of the thalamus, they initially branch extensively and try to innervate many targets, at which point Hebb's law takes over. Here's what happens:

- The axons going to LGN targets from ganglion cells that are far away on the retina won't fire together frequently because their activity is not as correlated as ganglion cells that are adjacent.
- Adjacent ganglion cells will not only be correlated with each other, but, but also with target cells they happen to innervate together, because when those ganglion cells fire at the same time, it creates a strong input signal to the target LGN cell.

Thus, axons of adjacent ganglion cells end up with high weights onto the same targets, while distant ganglion cells do not get weight increases when they innervate the same target.

As long as the initial projection from the retina to the LGN is approximately ordered (by chemical gradients, for example), Hebb's law guarantees that the final wiring will form a good map. This pruning and refining mechanism is so important in the visual system that, during development, before there are even photoreceptors in the retina, waves of organized, self-induced, spontaneous firing constantly move across the retina to produce an internally generated correlated firing.

When the thalamus projects to the cortex, a similar process occurs. Chemoaffinity mechanisms get the thalamic axons into the appropriate neocortical region where they compete for synaptic space on postsynaptic dendrites of cortical neurons. Hebb's law stabilizes the synapses from adjacent thalamic cells while removing uncorrelated inputs.

Environmental effects: Nature versus nurture

After the overall structure of the brain is set up under genetic control, as explained in earlier sections of this chapter, its fine details are established by activity within the nervous system itself. During embryonic development, activity-dependent modification of synapses is embedded in a highly competitive process in which neurons are first overproduced and then later the losers die off. After gestation, similar competitive processes occur for learning, but instead of killing off neurons on a large scale, the effects are mostly to change relative synaptic weights in an established system.

One unique aspect of human development compared to other mammals is that the production of neurons in humans continues for months after birth, followed by pruning that lasts up to a year. At that point, a relatively more stable neuron count is reached.

Most people don't think of "experience" as something that occurs during embryonic development. Yet the nervous system is active, even using unique, self-stimulation mechanisms to produce activity in the absence of direct sensory stimulation. Every mother also knows that fetuses kick and generate other motor responses. Tests have shown that at birth, infants are already familiar with their mother's voice from hearing it in the womb, for example. This is possible because neurons in newborns auditory cortices already respond better to the speech sounds of their mother in her language than to other voices in that language or to the mother speaking a different language she did not use while they were in the womb.

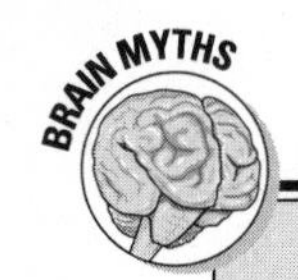

Epigenetics

After molecular biologists James Watson and Francis Crick discovered the structure of DNA, including the obvious way by which a double, complementary helix structure could reproduce, Crick elaborated the so-called central dogma of molecular biology. According to this dogma, DNA makes RNA, and RNA makes proteins. Information goes only one way, from DNA to proteins, never the other. However, this is not the case.

The expression of DNA itself is highly regulated, which allows the same set of chromosomes to produce hundreds of different cells types in an organism. Much of the DNA serves this regulatory function rather than producing, via RNA, proteins that are used for some structure within the cell.

The process of changing what portion of a cell's DNA is expressed involves environmental factors that affect the nucleus. This change in DNA expression is called epigenetics. Epigenetics works by suppressing gene expression from the DNA without altering the sequence of the silenced genes.

Taking the Wrong Path: Nervous System Disorders of Development

The fact that every human genome is a new, unique combination of genes derived from two parents means that each conception is an experiment. The first part of the experiment is whether this new genome can control the development of a viable fetus that will progress to full term and birth.

The common misconception about genetics is that single genes specify single traits independently. But this idea belies the complexity of the control process that generates viable multi-cellular organisms. A considerable portion of the genome consists of genes that regulate the timing and level of expression of other genes. A mutation that disrupts only one of many gradients specifying location or timing in the developing embryo can lead to malformation of vital organs or tissue structures and early death resulting in miscarriage.

Some of the more well-known genetic disorders include sickle cell anemia, cystic fibrosis, Tay-Sachs disease, and hemophilia. All of these are *autosomal recessive*, meaning that the inheritor must receive a defective gene from both parents to have the disorder. There are also *autosomal dominant* genetic disorders, such as Huntington's disease, in which a single dominant gene inherited from either parent produces the disorder.

In addition to the disorders that can be traced to particular genes, there are many genetic disorders for which no single gene has been identified or in which it is not even known that the disorder is genetic. Many disorders have multiple genetic causes with different levels of severity and susceptibility to environmental effects. For example, both autism and schizophrenia show high inheritance, that is, the tendency to occur in families and particularly in children whose parents suffer from the disorder. However, both disorders appear to have multiple genetic causes due to genes on different chromosomes.

Moreover, autism and schizophrenia also do not have 100 percent concordance even in identical twins having the same genome. This means that the genetic constellations that underlie these disorders must create (in some cases very high) susceptibilities to the disorder, but there still must be an environmental factor, such as slight differences in diet or social milieu. Environmental differences between identical twins exist even in the womb because the slightly different positions of twins in the womb exposes them differently to substances from the mother's circulation such as circulating levels of the stress hormone cortisol, as well as other unknown factors.

These variations also suggest that there may be positive feedback mechanisms at work in disorders such as autism and schizophrenia, which some people manage to break or never enter, while others with the same genetic susceptibility plunge down the disease spiral. Although the pendulum has swung strongly towards regarding disorders such as these as organically caused, different people clearly handle the genetic "deck they are dealt" differently with very different outcomes. This is a major mystery for 21st century neuroscience to explore.

Currently, scientists understand something about the function of about 50 percent of the 20,000 human genes. The search for genetic mutations that cause developmental disorders means that one must find one genetic mutation out of about 3 billion base pairs in the human genome, amid variation from person to person of about 0.1 percent, or approximately 1 million base pairs. Human genes average about 3,000 base pairs, although there is considerable variation, with some genes consisting of several million base pairs.

Looking for genetic developmental errors in mutant mice

In the last few decades, using mice has greatly aided the search for the genetic basis of genetic disorders. Not only do mice share many genes in common with humans, they're small and breed and mature quickly. As a result, researchers can rapidly test the effect of DNA mutations (a change in the *genotype)* on large numbers of animals (the result of the mutation on the form or behavior of the final animal, called the *phenotype)*.

"Junk" DNA

Many molecular biologists were shocked when it was originally reported that a considerable percentage of DNA — current estimates put this amount at 98 percent of the human genome — is never translated into any protein. The term *junk DNA* began to be applied to such sequences, implying that this DNA was in some sense vestigial, having once had a function that is now no longer needed. The situation turns out to be more complicated. Much of the genome is regulatory, such as coding for translational and transcriptional regulation of protein-coding sequences, rather than coding for protein directly. So some of the DNA once termed "junk" may be performing this function.

Another common situation is the existence of *introns* and *exons*. The gene for a particular protein may have multiple coding regions (exons) interrupted by regions that are not coded (introns). Both introns and exons are initially coded in the messenger RNA that's used to make the final protein, but the intron segments are removed by RNA splicing processes in the cell before the manufacture of the protein. However, much of the genome really is "junk" in that it consists of repeats of one to three base pairs over and over again.

To conduct these studies, researchers can deliberately induce specific mutations, typically by inserting a new DNA segment in the mouse embryo. In some cases, the intrinsic gene is also silenced. The technique called *knockout* involves the removing or suppressing a particular gene or gene sequence. For example, if scientists suspect a mutation in a particular protein as the cause of a human genetic effect, they can selectively make specific amino acid substitutions at particular positions in the protein in different lines of mice. This procedure can reveal which region of the protein amino acid substitutions cause genetic defects and therefore implicate protein folding errors, changed catalytic activity or receptor binding or other metabolic problems associated with the altered amino acid.

One of the earliest and most well known cases in which a mutation was associated with a developmental anomaly was the reeler mouse (so-called because of its tendency to spin around unevenly while moving). This mutation arose spontaneously and was reported in the early 1950's. The phenotype showed an abnormally small cerebellum and disrupted neocortical organization due to a defect in *reelin*, an extracellular matrix protein used to guide cell migration (see "Migrating along radial glial cells and differentiating," earlier in this chapter).

Knockout mice, in which genes have been turned off to determine their function by their absence, have been generated to study many human disorders such as diabetes, cancer, heart disease, arthritis, aging, obesity, Parkinson's disease, and even substance abuse.

Degeneracy and the genetic code

DNA consists of four possible bases at each location: adenine (A), cytosine (C), thymine (T), and guanine (G). A triplet of three consecutive bases is called a *codon.* The number of uniquely possible codons is 64 (four possible bases at each position times three positions). However, DNA actually generates only 47 distinct amino acids, so some amino acids are coded by more than one codon. For example, GAA and GAG both specify the same amino acid (glutamic acid). This redundancy is called *degeneracy.*

What this means is that mutations can change some bases in the DNA with no effect on the amino acid specified. In addition, many amino acids in some proteins can be switched without significantly affecting how the protein folds or its function. Because of this, researchers looking for the mutation that causes a genetic defect have to sort through considerable variation in DNA and amino acid sequences that aren't the cause.

Sometimes the search for the genetic basis of a disorder is helped if the unknown disease-causing gene happens to be on the same chromosome and close to a gene that specifies an easily observable trait whose genetic basis is known. This means that the chromosomes of people with the observable phenotype can be selectively examined to look for the nearby chromosome location that has the disease mutation.

Environmental effects on development of the human brain

The embryo is particularly susceptible to foreign chemicals because its cells are responding to chemical gradients and cell surface markers to migrate and differentiate to produce tissue form and types. Many things in the environment that have little effect on adults have profound effects on embryonic development. The general term for substances that cause birth defects is *teratogen.* These include

- Environmental toxins such as lead from paint and organic mercury that cause reduced brain size and mental retardation
- Ionizing radiation such as fallout from atomic weapons and x-rays
- Infections such as cytomegalovirus (CMV), herpes virus, rubella, and syphillus
- Prescription drugs such as thalidomide, an effective tranquilizer that unfortunately also resulted in gross limb deformities, and methotrexate, an anticancer drug

- Overexposure to non-prescription substances like alcohol (fetal alcohol syndrome) and caffeine
- Metabolic imbalances and deficiencies such as diabetes and folic acid deficiency

The Aging Brain

The brain develops by increasing neuron numbers and connectivity during embryology and early infancy. During this time, synaptic pruning also occurs, resulting in the reduction of redundant synapses and cells. The process of axon myelination, in which axons are wrapped with myelin sheaths to make action potential conduction faster and more reliable, continues until nearly the end of adolescence. The last parts of the brain to be fully myelinated are the frontal lobes, the areas of the brain most needed for abstract thought and judgment.

Normal adulthood is characterized by relative stability in cell numbers and structure, with learning continuing to modify synaptic strengths and continued replacement of neurons in a few loci such as the olfactory lobe and hippocampus. Neuron numbers start to decline with age, however, due to additional pruning, the accumulation of toxins and metabolic errors, and possibly genetic factors that program cell life.

Aging is a complex process that includes mechanisms that range from the sub-cellular level to the whole organism. For example, the telomeres at the end of chromosomes inside the nucleus of cells may only be capable of a finite number of divisions. Each time a cell divides, the telomeres shorten until they are too short for the cell to divide. The cell then may become inactive or it may die. At the whole organism level, joints wear out and blood vessels harden or get clogged with deposits. Any compromise of any of the interrelated tissues or organs of the body can adversely affect all the others.

Some theories of aging regard it as the accumulation of more or less random degeneration in multiple tissues that ultimately leads to death of the whole organism. Other theories point to specific phenomena such as telomere shortening to suggest that cell life and therefore aging is actually programmed.

Regardless of the mechanism, it is clear that most athletic and raw cognitive ability functions peak in the mid- to late-twenties. Cognitive capability, for a variety of reasons, continues to increase beyond this, however. How mental ability continues to increase after the peak of physical ability is accomplished by a change in strategy.

Living long and well: Lifespan changes in brain strategy

In cognition, the accumulation of knowledge and experience tend to compensate in later years for slightly slower reactions and short-term memory capacity. One distinction made by intelligence researchers is between fluid and crystallized intelligence.

Fluid intelligence concerns physical aspects of brain processing, particularly speed, that are much like similar athletic abilities, and which probably peak at about the same age. These attributes include recall speed, working memory size, and perceptual processing speed — the kinds of attributes that you would like to see in a jet fighter pilot or race car driver.

Crystallized intelligence is associated with knowledge, particularly with strategies for using knowledge. This type of intelligence continues to increase after the physical, fluid intelligence peak. Much of it has to do with recognizing patterns and having good intuition about situations. It is well documented, for example, that successful problem solvers spend relatively more time in the initial stages of understanding a problem and finding similarities to problems experienced before than poor problem solvers. Poor problem solvers tend to rush into the first solution scheme that comes to mind, and, although they start faster, pay later in time by pursuing bad options.

Many studies in the last few decades have found that intelligence and cognitive ability do not have to inevitably decline with age, at least in most people. Moreover, some loss of cognitive function appears to be restorable. Factors that appear to help maintain cognitive function include:

- **Intellectual practice:** Intellectual practice is crucial for even maintaining, let alone developing, cognitive skills during aging. The metaphor that the brain is like a muscle that has to be exercised is perhaps not so far off the mark. Problem-solving abilities thrive best when people are routinely, moderately intellectually challenged.
- **Good health and nutrition:** Is equally clear that good health and diet contribute to the maintenance of intellectual abilities. High blood pressure, diabetes, and substance abuse kill neurons. There is interesting recent evidence that exercise itself, such as running, increases brain rhythms associated with concentration and memory. Exercise may also stimulate the production of new neurons and the removal of waste or damaged neurons in some parts of the brain.

- **Nutritional and other dietary supplements:** Many supplements, like choline (related to the neurotransmitter acetylcholine), are also touted to improve brain function, at least in the short term.

Short-term boosts in cognitive function may derive from many stimulants, such as caffeine. Substances that produce short-term gains tend to have side effects, however, particularly when you ingest them in high doses. As yet, there appear to be no intelligence pills that are safe and effective. Of course, if you have vitamin *deficiencies* of some kind, taking supplements to improve overall function in the body might also improve cognitive abilities by improving health and alertness and reducing discomfort.

Accumulating insults: Aging-specific brain dysfunctions

Aging is a major risk factor for many neurodegenerative diseases, including Alzheimer's disease, Parkinson's disease, and vascular disease that affects the brain, just as it is for arthritis affecting the joints. Many of these disorders have both a genetic underlying propensity and some environmental trigger.

Alzheimer's disease

Alzheimers' disease used to be colloquially known as *senile dementia.* It was so commonly observed in the aged and so rarely otherwise that it was thought to be a more or less inevitable consequence of aging. Now we know that Alzheimer's is a specific disease associated with the build up tau proteins and extracellular debris such as neurofibrillary tangles. The probability of having this disorder increases greatly with aging, but Alzheimer's is not solely a disease of the elderly. Nor is it inevitable as you age. There are some very young Alzheimer's victims and some very old people without Alzheimer's.

The neurons that die in the initial stages of Alzheimer's disease are primarily *cholinergic* (they use acetylcholine as a neurotransmitter), but other neurons die as well. Neuronal death is particularly marked in the hippocampus early in the disease course. Because this structure is vital for transferring memories from short to long term (refer to Chapter 13), the typical progression of Alzheimer's is first loss of episodic memory ("I can't remember where I put the keys"), followed by loss of semantic memory ("I don't know where I am, and I don't recognize anything"), to almost complete loss of cognitive function and finally death from massive neuronal death that eventually compromises bodily maintenance functions.

Parkinson's disease

Parkinson's disease is associated with death of dopaminergic cells in a specific basal ganglia nucleus called the *substantia nigra*. The death of these cells interferes with the patient's ability to make voluntary movements or voluntary corrections during walking such as stepping over an obstacle. Researchers have had difficulty telling whether this disease is due to a genetic deficiency that simply takes a long time to play out or a combination of a genetic susceptibility plus an environmental trigger. (You can read more about Parkinson's as a motor control disease in Chapter 10.)

In a famous set of clinical cases from the 1980s, several heroin addicts injected themselves with a street-made synthetic heroin that turned out to be a substance (MPTP) that was highly toxic to the substantia nigra and produced Parkinson's symptoms in people much younger than those in which it normally occurs. While unfortunate for these patients, the discovery that MPTP could induce the disease (which could be done then in laboratory animals) allowed great progress to be made in evaluating treatments in animal models.

Autoimmune diseases

A number of neuronal degeneration disorders whose incidence increases markedly during aging are believed to be autoimmune disorders. An autoimmune disorder occurs when antibodies are made that attack body tissues rather than foreign invaders. Many autoimmune disorders are thought to occur as a result of viral infections by which some part of the viral protein coat is similar to (or evolved to mimic) a tissue in the body.

Multiple sclerosis is one of the most common autoimmune diseases of the nervous system. In multiple sclerosis, the myelin sheaths that wrap around the axons of the brain and spinal cord are attacked. This leads to reduced conduction and eventually failure to conduct action potentials along these axons, with general inflammation and scarring in nerve tracts.

The first sign of the disease is often muscular weakness. This may proceed to total paralysis and cognitive disability, although the course of the disease is highly variable. Although technically not a disease of aging, multiple sclerosis usually appears first in early adulthood, and it is more common in women.

Strokes

Aging increases the risk of strokes. A stroke is an interruption of the normal blood supply to the brain. (This blood supply is extensive because the brain constitutes about 20 percent of the metabolism of the entire body.) There are two basic types of strokes:

- **Ischemic strokes** in which vessel blockages produce loss of nutrient and waste transport.
- **Hemorrhagic strokes** in which blood vessels leak blood into the brain.

In both cases, a temporary disruption of brain function occurs in the affected region, followed by neuronal death, followed by some recovery due to plasticity and relearning, often aided by specific rehabilitation.

Many strokes go unnoticed because they produce few noticeable symptoms. A common pattern is for a victim to experience a long series of small strokes over several years that slowly compromise brain function, followed by a much larger stroke that causes treatment to be sought. Brain imaging then may detect the existence of the previous silent strokes.

Tumors

Tumors (neoplasms) occur in the brain as elsewhere in the body. The effects of brain tumors depend on their location and extent. Initial symptoms may range from nausea to muscular weakness to vision problems.

One relatively common brain tumor type with a particularly poor prognosis is a *glioma,* a tumor derived from glial cells in the brain. Gliomas are rarely curable, in part because the strong blood-brain barrier isolates the brain from the immune system so that even if almost all the tumor is destroyed by surgery, chemotherapy, or radiation, a few remaining cells can reinitiate the tumor and spread it to multiple brain areas.

It is also possible to have benign brain tumors whose prognosis is considerably better than gliomas. These are usually treated primarily by surgical removal, sometimes followed by postoperative radiotherapy or chemotherapy.

There are also secondary tumors of the brain. These tumors have invaded the brain from cancers that originated in other organs. The mechanism is that cancer cells from the primary tumor enter the lymphatic system and blood vessels and then circulate through the bloodstream to be deposited in the brain. Secondary brain tumors occur often in the terminal phases of patients with incurable metastasized cancer that originated elsewhere in the body. These are actually the most common cause of brain tumors.

Chapter 17

Neural Dysfunctions, Mental Illness, and Drugs That Affect the Brain

In This Chapter

- Examining the organic causes of mental illness
- Categorizing types of mental illness
- Looking at the pharmaceuticals used to treat mental illness

Plenty of things can go wrong with a system as complicated as the brain. One is mental illness, a very inadequate term for a spectrum of disabilities that includes various forms of learning disability (retardation), specific syndromes that may or may not be associated with retardation (Down syndrome verses autism), and poorly understood mental dysfunctions such as depression, bipolar disorder, and obsessive-compulsive disorder. Mental illnesses are among the most challenging and expensive health problems in the U.S. and worldwide.

Understanding and treating neural dysfunctions is one of the most active areas in neuroscience. In this chapter, I explain the current thinking on the causes of mental illness, list some of the more common or debilitating types, and look at drugs designed to treat them.

Looking at the Causes and Types of Mental Illness

Research into mental illness has shown that many types of mental illness are associated with neurotransmitter abnormalities that have a genetic base or were caused by some life experience. These illnesses or disabilities often arise from genetic causes, environmental causes, or some interaction between the two. For example, the genetic defect that results in Down syndrome is well known, and schizophrenia and autism both show high but complex heritability in that they can be caused by multiple genetic defects and are also influenced by the environment.

The simplistic idea that single gene defects "cause" single genetic syndromes is giving way to a more sophisticated idea that particular genetic variations produce susceptibilities to various kinds of retardation, syndromes, and mental illness, but the actual manifestation includes interactions with other genes and the environment, including the environment in the womb. For example, extreme stress on the mother that modulates circulating hormones and substances she ingests may affect some susceptible fetuses much more than others.

The modern emphasis on medical causes of mental illness doesn't mean that life experience, such as chronic or acute stress, can't produce mental illness in an otherwise "normal" brain. What happens, however, is that life experiences that cause mental illness often result in organic effects in the brain, such as chronic neurotransmitter imbalances or high levels of stress hormones. These may be treated most efficiently, currently, by using pharmaceuticals to alleviate the neurotransmitter imbalance, even if some sort of traditional therapy is also used.

From the couch to the pharmacy: A shift in treatment

From their beginnings, psychology and psychiatry have existed as distinct fields from disciplines like neurology. The main reason for this division is the former's fundamental belief in the causality of mental states. All psychoanalytic traditions, regardless of the school of therapy to which they belong, believe, for example, that therapy involves reversing the effects that some sort of experience or mental outlook (such as repressed feelings, subconscious urges, forgotten trauma, or inappropriate reinforcements that set up bad behavior) had on an otherwise normal brain. To overcome these experiences, therapy seeks to uncover memories that cause difficulties or to change behavior with new and more appropriate reward/punishment schemas.

Today, however, understanding that many mental illnesses are associated with neurotransmitter abnormalities, many psychological sub-disciplines, as well as the field of psychiatry in general, tend to view mental illness more as an organic brain problem to be treated with pharmaceuticals. In this view, therapy is used to assess the effectiveness and progress of the pharmacological treatment rather than as the primary treatment tool itself.

Mental illness was once considered to be caused by things solely outside the brain. Historically, it was often attributed to some divine (or demonic) intervention. And during the early and mid-20th century, schizophrenia and autism, for example, were thought to be the result of maladaptive child rearing ("refrigerator mothers," for example, in autism). Treating mental illness, then, focused on reversing the effects these outside elements had.

Genetic malfunctions

Sequencing the human genome (and the other genomes of other animals, particularly mice) has produced enormous advances in understanding how a number of brain dysfunction syndromes depend on particular genetic substitutions, deletions, or additions. Following is a list of several well-known genetic disorders that have serious cognitive or neurological effects:

- **Down syndrome:** Down syndrome is also called *trisomy 21* because it is caused by the presence of an extra 21 chromosome or part of it (humans have 48 chromosomes numbered 1 through 48). Down syndrome occurs in more than 1 in 1,000 births, and it is more common in children born of older parents due to the statistically greater chance of their having had this particular random chromosome damage. The syndrome is associated with not only moderate to severe mental retardation, but complications in other organ systems besides the brain which reduce the life expectancy of those with Down syndrome. Those who live into their 50s have a significantly increased risk of early Alzheimer's disease.
- **Fragile X syndrome:** Fragile X syndrome, so called because it results from a mutation on the X chromosome, is the most common inherited cause of intellectual disability. It is also the most commonly known single gene cause of autism. Fragile X syndrome is characterized by mental retardation and a number of noted physical, emotional, and behavioral features. Physical features include an elongated face, large protruding ears, flat feet, and low muscle tone. Fragile X individuals exhibit social anxiety, including particularly gaze aversion. Fragile X syndrome is neurologically associated with reduced function in the prefrontal regions of the brain.
- **Rett syndrome:** Rett syndrome is a developmental brain disorder characterized by abnormal neuronal morphology and reduced levels of the neurotransmitters norepinephrine and dopamine. Physical features include small hands and feet and a tendency toward microcephaly. Behavior traits include repetitive hand wringing, poor verbal skills, and a tendency to have scoliosis. The syndrome affects females almost exclusively (affected males typically die in utero).

- **Williams syndrome:** Williams syndrome is rare neuro-developmental disorder characterized by mental retardation, except for strong language skills. Individuals with Williams syndrome tend to be highly verbal and overly sociable. They have an "elfin" facial appearance with a low nasal bridge. This syndrome is caused by a deletion of a number of genes on chromosome 7.
- **Autism:** Autism is a spectrum disorder with multiple genetic causes and a range of characteristics in different individuals, ranging from severe retardation to slight social ineptitude. Asperger's syndrome is typically included in the autism spectrum as autism without significant language delay or dysfunction. Autistic individuals may have above-average intelligence and display unusual aptitude in certain technical or artistic areas. Many savants (people with extraordinary calculating, memorization, or artistic skills) are autistic. In its severe form, autism is severely debilitating, characterized by a total inability to engage in social interactions, repetitive behaviors such as continuous rocking, and extremely poor language abilities. Although it is clear that autism has multiple genetic causes due to its high heritability, the mechanisms of its effects remain poorly understood.

Developmental and environmental mental illness

Mental illness can clearly occur in a genetically normal brain which has suffered organic damage during development or later. It can also arise from trauma or stress that leads to indirect changes in the brain from factors such as chronic stress or sleep deprivation. Well-known environmentally generated brain dysfunctions include the following:

- **Fetal alcohol syndrome:** Fetal alcohol syndrome develops when the mother drinks excessive alcohol during pregnancy. Alcohol crosses the placental barrier and can damage neurons and brain structures leading to cognitive and functional disabilities such as attention and memory deficits, impulsive behavior, and stunted overall growth. Fetal alcohol exposure is a significant cause of intellectual disability, estimated to occur in about 1 per 1,000 live births. It is associated with distinctive facial features, including a short nose, thin upper lip, and skin folds at the corner of the eyes.
- **Maternal stress:** If a mother is highly or chronically stressed while pregnant, her child is more likely to have emotional or cognitive problems such as attention deficits, hyperactivity, anxiety, and language delay. The fetal environment can be altered when maternal stress changes the mother's hormone profile. It is thought that this occurs through the hypothalamic-pituitary-adrenal (HPA) axis via the secretion of cortisol, a stress hormone that has deleterious effects on the developing nervous system.

- **Post-traumatic stress syndrome (PTSD):** PTSD is a severe anxiety disorder that develops after psychological trauma such as the threat of death, as in war, or a significant threat to one's physical, sexual, or psychological integrity that overwhelms the ability to cope, as in sexual assault. Traumatic events cause an overactive adrenaline response, which persists after the event, making an individual hyper-responsive to future fearful situations.

  PTSD is characterized by low cortisol and high catecholamine secretion characteristic of the classical fight-or-flight response. These hormones divert resources from homeostatic mechanisms such as digestion and immune responses toward those needed for immediate, intense muscular exertion. Extreme or chronic stress can eventually damage the brain as well as the body. Some evidence shows that desensitization therapies, in which the PTSD sufferer re-experiences aspects of the stressor in a controlled environment, can mitigate some of its effects. Such therapy, if successful, may be superior to generic anti-anxiety medication that may deal only with the symptoms, rather than the cause, of the disorder.

Mental illness with mixed genetic and developmental components

The most common and well-known types of mental illness, such as depression and schizophrenia, have complex relationships between multiple pathways of genetic susceptibility and environmental triggers. Despite their extraordinary efforts to do so, scientists still don't understand the fundamental neural mechanisms that result in these mental disorders. However, significant progress has been made in the last two decades, using a combination of brain imaging, genetics, and *systems neuroscience* (modeling the brain as a dynamic, balanced system) to understand abnormalities of neurotransmitter function in the affected brains. With better understanding of the mechanisms behind these disorders comes better research approaches to determine how best to alleviate them.

Feeling blue: Depression

Depression, which affects nearly 15 percent of the population, is the most serious form of mental illness in terms of total cost. Some sources estimate the direct and indirect costs at over $50 billion yearly in the United States alone.

Depression is poorly understood, but the term clearly designates many different syndromes and diseases, some of which may have a strong genetic basis, while others may not. For example, there are many cases in which people have been without any significant depression their entire lives but have been plunged into depression by a single traumatic experience, such as death of a child or spouse. Depression also appears to result from chronic lower-level stress. Depression also develops in some individuals and families with high hereditability in the absence of any identifiable environmental trigger.

Following are some theories about depression and therapies based on them.

Monoamine hypothesis: Not enough serotonin

The monoamine hypothesis suggests that depression results from a deficit in the neurotransmitter serotonin. Most anti-depressants (particularly recent ones like Prozac) are designed to elevate serotonin levels; many also tend to elevate the levels of norepinephrine and dopamine.

However, depressed people do not show abnormally low serotonin levels, and artificially lowering these levels in people without depression doesn't result in depression. Thus, although elevating serotonin levels with drugs such as SSRIs (selective serotonin reuptake inhibitors, like Prozac) may alleviate depressive symptoms in some people, it's not at all clear — and in fact unlikely — that doing so creates a normal brain state in these people by restoring an intrinsic serotonin deficiency.

The anterior cingulate cortex: Blame it on the ACC

Brain imaging studies have pointed to the anterior cingulate cortex as a place in the brain whose activation might be responsible for some types of depression. The ACC is activated by pain, the anticipation of pain, and negative experiences generally. It also exhibits higher activity levels in depressed people than in non-depressed people. Therapies used to address an over-active ACC include

- **Ablation:** In some clinical trials, parts of the ACC were ablated (removed) in order to try to relieve intolerable pain in patients who were terminally ill. Many of these patients reported that they could still physically sense pain, but the sensation was no longer distressing, similar to the effects of some analgesics (I can personally attest to this experience when I was given Demerol for a separated shoulder).
- **Deep brain stimulation (DBS):** DBS, which partially inactivates the ACC in severely depressed patients, involves permanently inserting a small electrode into a particular brain area and implanting a stimulating electronic box much like a heart pacemaker that passes current pulses through the electrode. Some of these patients experienced immediate relief from pain as soon as the stimulation current was turned on.

The use of DBS for depression was preceded by a much more common use for Parkinson's disease patients. DBS stimulation of the *subthalamic* nucleus has produced immediate symptom relief in thousands of such Parkinson's patients (the subthalmic nucleus is part of the neural circuit in the basal ganglia that also includes the substantia nigra, the brain areas primarily affected by Parkinson's disease).

Shock treatments, negative feelings or emotions, and memory

Many studies have shown that people tend to interpret experience more negatively when they are depressed. A class of theories that might usefully be termed *downward spiral theories* suggests a feedback process in which negative feelings or emotions create negative memories, creating further negative feelings or emotion, eventually resulting in inescapable depression.

If this is the way depression works, the therapeutic strategy is to break this feedback cycle. Two types of neuronal-modulation therapy have been used to this effect: electroconvulsive therapy (ECT) and transcranial magnetic stimulation (TMS).

- **Electroconvulsive therapy (ECT):** ECT involves passing electric currents through the brain via scalp electrodes. The purpose is to induce a transient seizure that temporarily interrupts brain activity and causes retrograde memory loss. This memory loss typically spans the previous several months, some of which is recovered over the next days and weeks. There can also be *anterograde memory loss,* reduced ability to form new long-term memories, for several weeks.

  ECT was used frequently during the latter half of the 20th century for cases of intractable depression that involved significant risk of suicide. It was considered relatively non-invasive. However, its use fell out of favor for several reasons: a general dislike of inducing seizures in patients; the fact that there were frequent reports of lasting cognitive deficits including, but not limited to, memory loss; the fact that its effects often lasted only for a few months; and the availability of alternative pharmacological treatments that showed promise in some patients.

- **Transcranial magnetic stimulation (TMS):** In the last decade, TMS has been used in a somewhat similar manner to ECT. TMS involves creating a high-field magnetic pulse over a particular brain area via an external coil through which a high current is passed for a number of milliseconds. This magnetic field pulse produces currents inside the brain underneath the coil. Although the neural details of exactly how the electric currents affect the brain are not entirely clear, TMS does shut down brain activity in the affected area transiently without inducing a seizure (although it can produce seizures, especially if the stimulation is bilateral — that is, involving areas of both sides of the brain at the same time).

  Some clinicians have claimed success in treating depression with TMS. While the jury is still out because no large, appropriately randomized clinical trials have occurred, the technique appears to be much more benign than ECT.

  TMS was originally a research tool that was used to answer questions about whether processing in a particular brain area was necessary for perception or motor behavior. For example, researchers presented stimuli requiring subjects to make a judgment about movement while creating TMS pulses over motion detecting areas of visual cortex. When the TMS pulse compromised the subject's ability to do something, it generally validated brain imaging experiments that showed activity in the same area at a particular time associated with doing the same task.

Experiencing seizures in the brain: Epilepsy

Epilepsy is characterized by seizures in the brain. Seizures are incidents of hyper-synchronous neural activity during which normal, controlled brain function is severely compromised. Epilepsy has multiple causes, ranging from genetic to developmental abnormalities that may have environmental contributions. Treatments for epilepsy include

- **Pharmaceuticals:** Drugs used to treat epilepsy are called *anticonvulsants;* currently, 20 are approved by the FDA. Most of these are aimed at increasing GABA transmission (GABA is the most important and ubiquitous fast inhibitory neurotransmitter in the brain). About 70 percent of patients experience a reduction of seizures with one or more of these drugs — which leaves about 30 percent of epileptics with no improvement from any approved drug. Most anticonvulsants also have significant side effects.
- **Surgery:** Surgery is the major treatment possibility for patients whose seizures cannot be controlled by drugs or who cannot tolerate the side effects. In epilepsy surgery, the surgical team attempts to locate the focus of the seizure, which is the damaged area of the brain where the seizure usually starts, and remove it. Surgery is very effective where the seizure focus can be located and removed, but there are many cases in which this is not possible.

The search continues for better drugs and better imaging techniques to locate where the seizure originates.

Disordered and psychotic thoughts: Schizophrenia

Schizophrenia is a mental disorder in which thought is disordered and does not reflect reality. Schizophrenia is associated with significant social dysfunction and disability that, in severe cases, requires hospitalization.

Symptoms of schizophrenia typically appear in young adulthood, and they fall into two general categories:

- **Positive symptoms** mean active behaviors or processes such as hearing voices or trying to escape imagined people following the sufferer. These voices sometimes command sufferers to commit inappropriate acts and are associated with paranoia and delusional beliefs.

  Brain scans of schizophrenics during such auditory hallucinations show activity in their auditory cortex, suggesting that some internal source in the brain is generating activity in auditory areas that the schizophrenic cannot distinguish from actual hearing.

- **Negative symptoms** refer mostly to withdrawal and failure to engage in social interaction. Negative symptoms include flat affect and emotion, loss of motivation, social withdrawal, *anhedonia* (the inability to experience pleasure), and lack of attention to hygiene and routine activities of life.

Although positive symptoms can present a larger management challenge when patients act out delusional ideas, negative symptoms tend to contribute more to poor quality of life.

The genetic basis for schizophrenia

Because of the typical age of onset — young adulthood, the time when myelination in the frontal lobe is completed (the last major stage of axonal myelination during development; refer to Chapter 16) — schizophrenia is typically thought of as a frontal lobe disease that, due to its heritability, has a strong genetic and, therefore, organic basis.

One of many data arguing for the genetic basis of schizophrenia is the fact that, if one identical twin has schizophrenia, the other has a 40 percent chance of developing the illness, even if they're raised separately. If schizophrenia were due to random environmental effects, then the chance of both identical twins having it would be far less than 1 percent.

Treatment options

Pharmacological treatments for schizophrenia have enormously improved the lives of some sufferers in the last decades. This is particularly the case for alleviation of positive symptoms, which are generally the most difficult to manage. For unknown reasons, negative symptoms do not respond well to most current medications.

Most schizophrenics are treated with antipsychotic medications such as clozapine, quetiapine, risperidone, and perphenazine. What's not clear is why some of these agents work better in some patients than others for positive symptoms, and why none of them really work well with respect to negative symptoms. The prescription issue is complicated by the fact that several of these drugs have serious side effects. Interestingly, among the drugs that alleviate some symptoms are agents that increase acetylcholine levels in the brain. Nicotine, as from cigarettes, is a nicotinic agonist, and cigarette smoking, perhaps as a form of self-medication, is highly prevalent among schizophrenics.

Obsessing about OCD

Obsessive compulsive disorder (OCD) is an anxiety disorder characterized by intrusive thoughts that lead to repetitive behaviors in order to alleviate the anxiety related to the thought. Typical symptoms include excessive washing, particularly hand washing, repeated checking for something undone or missing, ritualistic adherence to certain procedures, hoarding, preoccupation with sexual or religious thoughts, and irrational aversions, such as extreme fear of germs. In extreme cases of OCD, the sufferer can be paranoid and psychotic. Unlike schizophrenia, however, OCD sufferers are typically aware of their obsessions and distressed by them. It has an incidence of about 2 percent.

OCD has been linked to an abnormality in the serotonin neurotransmitter system and is sometimes successfully treated with SSRIs (selective serotonin reuptake inhibitors). Mutations in genes linked to serotonin have been identified in some groups of OCD sufferers, but a strong overall genetic link has not been identified, and environmental factors may play a role in producing the disorder.

The Promise of Pharmaceuticals

The brain is a computer that uses electrical current flow to perform computations within neurons and chemical neurotransmitters to communicate between neurons (notwithstanding a minority of electrical synapses). Across the synaptic cleft from where the neurotransmitters are released, many different receptor types can exist for each neurotransmitter. The nervous system is finely balanced between excitation and inhibition at multiple levels, from the single neuron to the entire brain.

Nearly a century of work in systems neuroscience (gross anatomy and electrophysiology used to study the brain as a computational system) has provided a very detailed picture of the numerous neurotransmitter systems in the brain and the circuits in which they are used. These include the main fast excitatory neurotransmitters glutamate and acetylcholine, fast inhibitory neurotransmitters GABA and glycine, and numerous neuromodulators such as catacholamines and neuropeptides.

Most mental disorders are associated with some neurotransmitter imbalance, even if the imbalance was not the original cause. Most treatments for mental disorders, outside conventional psychotherapy for relatively mild mental problems, involve drugs that directly or indirectly affect the function of one or more neurotransmitter systems. Most of these drugs and supplements act by mimicking neurotransmitters. In the following sections, I explain how these drugs work.

Typical and atypical antipsychotic medications

The most effective drug treatments for schizophrenia are antipsychotic medications that reduce positive symptoms (few drugs alleviate negative symptoms). Antipsychotics typically suppress dopamine and sometimes serotonin receptor activity.

It was originally theorized that schizophrenia was caused by excessive activation of a particular type of dopamine receptor, the D2. Drugs that block D2 dopamine function reduced psychotic symptoms, while amphetamines, which cause dopamine to be released, worsened them. This led to the use of what are called *typical antipsychotic medications*, which include chlorpromazine, haloperidol, and trifluoperazine.

However, several newer antipsychotic medications, called *atypical antipsychotic medications,* are also effective that do not target the dopamine D2 receptor. Instead, these agents enhance serotonin function with much less blocking effect on dopamine. Atypical drugs include clozapine, quetiapine, risperidone, and perphenazine.

There has been recent interest in whether abnormally low numbers of NMDA glutamate receptors are involved in schizophrenia (postmortems of the brains of those diagnosed with schizophrenia show fewer of these receptors than exist in a normal brain). NMDA receptor-blocking drugs such as phencyclidine and ketamine have also been shown to mimic schizophrenic symptoms (the hallucinogen LSD has effects similar to ketamine).

Drugs affecting GABA receptors

When neuronal activity is excessive (as it is in epilepsy, schizophrenia, and depression), treatment strategies that aim to either reduce excitation or increase inhibition in the brain can be effective.

For epilepsy, the strategy is to increase inhibition of neuronal activity, and this is attempted through drugs called *anticonvulsants,* most of which aim to increase GABA transmission (GABA is the key inhibitory neurotransmitter in the brain). Currently 20 anticonvulsant drugs have FDA approval. Some notable examples include these (trade names are in parentheses): phenytoin (Dilantin), carbamazepine (Tegretol), and clonazepam (Klonopin).

Drugs affecting serotonin

Selective serotonin reuptake inhibitors (SSRIs) are a relatively recent class of drugs used to treat depression and anxiety disorders. They work, as their name suggests, by inhibiting, or slowing down, the reuptake of serotonin by the presynaptic terminal so that the concentration of serotonin stays higher in the synaptic cleft, activating the post synaptic receptors longer. Increasing serotonin that is naturally released appears to produce better results than increasing the concentration systemically.

Some currently used serotonin enhancers include the following (trade names are in parentheses): fluoxetine (Prozac), paroxetine (Paxil), and sertraline (Zoloft).

Drugs affecting dopamine

Dopamine is a catecholamine neurotransmitter that has five known receptors types (D1, D2, D3, D4, and D5). Dopamine is produced in the substantia nigra and the ventral tegmental area.

Attempts to supply dopamine to make up for its loss in the substantia nigra, as in Parkinson's disease, failed because dopamine does not cross the blood brain barrier if administered into the blood stream. However, its precursor in the pathway for its synthesis in cells, L-dopa, does. This treatment can mitigate Parkinson's symptoms for several years but eventually becomes ineffective. Moreover, in the course of treatment, patients may develop motor control syndromes called *dyskinesias.* The ineffectiveness is probably because of the loss of receptors for dopamine or the death of the cells dopamine normally activates.

Dopamine is also the reinforcement neurotransmitter for the reward system of the brain. It is released in the prefrontal cortex in response to food, sex, and neutral stimuli to which pleasurable stimuli have become associated. Drugs such as cocaine, nicotine, and amphetamines lead to an increase of dopamine in the brain's reward pathways and can hijack and overwhelm the natural reward system, leading to addiction.

Many anti-anxiety medications target dopamine receptors, particularly the D2 receptor, as discussed previously.

Some natural psychoactive substances

Naturally occurring psychoactive substances have been used in religious rituals and shamanism for thousands of years. Notable examples include Peyote (mescaline) and psilocybin, which include chemicals that can activate serotonin receptors and produce euphoria and hallucinations in large enough doses. Cannabinoids found in marijuana activate receptors called CB1 and CB2 that are involved in the brain's pain and immune control systems. Naturally occurring cholinergic antagonists from plants such as the deadly nightshade and mandrake also are hallucinogenic.

Part V
The Part of Tens

In this part . . .

Researchers know almost infinitely more about the nervous system now than they did even 15 years ago. They know what most parts of the brain do, to what other parts they are connected, and how all of this depends on the operation of individual neurons. And these discoveries are just the first steps toward an amazing future in which it will be possible to cure many brain disorders and to augment the brain far beyond its normal capability.

This part covers, in a very few pages, the overall organization of the brain and how technology is likely to change it in the next decade. Almost nothing in previous human experience can compare to what's about to happen!

Chapter 18

Ten Crucial Brain Structures

In This Chapter

- Areas that control movement
- Areas that mediate consciousness and thought
- Areas that impact the senses

The brain consists of many distinct functional areas. Some of these are anatomically distinct; others are not. Because different brain areas are responsible for particular functions, destruction of that area tends to lead to a severe loss of that function. The areas I list in this chapter are those that have a known crucial importance for cognitive function, are the subject of considerable current research, and are specifically affected by significant brain diseases such as Alzheimer's and Parkinson's diseases.

A list of ten brain areas is small compared to the hundreds of named central nervous system structures and axon tracts. Nevertheless, you could probably pass yourself off as a neurosurgeon if you could name just these ten. So here they are: ten brain areas that frequently make the news (but, please don't attempt any neurosurgeries until you've thoroughly read Volume II).

The Neocortex

When you look at a human brain, most of what you see is neocortex. This mammalian feature dwarfs the rest of the brain, which it totally covers and nearly encloses. The neocortex has two remarkable properties with respect to the rest of the brain:

- It's huge, containing the largest proportion of the neurons in the entire brain with the possible exception of the cerebellum.
- It's surprisingly uniform in architecture, consisting of the same six-layered structure with the same neuronal types and circuits throughout.

The neocortex is at the top of the processing hierarchy for all the senses, as well as for control of behavior. To read more about the neocortex's role in behavior, head to Chapters 13 and 14.

The Thalamus, Gateway to the Neocortex

The thalamus is the gateway to the neocortex. In each cerebral hemisphere, the thalamus is very near the center of the brain, passing information to and from various areas of neocortex. It is like the hub of a wheel that is the concentrator and distributor of all forces.

The olfactory system is the only sensory system in which there is a projection from the periphery (the olfactory bulb) directly to neocortex. All other senses involve some peripheral processing, followed by a projection to a specific area of the thalamus, which then projects to a primary area of neocortex for that sense. In the olfactory system, the neocortex (orbitofrontal cortex) projects to an olfactory region of the thalamus, which projects back to orbitofrontal cortex.

Because the thalamus receives inputs from all the senses and from the motor control system and the reticular formation (responsible for alertness and attention), it is like an orchestra conductor apportioning activity among the various instruments.

The thalamus is particularly important in making a fast, efficient link between the frontal lobes and the parietal, occipital, and temporal lobes. The information shuttled between these areas relates to attention, awareness, and consciousness.

Because of its central role in the senses, you can find lots of information about the thalamus in the chapters in Part II.

The Pulvinar

In the search for brain areas that appear to control lots of other brain areas, neuroscientists are frequently led to the *pulvinar*, a nucleus in the most posterior region of the thalamus. This nucleus, which has widespread connections with all of the neocortex, is involved in attention, particularly visual attention and eye movements. The pulvinar appears to integrate vision and goal pursuit so as to link appropriate visual stimuli to context-specific motor responses, while managing visual processing to ignore irrelevant visual stimuli. For more on the pulvinar, head to Chapter 13.

The Cerebellum

The cerebellum is one of the oldest parts of the vertebrate brain. It is also one of the largest in terms of the number of neurons, with some estimates placing the number of neurons in the cerebellum at more than the entire rest of the brain, including the neocortex.

The function of the cerebellum is to modulate and coordinate motor behavior. It performs this task by detecting errors between what is "programmed" by the frontal lobes for a particular movement and what is actually executed (such as trying to take a normal step and stepping in a hole). The cerebellum is also responsible for learning and enabling rapid motor sequences. When you first learn to ride a bicycle, for example, you have to think about everything you're doing, but after learning, you don't think about those details because the process has become programmed within your cerebellum.

The cerebellum is also involved in cognition in terms of thinking about yourself or other people or things moving. For example, when you play chess, you have to imagine how the pieces can legally move and which of opponent pieces they'll interact with when they do so. The cerebellum is activated during this thinking.

Based on all this info, you might think that damaging the cerebellum would lead to profound dysfunction or even death. However, the main effect of damage to the cerebellum is to make people clumsy. There are, in fact, several well documented cases of people born without a cerebellum that lead almost completely normal lives.

Read more about the cerebellum and its role in movement in Chapters 8 through 10.

The Hippocampus

The hippocampus is part of the circuit that makes memories. This structure receives inputs from the entire neocortex and projects back out to the same areas.

The storage in the hippocampus is temporary, however. The hippocampus can play back a sequence of events in context and activate the cortical areas that were activated by the event itself. This playback occurs typically during sleep, especially during REM sleep. The result of the playback is that the memories that were stored for a short term in the hippocampus cause long-term storage back in the neocortical areas that were activated during the original episode.

The hippocampus evolved before the neocortex, allowing vertebrates older than mammals, such as lizards and birds, to learn from experience. The continued development of the hippocampus and its enabling of complex, learning-dependent behavior may have spurred the later development of the neocortex.

You can find out more about the hippocampus and its role in memory and intelligence in Chapters 12, 13, and 15.

Wernicke's and Broca's Areas

Wernicke's and Broca's areas are two areas instrumental in language. Wernicke's area, at the border between the superior temporal lobe and the parietal lobe, functions as a high order auditory association area.

The function of Wernicke's area is to process speech. To do this, it has to interact with other areas of the brain that store information about things (semantic networks), particularly in the temporal lobe. Wernicke's area also has to help extract meaning from sentences according to grammar. In doing this, it interacts with other brain areas also, including Broca's area.

The language functions of Wernicke's area are carried out only on the left side of the brain of 95 percent of right-handers and a majority of left-handers. Damage to Wernicke's area results in reduction in the ability to comprehend and produce meaningful speech. The corresponding area on the right side of the brain processes tone of voice indicating irony, humor and other aspects of prosody.

Broca's area is located in the frontal lobe just anterior to the areas of primary motor cortex responsible for controlling the tongue, lips, and other speech articulators. Damage to Broca's area results in difficulty in producing speech. Severe damage to Broca's area can also reduce some aspects of speech comprehension as well.

The identifications of Broca's and Wernicke's areas in the late 1800s were one of the first unequivocal instances of localization in the brain of specific functions. Moreover, both of these localizations showed that the left and right sides of the brain were not functional mirror images but were part of a pattern of hemispheric specialization with the left side of the brain processing most language and the right side most spatial manipulation capabilities.

To read more about these areas, head to Chapter 13.

The Fusiform Face Area

The fusiform face area (FFA) is a region of the medial temporal lobe that underlies our ability to recognize faces. It may also function in our ability to recognize purpose in some kinds of behavior.

Here's how it works: Visual stimuli is processed in a series of stages in the cortex in which the neurons become more selective about the features of a stimulus but less selective about its position.

At a certain point, the processing sequence splits into two streams, called the *what* and *where* (or how) streams. The where stream proceeds into the parietal lobe and involves the use of vision to manipulate objects and to navigate in the world. The what stream cascades through several areas of the infero-temporal lobe, which have neurons that are very selective for specific complex shapes such as ellipses and stars. The complexity of shapes selected for becomes greater and greater as one proceeds along the infero-temporal lobe until the anterior pole is reached.

Close to the end of the ventral visual pathway processing stream is the FFA. Here are the cells that selectively respond to faces, hands, and other biologically significant stimuli.

There are several well-documented case studies of patients with damage to FFA. When damage includes particularly the right FFA, patients lose the ability to recognize faces, even their own, though they may be able to distinguish other similar looking objects like cars or animals.

For more on the fusiform face area, head to Chapters 5 and 13.

The Amygdala

Many functional similarities exist between the amygdala and the hippocampus. However, whereas the hippocampus is extensively connected to the neocortex, the amygdala has extensive connections with low-level sensory systems through the thalamus. The amygdala also has connections directly from the hippocampus.

The amygdala functions as a memory system for emotionally salient events. However, unlike the hippocampus, which works with the neocortex to reproduce a cortical representation of some event of which we can become conscious, the amygdala produces unconscious autonomic responses to remembered stimuli, or stimuli resembling those, that result in feelings of fear, disgust, or apprehension. The function of the amygdala is generally to foster a withdrawal reflex from situations and events resembling those that in the past may have caused or almost caused pain. This system works with the instinctive system all mammals have that evokes fear at the sight of snakes, spiders, blood, and expressions of pain or fear on the faces of others.

There have been well-documented cases of damage to the amygdala resulting in the person lacking concern for the suffering of others, including extreme cases in which those with amygdala damage inflict suffering without remorse or empathy.

Read more about the amygdala in Chapters 7 and 14.

The Lateral Prefrontal Cortex

Working memory is the representation, in our mind, of what is salient about the current situation. The main brain area responsible for working memory is the lateral prefrontal cortex. This area of the neocortex receives inputs from the rest of the neocortex, the thalamus, and the hippocampus, and makes reciprocal connections back to many of these areas. Little is known about functional subdivisions within this large area of the prefrontal cortex, which is nonetheless essential for abstract thought.

The lateral prefrontal cortex is the seat of rationality and the part of the brain that most liberates us from purely instinctive behavior. By enabling the representations of things that are not actually in front of us at the present time, it permits planning and complex goal pursuit. The lateral prefrontal cortex is also essential for *episodic memory,* the memory for a particular event, in the context of the situation around that event, as opposed to memories of facts or general associations.

Damage to the lateral prefrontal cortex impairs the ability to act in a manner appropriate to current circumstances and to change goals or sub-goals in response to real world contingencies. Behavior becomes more stimulus-driven and stereotyped.

I cover the lateral prefrontal cortex in Chapter 14.

The Substantia Nigra (Basal Ganglia)

The basal ganglia are a complex, interconnected set of subcortical nuclei that control behavior at the level below the neocortex. The substantia nigra performs a crucial modulatory role in this system.

One reason that the basal ganglia have become relatively well known is Parkinson's disease, which is caused by a degeneration of dopamine producing neurons in the substantia nigra.

To read more about the basal ganglia and its role in movement, head to Chapter 10.

The Anterior Cingulate Cortex

In his famous treatise on man, Descartes located the soul in the pineal gland because it is the only brain structure that is singular rather than being bilaterally symmetric on the left and right sides of the brain. If I had to make the same guess about the soul's location, after passing the temptation to choose the pulvinar, I would more likely choose the anterior cingulate cortex (ACC).

The anterior cingulate is the anterior part of the cingulate gyrus, located just above the corpus callosum and below the neocortex. It is not neocortex but *mesocortex,* an earlier type of cortex that evolved as the top of the hierarchy of the limbic system.

The anterior cingulate is special because it seems to control neural processing throughout the neocortex, allocating this processing according to task demands. It is also implicated in subjective experience associated with consciousness, particularly consciousness of pain. The anterior cingulate is activated by pain and even the anticipation of pain. Electrical stimulation of the ACC can eliminate the perception of pain without removing the sensation of the stimulus that is painful.

The ACC is also activated when you struggle with a difficult task, particularly when you make errors. This activation is part of the ACC's role in allocating neural processing across the neocortex according to task demands. In this function, it works with and is extensively connected to the lateral prefrontal cortex. In this regard, you can think of the lateral prefrontal cortex as holding the content of thought and the ACC as selecting that content.

Chapter 19

Ten Tricks of Neurons That Make Them Do What They Do

In This Chapter

- Overcoming limitations caused by size and distance
- Fine-tuning processes
- Devising a development strategy unique to vertebrates

Neurons are cells. As cells, they contain components common to all animal cells, such as the nucleus and Golgi apparatus. However, neurons have other features unique to neurons, or at least, not common in other cells. These unique features exist because neurons are specialized for processing and communicating information. This specialization evolved because it allowed organisms to increase their survival chances by moving within their environment based on sensing things like food, toxins, temperature, and predators.

Neurons are so specialized compared to any other cells that the study of neurons and neural organization comprises many schools of neuroscience. If you're interviewing for a job in a neuroscience lab, you will impress the lab head by knowing about the ten neural cell attributes I cover here.

Overcoming Neurons' Size Limit

As information processors, neurons receive information from other neurons, perform computations on that information, and send the output of those computations to other neurons.

Neurons receive information via synapses. Most neurons have synapses on their cell bodies, and the first neurons probably were limited to such. However, synapses have a minimum size so that a normal sized cell could only have a maximum of a few hundred synapses on its cell body.

In order to have more synapses, cell bodies could grow larger, but doing so costs too much (a cubic change in volume translates to a square change in area). Neurons solved this problem by extending roughly cylindrical processes called *dendrites* such that a linear increase in length and volume produced a squared increase in area. Dendrites allow neurons to receive tens of thousands of synapses.

A single neuron is an enormously complicated computing device. Here, very briefly, are some key things to know about synapses and dendrites:

- **Synapses are either electrical or chemical.** Electrical synapses are simple and fast, but inefficient and inflexible. Chemical synapses are more powerful and flexible, but they're slower than electrical synapses. For the details on how electrical and chemical synapses work and the tradeoffs involved, head to Chapter 3.
- **The dendritic tree is a place for synapses that input to the neuron.** But dendritic trees do more than just provide space for synapses. They also allow synapses that are close to each other to interact nonlinearly. Synapses close to the cell body have larger effects on the current spreading to the cell body than synapses that are farther away. The dendrites' branching structure is also important because it provides numerous computational subunits at a level below that of the neuron itself, allowing tens of thousands of time-varying presynaptic inputs to interact in dynamically complex ways to produce a final output.

Getting the Biggest Bang for the Buck with Dendritic Spines

Neuronal dendrites are not merely smooth cylindrical surfaces between the branch points. Rather, they are covered with little mushroom-like appendages called *spines.* Spines are the site of much of the synaptic input to the dendrites, particularly excitatory input.

Here's the interesting thing about spines: They can appear and disappear and dynamically change their shape based on activity in both the presynaptic and postsynaptic neuron. In other words, spine shape changes appear to be a mechanism for dynamic changes in synaptic strength that underlie learning and plasticity in the nervous system.

The shapes of spines impact at least two mechanisms that affect synaptic efficacy:

- The length and diameter of the spine neck affect the amount of current that reaches the main dendritic shaft from the synapse at the spine head.
- The spine volume appears to affect ionic concentration changes associated with synaptic current, such that small volumes may cause currents to saturate at lower levels than higher volumes.

Ligand-Gated Receptors: Enabling Neurons to Communicate Chemically

Ligand-gated receptors are the protein complexes on the postsynaptic side of synapses. They connect the world outside the neuron with the world inside by allowing ions to move through channels in the membrane. The receptor is selective to particular kinds of messages from other neurons, based on the neurotransmitter released by the presynaptic neuron. The ion channel that is opened when the receptor binds a neurotransmitter is the message that is communicated to the rest of the cell as synaptic input current.

Ligand-gated ion channels known as *ionotropic* receptors are different from another type of receptor called *metabotropic* receptors. The metabotropic receptor functions by having a ligand binding site on the exterior of the membrane, like the ionotropic receptor, but it has no channel in its receptor complex. Instead, ligand binding causes the release of an intracellular messenger that was bound to the interior side of the complex, that activates (usually by opening) a channel somewhere else on the membrane (usually nearby).

Ligand-gated ionotropic ion channels usually mediate fast synaptic transmission underlying behavior. Metabotropic receptors are typically slower and mediate modulatory or homeostatic responses, although there are a few cases of fast metabotropic receptor channels. They make it possible for a single neurotransmitter to have different effects in different postsynaptic neurons.

Getting Specialized for the Senses

During the evolution into multicellular life forms, neurons became specialized for sensing aspects of the environment. This specialization was the result of the development of specialized membrane receptors or intracellular organelles in single cells. Neural sensory receptors are cellular transducers that respond to energy, forces, or substances in the external or internal environment and convert the detection into electrical activity, often through modulating the release of a neurotransmitter. Here are some examples:

- Photoreceptors in vertebrate eyes have structures derived from cilia that capture light photons. Capturing these photons causes a G-protein intracellular cascade that closes sodium-permeable ion channels, hyperpolarizing the cell, and reducing the release of glutamate, the photoreceptor neurotransmitter.
- In the ear, auditory hair cells also have cilia that, when bent, open ion channels that depolarize the cell and cause action potentials in the auditory nerve.

Computing with Ion Channel Currents

Neural membranes, like those of most animal cells, are virtually impervious to the flow of water and most ions. Neural membranes differ from those of other cells, however, in that they have many different ion channels that can be activated by ligands or voltage.

When ion channels are open and allow sodium ions to flow through, the neuron is excited. It is inhibited when potassium or chloride channels are open. Neurons have thousands of ion channels of different types in their membranes that open in complex, time-varying combinations.

The neuron computes the interaction of all the ion flows of all the channels in the neuron. For this computation to occur, a certain number of excitatory inputs have to be simultaneously active, while a certain number of inhibitory inputs must not be active. Given 10,000 different inputs, the number of unique combinations a neuron can discriminate is an astronomically large number.

The structure of the neuron's dendritic tree is crucial in this computation. Researchers know that less current injected at distal synapses reaches the cell body than that injected at proximal synapses, and the time course of the distal current is slower. A branching dendritic tree yields many more synaptic sites on the many distal branches than on proximal ones, which partly makes up for the reduction in current magnitude from individual distal synapses. It does not make up for the slower time course of distal versus proximal synapses, however. Because of this, inputs to neurons that tend to dominate the post-synaptic neurons' fast activity tend to occur on proximal synapses, while distal synapses are primarily modulatory.

Keeping the Signal Strong across Long Distances

The most remote dendrites in a neural dendritic tree are rarely more than a few hundred micrometers from the cell body. Synaptic inputs at this distance are severely weakened by the time they reach the cell body by what is called *electrotonic* spread over the membrane. But even though the weakening is significant, there are enough synapses on the most distant dendrites to result into an effective input at the cell body.

It's a different story with the axons, however. Most cells have only one axon leaving the cell body, and this axon may travel a meter through the body and then branch into hundreds or thousands of axon terminals.

The single most important invention of the nervous system is the action potential. The action potential uses transient, voltage-gated sodium channels in the axonal membrane to create a voltage pulse that, by causing a chain reaction across adjacent cells, can send a signal to the most distant synapses without weakening. Each action potential pulse is essentially identical, and the shape of the action potential is more or less the same as at the terminal as it was at its origin.

The Axon: Sending Signals from Head to Toe

The axon is the device by which a neuron sends signals to muscles, glands, or other neurons. Axons are cable-like structures that have two functions: communicating action potentials from the cell body to the axon terminal and, at the axon terminal, releasing a neurotransmitter that binds to a membrane receptor on the postsynaptic cell and opens an excitatory or inhibitory ion channel.

To reach their postsynaptic targets, axons manage to grow long distances (a meter or more) and find those targets. Not only do axons grow long distances, but they are also able to find multiple targets in multiple brain areas and, at each target region, branch into the appropriate number of axon terminals and make contact at the right dendritic locations of the appropriate cells.

Sending a message from your head to your toe involves a two-axon link. A motor neuron in your primary motor cortex sends its axon down the spinal cord to synapse on a motor neuron that controls a muscle in your toe. That motor neuron in the spinal cord sends its axon down your leg, foot, and to your toe. The entire length of your body is traversed by two cells.

Speeding Things Up with Myelination

Axons are marvelous devices for conducting action potentials from a cell body to an axon terminal that may be more than a meter away. How quickly the action potential travels is roughly proportional to the diameter of the axon. Very fine axons may conduct action potentials at a rate of a few meters per second; larger caliber axons can conduct action potentials more quickly.

This scheme works well in invertebrates, which have a small numbers of neurons, relatively speaking. Natural selection has produced some very large-caliber axons in some invertebrates where high speed is needed. The most famous example is the squid giant axon, which can be as much as one millimeter in diameter, big enough to see with the naked eye! The squid giant axon mediates the escape reflex by activating the siphon jet.

This squid giant axon is so large that early electrophysiologists studying the action potential used them for their studies.

But in vertebrates with hundreds of millions or billions of cells in their nervous systems, axons a millimeter in diameter aren't going to work. The solution? Myelination, which allows small-caliber axons to conduct action potentials very rapidly.

In myelination, certain glial cells wrap many times around almost all the axon, but they leave gaps at regular intervals. These intervals are called the *nodes of Ranvier,* which is where all the voltage-dependent sodium channels are concentrated.

Myelinated axons conduct axon potentials in such a way that the action potential jumps from node of Ranvier to node. This process allows axons only a few micrometers in diameter to conduct action potentials at speeds of up to 100 meters per second.

Neural Homeostasis

Complete genetic coding outlining the entire structure of the nervous system is possible for an animal, like an invertebrate, that has a few hundreds or thousands of neurons. In fact, in many invertebrates, every neuron and the majority of synapses are genetically specified and identical from animal to animal. But in vertebrates, which have millions or billions of neurons and only about 20,000 genes, such a development scheme isn't feasible.

Therefore, in vertebrates, the nervous system is developed by a process that specifies general rules for connectivity but lets stimuli from the environment and random events within the developing nervous system compete for synapses. In other words, synaptic and neural fitness are locally determined rather than specified by some outside, master design. This is certainly a valuable construction procedure for a system that has to build itself.

One of the principles that allows this system to work is homeostasis, in which each neuron maintains a certain balance of excitatory and inhibitory inputs, and a minimum number of active output synapses in order to survive; otherwise, it dies. Competition for a limited number of input synapses is a mechanism that makes neurons in a given area respond to different input combinations, including new ones that are acquired during learning.

Some neuroscientists have suggested that neurons actively seek to have variability in their responses over time. This may make for more robust neural circuits that can tolerate variation in their inputs, such as from noisy neural sensors. Neural noise analysis is a growing area in systems neuroscience.

Changing Synaptic Weights to Adapt and Learn

The effort to uncover principles of learning and memory used to be called the *search for the engram,* the memory trace in the brain that constituted a memory. Overwhelming evidence now shows that learning in the nervous system occurs because neurons modify the strength of the synapses between them. This synaptic modification allows small neural circuits to become highly selective for stimuli that have been learned.

Learning is a less extreme version of changes in neural circuits that occur during embryonic and early postnatal development. During development, an exuberance of synapses and neurons form and compete for synaptic connections. Unfit synapses are pruned, and neurons that don't emerge from synaptic pruning with enough connections die off entirely.

This situation is the basis of many critical periods in development in which certain types of experience, such as having useful visual output from both eyes, is necessary for some cortical connections to form. After the critical period, the winning synapses and circuits are stabilized, and later experience can't produce large-scale neural rewiring.

After development, synapses in a few selected areas of the brain are still malleable, but not everywhere, and neuronal death caused by the lack of appropriate synapses either stops or proceeds very slowly. In addition, synaptic alteration occurs in a much more restricted manner. Learning in the hippocampus is unique because it consists first of changes in synaptic strength, followed by growth of new neural connections, and then new neurons.

This doesn't mean that the adult nervous system lacks adaptation mechanisms. The eye adapts to dim versus bright light by using rods rather than cones, and people manage to sleep on rocking boats and in noisy environments. And it is unclear how much most of the adult nervous system is capable of changing, given the right trigger. It could be far more than initially thought.

Chapter 20

Ten Amazing Facts about the Brain

In This Chapter

- Uncovering details about the brain's physical structure
- Clearing up common misunderstandings on how the brain functions
- Some brain esoterica just for fun

There is a lot about the brain we don't know, which isn't too surprising given that what we *do* know about the brain indicates that it is the most complicated structure on earth (or anywhere else). The brain is amazing because of what it is (a switchboard of billions of neurons), and because of what it does (the memory and artistic skills of some savants almost defy belief). In this chapter, I go over 10 attributes of the human brain that, when you think about it, are pretty amazing.

It Has 100 Billion Cells and a Quadrillion Synapses

Your brain is built from an enormous number of cells, on the order of 100 billion. Each of these cells makes about 10,000 connections, called *synapses,* with other neurons, yielding about a quadrillion connections. In fact, your brain has enough neurons and neuronal connections that theoretically it could store each and every experience in your life, including all the visual, auditory, tactile, and other sensations associated with those experiences. The number of distinct brain states expressed this way far exceeds the number of atoms in the known universe.

Most of the cells in your brain are in the neocortex and the cerebellum. The cerebellum uses lots of cells to allow high precision in coordinated movement. The neocortex uses lots of cells for high precision in sensory discrimination and for planning complex behavior.

Some people equate the brain and how it works and stores information with microprocessors. They say that the 100 billion neurons in the brain are like the transistors in a microprocessor. In truth, even the most advanced microprocessors, which have a few billion transistors, can't hold a candle to the power of the human brain. Why? Because neurons are more complicated than transistors. Digital transistors tend to be either "on" or "off" and connected to only a few other transistors. Each neuron in the brain, however, is like a computer itself, capable of an almost infinite variety of different states. Neurons also typically output to around 1,000 other neurons using about 10,000 connections.

In addition, while microprocessors work mostly in a series, processing a single instruction at a time, neurons work in parallel, processing billions of things at the same time.

Consciousness Doesn't Reside in Any Specific Area of the Brain

Many people misunderstand how the brain works because of the popular — but incorrect — notion that certain functions reside solely in particular parts of the brain. One part of the brain is thought to house your ability to taste, another your ability to see, another your ability to move your right hand, and so on. Given this perception, it's easy to see why people tend to think that consciousness resides in a particular area, and because we think of consciousness as the highest brain function, we also tend to put this "consciousness location" at the top of the brain hierarchy. Well, as I explain throughout this book, this just isn't how the brain works.

The brain doesn't have any consciousness neuron, and no particular brain area, by itself, serves as the seat of consciousness. Nor is consciousness just a function of brain size; otherwise, elephants would be conscious and humans not. No place in the brain receives the results of all the neural processing in the rest of the brain, so there's no "top" to the neural hierarchy, and nothing in the brain "looks at" images formed in other parts of the brain.

Instead, several areas of the brain are necessary for consciousness. Areas of the brain that seem to be necessary for and activated by consciousness include the thalamus, the prefrontal cortex, and portions of the parietal and medial temporal lobes. Still, these areas are not unique to the human brain, although the prefrontal cortex is larger (as a percentage of body weight) in humans than any other animal. In addition, damage to the reticular formation in the brain stem produces unconsciousness, but that brain area exists in non-mammalian vertebrates such as lizards and frogs.

It Has No Pain Receptors

Although the experience of pain is dependent on brain areas such as the anterior cingulate cortex, the brain tissue itself actually has no receptors for pain. The brain "experiences" the pain reported by receptors elsewhere in the body, mostly in the skin.

Because brain tissue has no pain receptors, surgery can be done on the brain with the patient fully awake (although tranquilizers and analgesics are typically given to reduce anxiety, and local anesthetics are used for the early phase of the surgery where the scalp skin is cut in order to remove a piece of the skull).

When you have a headache, then, it is not usually because something in your brain hurts, but because a pain message from somewhere in your body reaches the brain. For example, you can get a headache because you actually have mild indigestion or some other body pain of which you are not directly aware.

This doesn't mean, however, that brain dysfunction can't be felt as pain. Migraine headaches appear to be due to transient vascular problems in the brain that lead to abnormally high neural activity, which is felt as pain not because pain receptors are activated, but because some unknown brain circuit interprets that excessive activity as painful. Tumors and strokes may induce excessive brain activity that the brain similarly interprets as painful. In some cases, pain receptors may be activated but referred to the wrong place (as when trigeminal pain receptors are activated).

Another example is the pain associated with looking directly at a very bright light. There are no pain receptors for bright light in the eye, but the brain interprets something about the firing of ganglion cells going from the eye to the brain as indicating that the light level is high enough to be damaging, and it's felt as pain.

Cutting the Largest Fiber Tract in the Brain Produces Few Side Effects

The largest fiber tract in the nervous system is the corpus callosum, which connects the left and right brain hemispheres. This fiber tract contains about 200 million axons (slightly more in women than men).

This tract has been severed surgically many times to stop epileptic seizures from spreading from one cerebral hemisphere to the other, becoming amplified, spreading back, and producing a whole brain *grand mal* seizure.

When this procedure was apprehensively tested on a few patients, the results were remarkable. Several of the patients experienced a significant reduction in seizures, and none of the patients showed any obvious neurological deficit from the surgery. Later, careful experiments by Roger Sperry and Michael Gazzaniga in these split-brain patients showed that there were side effects, such as the fact that their two cerebral hemispheres operated independently and had quite different capabilities. Language and the ability to express abstract thought, for example, is almost totally localized in the left hemisphere in most people, whereas the right hemisphere dominates the processing of visual space. If stimuli were shown in the left visual field, which is processed by the right hemisphere, these patients could say nothing whatsoever about what was shown, but they *could* point with their left hand (also controlled by the right hemisphere) to a picture of the previously shown object.

Einstein's Brain Was Smaller than Average

A weak but statistically significant correlation exists between brain size and intelligence in humans. In a nutshell, there is a general consensus that animals capable of what appears to be intelligent behavior, such as apes and dolphins, have larger brains than less intelligent animals, such as lizards.

Despite the correlation, however, variability in humans is enormous. Average human brain weight is about 1,500 grams (about 10 percent less than the estimate for Neanderthal brains, by the way). Einstein's brain weighed in at about 1,230 grams, which is about 18 percent below average. He did, on the other hand, have somewhat wider than normal parietal lobes, which may have contributed to his mathematical ability.

Clearly, total brain size is only one attribute that underlies intelligence. Brain size may represent a potential capability, like height, that has advantages in some intellectual arenas but not others. The American autodidact Chris Langan, who has a very large head and scores at record levels on some IQ tests, is employed as a bar bouncer.

Adults Lose a Hundred Thousand Neurons a Day with No Noticeable Effect

Neuronal count peaks at about birth in humans; then, over the lifespan, about 10 percent of this is lost, which amounts to losing several hundred thousand neurons a day, more or less. Although some may argue male behavior during midlife crises is proof of the negative impact losing so many neurons a day can have, the fact is that this loss shouldn't compromise mental capacity that much.

The loss of so many neurons that occurs in normal aging without any apparent mental compromise is one reason most neuroscientists don't believe that single neurons store single memories. If they did, death of individual neurons would produce sudden, irretrievable loss of specific memories, which doesn't happen.

Instead, information is spread out over many cells and synapses. When neurons in the brain die, it almost certainly weakens but does not erase any memory they were involved in storing. Other neurons in the network undoubtedly compensate to re-strengthen the memory. It may be that senility occurs when so many neurons are lost that not enough are left to compensate for the ongoing neural death.

Pound for Pound, It Takes a Lot of Energy

The brain has the highest metabolic activity per mass of any part of the body. It constitutes about 5 percent of your body mass but consumes about 20 percent of its energy. This also means that about 20 percent of the total blood flow in the body is coursing through the brain, supported by blood vessels whose length, including capillaries, is hundreds of miles.

Is bigger always smarter?

One thing to consider when thinking about the correlation between brain size and intelligence is the relationship between brain size and body weight. Elephants have larger brains than humans, but humans have a much higher ratio of brain size to body weight. Large bodies sometimes require larger brains for reasons that have nothing to do with intelligence: for example, neurons in large bodies need long axons, which requires larger neural cell bodies and supporting glia. This kind of scaling would not in itself cause an increase in intelligence.

In addition, some animals with very small brains exhibit surprisingly intelligent behavior. Birds are one example. Birds may have to have small brains because they fly, and excess weight is a big problem for flyers. It may be that birds' brains are in some sense more efficient than mammal brains. Octopuses also exhibit some extraordinarily complex behavior despite having very small brains. These invertebrates are so different in many ways from vertebrates, and especially mammals, that it is hard to know exactly how to compare their capabilities with ours.

The high energy use by the brain has been taken advantage of in brain imaging. Brain scan techniques like fMRI (functional magnetic resonance imaging) actually detect either the increased blood flow in brain areas that are highly active or the level of blood oxygenation in active versus less active areas.

The high metabolic demand created by big brains has been used to argue that animals with large brains, such as dolphins, must be highly intelligent, even though their intelligence may be fundamentally different from human intelligence; otherwise, there would be no fitness advantage of supporting such a large brain, and the forces of natural selection would have favored smaller rather than larger brain size.

It's a Myth That We Use Only 10 Percent of Our Brains

You've certainly heard it: Humans use only 10 percent of their brains, but there is no scientific basis for this assertion.

The origins of this myth are obscure. Some researchers have traced the idea to misattributions of the words of Albert Einstein, Dale Carnegie, or William James (a Harvard psychologist). Others use the following to bolster the idea:

- **The fact that some people who, through some developmental injury or anomaly, have brains 10 percent the size of the average adult human but appear normal.** The fact that the brain can developmentally correct for some such injury or anomaly doesn't mean that these people's brains and functionality are completely equivalent to those with non-injured brains. Nor does it mean that all the neurons in non-injured brains aren't actually used.
- **The existence of savants, people with extraordinary calculating or artistic skills that seem beyond normal human capability.** The idea here is that savants use some percentage of the brain for these skills that the rest of us don't. That may be the case, but most savants show profound disabilities in other areas. So how would you equate the novel ability to memorize phone books while simultaneously being unable to conduct many normal life functions such as dressing oneself with the idea that savants use more of their brains than people who don't possess extraordinary ability in any particular area but who engage in full and meaningful lives?

The assertion that we use only 10 percent of our brains is not only untrue, but meaningless. Consider what it would mean if the statement were true: We could remove 90 percent of the brain with no noticeable difference. This is certainly not the case. Every known case of such extensive brain damage shows profound incapacity. Or maybe we could kill 90 percent of the neurons throughout the brain with no effect if we killed only the redundant ones. Although an interesting idea, it's currently technically impossible to test. There is no way, therefore, to assert the truth of the statement from any evidence, or to determine whether the percent we really use is 20 percent, 89 percent, or anything else.

You can't simply count neurons to estimate level of function. For example, by some estimates, the cerebellum has as many neurons as the entire rest of the nervous system, but if the cerebellum is destroyed, the primary result is people who are somewhat clumsy and have slightly slurred speech. The fact that you can function without a cerebellum doesn't mean you don't "use" the neurons in it when you have one.

Brain Injuries Have Resulted in Savant Skills

Savants are people who have extraordinary skills, usually in a limited domain like music, mathematical calculation, or art. Many savants have autistic characteristics. The brains of most savants, to the extent determinable from imaging, are anatomically normal, but there are exceptions. The brain of Kim Peek, the model for the 1988 film *Rain Man,* was severely abnormal with, among other things, no corpus callosum. Some savants, however, are people who started life with what appear to have been normal brain structures and abilities, but who suddenly became savants after an injury. These, such as the two people identified here, are called *acquired savants:*

- **Orlando Serrell** is one such acquired savant. He possessed no special skills until, when he was ten-years-old, he was struck by a baseball on the left side of his head. After an initial period of headaches, he suddenly demonstrated the ability to perform complex calendar calculations and nearly perfect memory for every day since the accident.
- **Derek Amato** acquired musical savant skill after a head injury incurred when diving into a shallow pool. A few days after the head injury, he sat down at a friend's piano and was suddenly able to play and compose music, despite never having played before.

There are several cases in the clinical literature of people who have suddenly become accomplished painters after brain damage associated with left fronto-temporal dementia. The painting skills showed obsessive ability to represent realistic detail, but little symbolism or abstraction. Interestingly, these patients tended have severely deficient semantic memories (general world knowledge), but relatively intact episodic memories (memories for particular events). This is the opposite pattern to Alzheimer's disease.

Although not in the same category as these acquired savants, extraordinary memory recall elicited by brain stimulation during neurosurgery (such as done by Canadian neurosurgeon Wilder Penfield on epileptics) or hypnosis suggest that some untapped memory or skill repository may exist in all of us that remains latent unless released following some unusual stimulus. How we would have such latent skills and why we'd have them but never use them is a deep mystery in neuroscience.

Adult Brains Can Grow New Neurons

The central nervous systems of babies and non-mammalian vertebrates such as fish readily regenerate damaged brain areas and tracts following injury. This isn't the case with adults. Overall, we lose neurons as we age. In addition, after brain injuries such as strokes, even when there is some recovery of function, the area of the brain damaged doesn't heal; it still consists of scar tissue and vacant spaces filled with fluid.

However, the idea that the adult brain is fixed and incapable of neural regeneration is undergoing profound change. In the last quarter of the 20th century, study results began showing that some regions of the adult human brain sometimes grow new neurons. Particularly exciting was that discovery that one of the regions where new neurons can be added is the hippocampus, a medial temporal lobe area crucial for forming long-term memories, and that the growth of new neurons in the hippocampus is associated with this learning.

Another place that new neurons grow in adults is in the olfactory system; specifically, olfactory receptors are constantly being replaced. Researchers hypothesize that these receptors, in detecting odors, are exposed to toxins and that the way to maintain function is to replace them routinely.

Recent research has exploded on the capabilities of stem cells, which divide and differentiate into the different final cell types during an organism's development. Very recently, techniques have been found to convert adult, differentiated cells into stem cells, which offers the possibility of pharmacologically triggering neural regeneration where it might be needed.

Chapter 21

Ten Promising Treatments for the Future

In This Chapter

- Treatments involving genetic manipulations
- Treatments involving stimulating the brain
- Prosthetics that address neural dysfunction

Researchers are at the brink of a revolution in treatment of brain diseases. Genetic modifications may cure brain dysfunctions such as strokes, and the effects of aging and Alzheimer's may be reduced or reversed. In just a few years, stem cells may be transplanted to regenerate and cure brain disease. Pharmacological techniques may allow some cells to be reprogrammed into stems cells to correct for current and even future damage or disease. In addition, human brains may be augmented by increasing their size and computing power. Drugs may be developed that reinstitute learning capacity in the adult brain like that of a newborn.

In this chapter, I look at the current strategies (like deep brain stimulation) that show a lot of promise and cutting-edge technology (like neuroprostheses) that show a lot of potential.

Correcting Developmental Disorders through Gene Therapy

Genetic mutations, copying errors, unfortunate gene combinations, and environmental toxins can produce profound disorders that severely compromise human potential, even before birth. Among the most well-known developmental genetic disorders are Down, Rett, and Fragile X syndromes.

There are no surgical or pharmacological corrections for these conditions. However, there is increasing understanding of the function of much of the human genome, and rapid improvements in the early 21st century have been made that will enable practitioners to silence genes or add new ones.

One route to altering genes or gene expression is with the use of retroviruses. *Retroviruses* are RNA viruses that produce DNA in the host cell after they enter. This DNA is incorporated into the host's genome, after which it replicates with the rest of the host cell's DNA. Retroviruses can be engineered with sequences that knock out host genes or insert new genes into the host.

Any genetic anomaly that was detected early in pregnancy could, in theory, be treated by a specifically engineered retrovirus that functionally substitutes a "normal" gene for the mutated or defective one. Early phase clinical trials have already shown some success with gene therapies for some forms of hereditary retinal degeneration in adolescents.

Augmenting the Brain with Genetic Manipulation

Most medical interventions into the body or brain are attempts to fix something that is broken, rather than to improve upon what is considered "normal." This is changing, however. The genes that are the most interesting in this regard aren't the ones that make proteins for cell structures or metabolism, but the genes that regulate other genes.

For example, a small number of genes likely controls the relative size of the neocortex. What if the neocortex were bigger? There is little doubt that scientists will know, if they don't already, how to grow a larger human neocortex by 2020.

There are, of course, Frankensteinian implications associated with these kinds of augmentations. Even if genetic modification for intelligence augmentation were outlawed altogether in most developed countries, it's hard to imagine that someone, somewhere wouldn't try by. And a gene injected into a fertilized ovum produces a transgenic modification that will be in the germ line — that is, in all future generations.

And that doesn't even begin to address the ethical questions that would arise. If the transgenic cortex enlargement is done illegally, does that deny citizenship or other rights to the children born who have such manipulations? Will such individuals be discriminated against informally or legally? Might they discriminate against us? What if they are autistic savants, brilliant in some specialized areas by hopelessly socially retarded?

Correcting Brain Injury with Stem Cells

Stem cells are cells that exist during development (and sometimes afterward) that are undifferentiated and retain the ability to turn into specialized cells such as neurons, kidney cells, blood vessel wall cells, and so on. Research suggests that injecting stem cells into damaged tissues like the brain or heart causes the cells to differentiate according to the host environment into the appropriate tissues of that environment, sometimes repairing the damage.

One of the major obstacles to using stem cells to treat injury or disease has been obtaining the stem cells. They exist in adults but are rare and difficult to identify. They are plentiful in embryos, but harvesting embryos to get stem cells has obvious ethical difficulties. Moreover, stem cells obtained this way, like any foreign transplant tissue, may be rejected by the host.

Several technologies are coming to the rescue, however:

- The development of non-destructive tests to identify stem cells in the person with the tissue damage; once found, these stem cells be transplanted back into the person.
- Techniques that can reverse the differentiation in any adult cell, such as a skin cell, and pharmacologically turn it into a particular type of stem cell, such as a neural stem cell.

A number of laboratories are attempting to restore vision caused by photoreceptor death by using stem cells to replace the photoreceptors; others are inserting DNA to rescue photoreceptors or turn other cells into photoreceptive cells. There also have been mixed reports of symptom relief from injecting either stem cells or non-stem dopamine-releasing cells into the substantia nigra of Parkinson's patients. Other problems to be solved include ensuring the proper integration of stem cells into the host such that they restore function properly.

Using Deep Brain Stimulation to Treat Neurological Disorders

One of the most exciting recent developments in the treatment of several neurological disorders is the use of deep brain stimulation (DBS) to substantially alleviate the symptoms of the disorder. So far, most targets for DBS have been in the basal ganglia to treat movement disorders. In DBS, one or more electrodes is permanently inserted into the basal ganglia target nucleus, and an implanted electronic device akin to a cardiac pacemaker passes current pulses through the electrode(s) to stimulate neurons in the ganglia. At this point, researchers aren't totally clear whether the main effect of the stimulation is general excitation, general inhibition, or the production of some beneficial pacemaking activity by causing many cells to fire synchronously.

The results of DBS in many patients have been dramatic. These electronic devices can be turned on an off at will. Some Parkinson's patients, for example, can be seen to exhibit the typical stooped posture and shuffling gate after the surgery with the device off, but, as soon as the current is turned on, they are able to walk and engage in sports like basketball.

FDA approval has also been given for DBS to treat pain and major depression. Some experimental DBS treatments have been done for obsessive-compulsive disorder and Tourette's syndrome. The field is so new that the long-term effects and possible side effects are currently unclear, but in many patients, the symptom relief has been far better than any drug treatment with considerably fewer side effects.

Stimulating the Brain through TMS and tDCS

There are two very new types of brain stimulation: transcranial magnetic stimulation (TMS) and transcranial direct current stimulation (tDCS).

- **TMS** involves creating a very short high magnetic field pulse via a coil outside the head over the brain area to be stimulated. TMS appears to produce relatively localized, transient electrical currents that disrupt neural activity. TMS has some resemblance to electroconvulsive therapy (ECT) except that ECT clinicians normally try to induce a seizure, whereas in TMS seizures are usually avoided. Although TMS was initially used to indicate what brain area is involved in execution of a task, it has recently been used to treat mental disorders, like depression, by repetitively stimulating brain areas.
- **tDCS** is very new at the time of this writing. It involves injecting a small direct current between two electrodes on the scalp. Areas of the brain near the anode appear to be generally stimulated and areas near the cathode depressed. Reports have been made for results as wide ranging as improving math scores on standard tests to alleviating schizophrenic symptoms. If these results hold up, tDCS may be a powerful, yet simple tool for minimally invasive modulation of brain activity.

Using Neuroprostheses for Sensory Loss

The loss of a major sense such as vision or hearing is one of the most disabling of all nervous system disorders. Most vision and hearing losses occur from damage to the peripheral receptors or receptor organs — the eye in the case of vision and the inner ear in the case of hearing. While many people are familiar with the use of prosthetics for limb loss, not many are aware of *neuroprostheses*, prosthetics designed to address nervous system disorders.

One of the most successful applications of neuroprosthesis is the artificial cochlear stimulator, used to address hearing loss. Cochlear stimulators consist of a microphone, frequency analyzer, and transmitter outside the head, and a receiver and cochlear stimulator inside. Worldwide, the number of cochlear implants is approaching 200,000. In many cases, these implants allow the wearer to converse normally in person or on the telephone.

The situation for artificial vision is much more difficult. One reason is that the visual system is a more complex system (one million ganglion cell axons versus 10,000 auditory nerve fibers). Another is that the cochlea provides a uniquely appropriate physical interface for relatively simple, 20-electrode arrays in a cochlear stimulator. The visual system doesn't have a similar interface.

Early artificial vision devices have been located in either the retina or visual cortex.

- **Retinal devices:** Devices located in the retina have been used when the vision loss is caused by photoreceptor death but the rest of the retina, including the output retinal ganglion cells, is intact. Current devices have allowed some blind patients to see a few blobs of light, but they haven't yielded high enough acuity for reading.
- **Cortex devices:** Some visual prostheses have been implanted in the cortex. Some patients with these devices have been able to resolve 20 or 30 pixels, which is enough to detect, but not really identify objects.

Addressing Paralysis with Neuroprostheses

Spinal cord injuries and strokes have caused paralysis in millions of Americans. Although considerable research continues to try to discover how to neurally repair a spinal transaction, there continues to be no way to re-grow damaged axons and cells necessary to allow movement for most spinal cord injury victims.

Many neuroscientists believe electronically bypassing spinal cord injuries should be possible. The strategy would be to record the activity of command neurons in the primary motor cortex that are sending signals to move the muscles but which don't reach them because of the injury. Microelectrode arrays would record these signals, and electronic circuits would analyze and transfer these signals past the transaction to either the motor neurons controlling the muscles or to the muscles directly.

An alternative therapy is to record and analyze brain signals to control a prosthetic device such as an artificial arm, or the ability to move a cursor and click mouse buttons for computer control. Because opening the brain and installing recording arrays is highly invasive, this approach hasn't been tried extensively.

Building a Better Brain through Neuroprostheses

Gene modification carries the ability to create something beyond what is normal for a human, as well as for repairing defects. The same is true for neuroengineering, which involves interfacing the nervous system to electronic devices such as computers and the Internet.

Cochlear implants for deafness and deep brain stimulation are only the beginnings of new technology that will link the brain directly to computers. Microelectrode arrays with hundreds and soon thousands of electrodes can both record and stimulate assemblages of neurons at the single cell level.

Experiments have demonstrated that humans (and monkeys) can use electrode arrays implanted in their motor cortices to move computer cursors and artificial arms just by thinking about doing so. Imagine sending requests for information to the Internet by a wireless relay from a recording array in the brain and receiving the answer wirelessly back through a stimulating array. You could soon conceivably communicate with any person on earth just by thinking about doing so. You could also request and receive the results for any data search or calculation.

Given that the basic principles of nervous system stimulation used routinely for cochlear implants and experimentally for visual prostheses can deliver information into the brain, it's not at all beyond imagining a near future (a decade) in which these capabilities are available.

And once we connect the brain to electronics, it's not too far-fetched to imagine modifying the brain, genetically or with stem cells, to facilitate such connections. Sound like a B-movie science fiction plot, with modified people versus "natural" ones? Keep in mind that, unlike much science fiction that is based on made-up science, the techniques for brain-computer interfacing are already here, already being used, and rapidly improving.

Engaging in Computer-Controlled Learning

Learning is hard. We all spend at least 7 hours a day for 16 years just to qualify for a high school diploma — and this is if you're what's classified as *normal*. If you have dyslexia, autism, or some other learning disability, your struggle to achieve competence may be arduous.

Neuroscientists now understand much more about how the brain learns than even 10 years ago, and what they know is that learning in some contexts at some rates is much easier than in others. Computers are starting to be used to implement contexts for learning that accelerate it in children, people with learning disabilities, and older adults.

Highly proficient computer tutors are being embodied in *avatars,* computer simulations of teaching characters with whom the student interacts. Being taught by such avatars will be like having like Plato, DaVinci, and Einstein (to mention a few) all wrapped into one tutor, who also possesses extraordinary personal and communication skills. The infinitely patient and knowledgeable computer tutor would set challenges at the optimum level with respect to the student's current ability, and these challenges would be interesting but not intimidating. Learning could be embedded in games. Mathematics and science could be presented in an intuitive, interactive, visual manner.

Avatars are likely to become helpful companions for children, adults suffering dementia, and possibly even people afflicted with psychological disorders such as depression, schizophrenia, and autism. In research, an avatar system called *FaceSay* (`www.symbionica.com`) has shown some success teaching autistic children to monitor facial expressions. Neuroscience-derived avatars may also functions as models for some types of mental illness.

Treating Disease with Nanobots

Nanotechnology is the assembling of devices with molecular components to achieve nanoscale structures. Current conceptions are for nanoscale devices to be autonomous, with the ability to sense, move, obtain, or store energy, and to accomplish some programmed function. A common name used for the (currently conceptual) devices is *nanobots.*

Many schemes have been proposed by which nanobots would be injected into the body, travel to some locus of disease, and treat the disease. Treatments might include releasing drugs, killing cancer cells, or repairing blood vessels. All of this is very speculative right now, based on the current ability to make things like molecular-sized gears, and a few other components.

Nevertheless, should nanobots become capable of some of the functions mentioned previously, they would also have a serious impact on neuroscience. For example, nanobots could repair damaged axon tracts and reverse paralysis by some combination of mechanically creating a pathway and laying down cell adhesion or affinity molecules. Scavenger nanobots could conceivably remove neurofibrillary tangles and plagues associated with Alzheimer's disease. By interacting with functioning neurons directly, they could also comprise the interface between inserted electronic processing chips and brain activity. These potential nanobot functions are, however, by far the most speculative ideas in this chapter and are all likely much more than one decade away.

Index

• A •

A1 (auditory area 1), 112
ACC. *See* anterior cingulate cortex (ACC)
accessory optic and pretectal nucleus, 94
accommodation (vision), 84
accountability, 172
acetylcholine
  actions, 42, 43, 52
  neurotransmitter activity, 183
  receptors, dysfunctional, 147
achromatopsia, 99
acquired amusia, 203
action or thought first, question of, 170–172
action potential
  auditory processing, 107
  function, 331
  muscle cell, 144–146
  neuron, 50, 55–57
  neurotransmitter release, path to, 57–58
active sense of touch, 68
adaptation
  facilitation compared to, 262–263
  retinal, 87
adenosine triphosphate (ATP), 146
adrenal medulla, 183
afferent, 142
ageusia, 133
aggregate field theory, 12
aging, 81, 300–304
agonist, receptor cell, 145
alliesthesia, 134
all-trans retinal, 86
alpha motor neuron, 40, 41, 152
alpha rhythm, 193
Alzheimer's disease, 302
amacrine cell, 89–91
Amato, Derek (savant), 341
amblyopia, 100
amnesia, temporary, 217, 278
amplitude of sound, 108–109
amusia, 115, 203
amygdala
  function, 32, 33, 323–324
  hypothalamus activation, 184
  memory, role in, 211
  olfactory bulb projection to, 124
anatomical terminology, 25
AND gate, 265–266
anesthesia as artificial unconsciousness, 217
anhedonia, 313
animal nervous system, 11, 15, 40
anion, definition of, 53
anosmia, 133
anterior cingulate cortex (ACC). *See also* deep brain stimulation (DBS)
  ablation, 310
  depression, role in, 310
  function, 33–34, 256–258, 325
  memory, role in, 211–212
  pain, activation by, 80
  problems with, 258
anterograde memory loss, 311
anvil bone, 106–107
anxiety disorder, 309, 313–314
aphasia, 114, 238
Aplysia , 264
appetite regulation, 134
apraxia, 98, 148, 176
area 17, 30, 95–96
area V4, damage to, 99
artificial intelligence, 241
association cortex, 17, 236
astrocyte function, 60
ATP (adenosine triphosphate), 146
attended auditory input, 112
atypical antipsychotic medication, 315
auditory area 1 (A1), 112
auditory association area, 113
auditory canal, 104, 106
auditory hallucination, 312
auditory localization, 115–117
auditory processing. *See also* ear; sound
  description, 30, 36
  hearing aids, 110
  loss, 117–118
  pons involvement, 36
  relay stops, 110–111
auditory receptor, 105
augmentation, neurological, 22, 344
autism
  cause, theories on, 19, 288, 307
  genetics, 297
  learning, computer-controlled, 349
  males, preponderance in, 39
  mirror neurons, 174
  savants, occurrence in, 308, 341–342

autoimmune disease, 303
autonomic nervous system. *See also* homeostasis
components, 180
controlling, 183–185
description and function, 17, 42–43, 180–181
dysfunction, 186
autosomal recessive and dominant genetic disorder, 296
avatar, 349
awareness
consciousness compared to, 213, 214
as intelligence, 207
axon
bifurcation, 71
connection of brain areas, 290–291
description and function, 14–15, 48, 49, 331–332
hillock, 56
increase in, 228
myelination, 57, 332
regeneration, 60, 148
terminals, 50
tract, 14
azimuth, computing, 115–116

• B •

balance
alcohol consumption impairment, 139
control, 94, 153
loss of, 118
barn owl, auditory localization in, 117
baroreceptor, 183
basal ganglia
action initiation, 166–168
anatomy, 34
degeneration, 176–177
function, 34–35, 166–168, 325
behavior. *See also* mood; *specific behavior*
appropriate, brain areas for learning, 213
brain involvement, 17–18
control, 161, 208
hormonal influence, 40
patterns, 32
utilization, 246
Bell, Alexander Graham (inventor), 218
beta wave, 193
biceps-triceps system, 151–152
binocular disparity, 96
biological "clock." *See* circadian rhythm
biological variation in intelligence, 204–205
bipolar cell, 87–89
bitter taste, 129
blindness, 99–100, 218
blindsight, 220
blue-green color blindness, 99
bodily-kinesthetic intelligence, 200
boldfaced text, purpose of, 3
Boring, Edwin G. (psychophysicist), 127
brain. *See also* consciousness; hemispheres of the brain; intelligence; *specific brain structure and region*
aging, impact of, 300–304
augmentation, 22, 344
cells and synapses, number of, 335–336
computations, 14
crucial structures, ten most, 319–325
development, 39–40
energy consumption, 339–340
gender-based differences, 38–40
gross structure, developmental errors in, 19
input, 73–77, 86–87, 129–131
muscles, direct control of, 162
pain receptors, lack of, 337
parallel activity, 223
percentage used, 340
recordings, 44, 61–63
regions, 12
size and intelligence, 38, 338, 339
sound processing, 112
spinal cord and, transition between, 35–37
structural organization, 38
triune, 211
tumors, 304
variations, 37–40
weight, 9, 24
brain cancer, 60
brain damage
abilities, loss of specific, 203
amusia, 115
developmental disorders, 343–344
environmental factors, 20
frontal and parietal lobes, 176
functionality despite, 340
MT area, 97
recovery, 204
right side area 22, 114
stem cell therapy, 345
utilization behavior from, 246
"brain myths" icon, purpose of, 6
brain rhythm, 192–194
brain stimulation
deep brain stimulation (DBS), 21, 35, 345–346
transcranial magnetic (TMS) and transcranial direct current stimulation (tDCS), 278, 311–312, 346

brain studies
  cytoarchitectonics, 43
  imaging, 44–46
  tract tracing, 43–44
brainstem, 129, 130
Broca's area
  function, 114, 322
  language specialization, 238
Brodmann, Korinian (anatomist), 95

calcium
  channels, 57
  concentration, monitoring of, 63
  in neural membranes, 53–54
canonical circuits, 230, 231
Carnegie, Dale (writer), 340
carotid artery, baroreceptors in, 183
case-based intelligence, 255
cataplexy, 195
cataracts, 100
catecholamines, 53
cation, definition of, 53
caudal, definition of, 25
caudate, 34
caudate-putamen complex, 34
cell surface markers, 290–291
cellular specialization as intelligence, 207
central fissure, 28
central nervous system. *See also* brain; spinal cord
  components, 23
  movement controlled by, 17
central pattern generator, 40, 143, 156
central sleep apnea, 196
central sulcus, 27
cephalopods, intelligence in, 207
cerebellum
  anatomy, 34, 159
  description and function, 14, 37
  function, 157–159, 169, 321
  motor learning, 157, 169
  systems, 158
cerebellar cortex, 157–158
cerebral cortex. *See also* neocortex
  areas, 97
  development, 287–289
cGMP (cyclic GMP), 86
chapter organization, 2
chemical messengers, 290–291
chemical synapse, 328
cholinergic neuron, 183
cholinesterases, 145
chorda tympani, 129–131
chorea, 177
cingulate gyrus, 32
circadian rhythm. *See also* sleep
  control of, 31, 94–95
  jet lag, 196
  process, 187–188
circuits, neurological
  developmental errors in, 19
  role of, 13–14
circumvallate papilla, 126–127
CIT (constraint induced therapy), 80
closed-angle glaucoma, 100
closed-loop reflex, 150–153
cochlea
  anatomy, 108
  function, 105
cochlear implant, 21–22, 347
"cocktail party" situation, 219
cognitive function, maintaining, 301–302
coincidence detection, 271, 272–275
color blindness, 99
color perception, 85, 98
comas, paralysis compared to, 216
communication, importance of, 218
communication neuron, 13
comparator neural circuit, 152–153
complex cell, 231
computation neuron, 13
computational neuroscience, 265–266
computers
  consciousness in, 214
  logic gates, 265–266
conductive hearing loss, 117
cone, eye, 85, 99
consciousness, 336. *See also* awareness; thought
  assumptions about, 213–214
  awareness compared to, 213, 214
  communication, importance of, 218
  computers, 214
  description, 207
  epiphenomenon of, 161, 171
  language, relation to, 218–219, 239–240
  loss of, 217–218
  memory, role in, 219–222
  studying, 215–218
  types, 214
  unconscious processing, 219–222
"consciousness neuron," 161, 170–172
constraint induced therapy (CIT), 80
context, 276–277
conventions used in this book, 2–3

corpus callosum
cutting, impact of, 337–338
description, 27
cortex
thalamus, sensory relays from the, 233–235
vision prostheses in, 347
cortical map
auditory, 293
development, 291–295
Hebb's law, application of, 294–295
skin, 76–77
vision, 292
cortisol, 308, 309
Cotterill triangle hypothesis, 216
Crick, Francis (molecular biologist), 296
crosstalk, dorsal-ventral pathway, 98
crystallized intelligence, 301
culture
brain differences, impact on, 39
pain tolerance, influence on, 81
cuneate nucleus, 37
current, electric, 55
cyclic GMP (cGMP), 86
cytoarchitectonics, 43

• D •

D2 (dopamine receptor), 315
DBS. *See* deep brain stimulation (DBS)
deafness
causes, 117–118
cochlear implants, 21–22
examples, 218
decision-making, 17–18, 28. *See also* veto process
deep brain stimulation (DBS)
description and outcomes, 21, 345–346
targets, 35
degeneracy, genetic, 299
delta wave, 193, 194
dementia, senile, 302
demonic intervention, 307
dendrite
description and function, 14–15, 48, 49
spines of, 328–329
structure, 48
synaptic receptors, relation to, 49, 328
depolarization, 55
depolarizing bipolar cells, 89
depression
anterior cingulate cortex, low activity in, 257–258
pain, relation to, 77, 80
theories and therapies, 309–312
depth perception, 98
dermis, definition of, 68
deuteranopia, 99
developmental adaptations to the environment, 260
developmental disorders, 308–314, 343–344
diabetic retinopathy, 100
dietary supplements for cognitive function, 302
diffusion tensor imaging (DTI), 46
distributed coding, 130–131
divide-and-conquer approach to studying consciousness, 215
divine intervention, 307
dLGN (dorsal lateral geniculate nucleus), 91
DNA mutations, 297–299
dopamine, 177, 316
dopamine receptor (D2), 315
dorsal, definition of, 25
dorsal lateral geniculate nucleus (dLGN), 91
dorsal root
description, 41
ganglia, 71
dorsal stream, 97–98
dorsal-ventral pathway crosstalk, 98
Down syndrome, 307, 343–344
downward spiral theories, 311
The Dragons of Eden: Speculations on the Evolution of Human Intelligence (Carl Sagan), 211
dreams, 191–192
drugs. *See* pharmacological therapy
DTI (diffusion tensor imaging), 46
dyskinesia, 316

• E •

ear. *See also* auditory processing
anatomy, 104–105
auditory hair cells, 50
ringing in the, 118
eardrum, 104, 106
Economo, Constantin von (neurologist), 174–175
ECT (electroconvulsive therapy), 311
Edelman, Gerald (Nobel laureate), 214
Edinger-Westphal nucleus, 95
EEG (electroencephalography). *See* electroencephalography (EEG)
efferents, 142
efficiency in motor control, 143
Einstein, Albert (physicist), 338, 340
elderly persons, 300–304
electrical synapses, 328

electricity
brain imaging, role in, 45–46
in the nervous system, 14–15, 53–58
vision processing, 85
electroconvulsive therapy (ECT), 311
electroencephalography (EEG)
brain rhythms, 192–194
description, 44
sleep stages, 189–190
electroshock therapy, 278
electrotonic spread, 331
elevation, detecting, 116–117
11-cis retinal, 86
embryonic development, "experience" in, 295
emotion. *See also* mood; *specific emotion*
importance of, 199
intelligence, relationship with, 208–213
loss of, 212–213, 279
memory of strong, 209
olfaction, invoked by, 123–124
processing, 33
views of, 208
endogenous opioid, 78
endorphins for pain management, 78
end-plate potential, 145
enkephalin, 53
entorhinal cortex
function, 122
olfactory bulb projection to, 124
environment, adaptation to the, 260–261
environmental factors, 20, 299–300, 308–309
environmental input, 50–51
epidermis, 68
epigenetics, 296
epilepsy, 312, 337–338
epinephrine, 53
epiphenomenon of consciousness, 161, 171
episodic memory, 18–19, 235, 276–277
ERP (evoked response potential), 44
error, response to, 257
ethics of genetic manipulation, 22, 344
eukaryotic life, 10, 207
evoked response potential (ERP), 44
evolutionary adaptation to the environment, 260
examination of the brain. *See* brain studies
exercise for cognitive function, 301
exon, 298
"experience" in embryonic development, 295
extensor-flexor muscle pair, 151–152
eye. *See also* vision
brain, transmission to, 86–87
injury, 100
light entering, 84
movement. *See* saccade
photoreceptors, 50
phototransduction, 85–86
retina, function of the, 85

• F •

FaceSay (avatar system), 349
facial expression, 36
facilitation, adaptation compared to, 262–263
fast, excitatory neurotransmitters, 52
fast, inhibitory neurotransmitters, 52–53
fear, storage of, 33
feeling, loss of, 79–80
fetal alcohol syndrome, 308
FFA (fusiform face area), 98, 323
fiber tract, neurological, 337–338
fight or flight response, 42–43, 182–183, 184
filiform papilla, 126–127
fissure of Rolando, 28
flagella, synchronization of, 10–11
flavor detection, 119, 125
flexibility in motor control, 143
flexor-extensor pair, 142
fluid intelligence, 301
fluorescent dyes in optical imaging, 63
flutter, sensation of, 70
fMRI (functional magnetic resonance imaging), 45
foliate papillae, 127
food avoidance, 125
fovea, photoreceptors in the, 87
Fragile X syndrome, 307, 343–344
free nerve ending, 72, 73
"free will" of humans, 161
frequency of sound
information about, transferring, 108–109
measuring, 106
tonotonic maps, 112
frog
eye movements, 35–36
neocortex size, 245
frontal lobe. *See also* primary motor cortex
behavior, impact on, 29
damage, 252–253
description, 27
divisions, 29
function, 28–29, 162–164, 226
size and intelligence, 17–18
frontal operculum cortex, 132
function gene, 261
functional magnetic resonance imaging (fMRI), 45

fungiform papilla, 126–127
fusiform face area (FFA), 98, 323
Fustin, Joaquin (professor), 249

• G •

GABA (gamma amino butyric acid)
  actions, 53
  receptors, medications affecting, 315
Gage, Phineas (railroad worker), 212–213, 279
gait. *See also* locomotion
  alternating, 156–157
  definition, 155
gambling, excessive, 254–255
gamma amino butyric acid (GABA)
  actions, 53
  receptors, medications affecting, 315
gamma oscillation, 251, 252
gamma wave, 193, 194
ganglia. *See also* basal ganglia
  definition, 71
  pulses. *See* action potential, neuron
  retinal, 89–91, 188
  types and classes, 90
gap junction, 141
Gardner, Howard (psychologist), 200
gate theory, 79
gating function, 93
Gazzaniga, Michael (scientist), 240, 338
gender difference
  cognitive processing, 38–40
  pain tolerance, 81
gene, function, 261
gene therapy for developmental disorder, 343–344
genetic disorder, 307–308
genetic modification for intelligence augmentation, 22, 344
genetic mutation, 297–299
genotype, change in, 297
German measles, hearing loss and, 118
glaucoma, 100
glial cell
  development, 288–289
  function, 23
  radial, 288, 289
  types, 59–61
glioma, 304
global recovery from brain damage, 204
globus palladus, 35
glossopharyngeal nerve, 129
glutamate
  actions, 52
  receptors, 270–271
  vision processing, 85, 86, 89
glycine, 53
goal planning, 163–164
Golgi stain, 43–44
Golgi tendon organ, 139
gracile, 37
grand mal seizure, 337–338
gray matter, 14, 226, 227
grid cell, 276
growth-regulating hormone, 185
Guilford, J.P. (psychologist), 200
gut feeling, 254
gyrus and sulcus of the neocortex, 26–27

• H •

habituation, 263, 264
Haeckel, Ernst (biologist), 285–286
hair cell, 105, 107, 330
hallucination, 312, 316
hammer bone, 106–107
head injury as cause of blindness, 100
headache, migraine, 337
hearing. *See* auditory processing
hearing aid, 110
Hebb, Donald O. (psychologist), 293
Hebb's law, 293–295
hemispheres of the brain
  description, 27
  lateralization differences, 39
  specialization, language and vision, 236–239
  variations, 38
hemorrhagic stroke, 304
Heschl's gyrus, 112
hippocampus. *See also* learning; memory
  function, 32, 33, 321–322
  NMDA receptors, 269
  olfactory bulb projection to, 124
homeostasis. *See also* autonomic nervous system
  control, 31
  definition, 16
  importance of, 179
  intelligence, as primitive, 207
  neural, 332–333
homunculus map
  motor, 162–163
  properties, 28
  skin, 30, 76–77, 293

horizontal cell, 87–89
hormones
brain development, impact on, 40
gastrointestinal, 134
neurohormones, 185
neurotransmitters, evolution into, 53
"how to" pathway, 97–98
Hubel, David (scientist), 96, 231
human beings, first appearance of, 10
humor, inability to discern, 114, 238
Huntington's disease, 177
hyperpolarization, 55, 86
hyperpolarizing bipolar cells, 89
hypnosis, 279
hypogeusia, 133
hypothalamus, 31, 184

• I •

icons, explanation of, 5–6
illusion, visual, 101–102
image, right-left sorting of, 91–93
imaging studies of the brain, 44–46
impulse control, veto power over, 172, 213–214
incus, 106–107
inferior colliculus, 35–36
infero-temporal cortex, 98
information delivery, neuroprostheses for, 348–349
initial segment, 56
inner ear, 104, 105, 107–109
insomnia, primary, 195
instinct, 254
insula, 132
intellectual practice, 301
intelligence. *See also* anterior cingulate cortex (ACC); language; lateral prefrontal cortex; memory; orbitofrontal cortex
artificial, 241
augmentation, 22, 344
brain size, impact of, 38, 338, 339
case-based, 255
components of, 204–206
crystallized, 301
emotion, relationship with, 208–213
fluid, 301
human compared to animal, 18
levels, 207–208
nature of, 201–204
neocortex and frontal lobe size, impact of, 17
power for prediction, 200
rule-based, 199
specialization of, 201–203
types, 200
interaural intensity difference, 116
interaural time difference, 116
internal body function, movements that regulate, 138
interneuron, definition of, 230
interpersonal intelligence, 200
intracellular electrode recording, 62
intrapersonal intelligence, 200
intrinsic optical change, 63
intrinsically photoreceptive cell, 95
intron, 298
invertebrate brain, 224
ion, definition of, 53
ion channel, 86, 330
ion pump, 54
ionotropic receptor, 51, 52, 329
ipsilateral, definition of, 111
ischemic stroke, 304
italicized text, purpose of, 2

• J •

James, Williams (psychologist), 184, 340
"junk" DNA, 298

• K •

Kandel, Eric (neuropsychiatrist), 264
Kanizsa triangle, 101, 102
Keller, Helen (activist), 218
kinesthesis receptor, 73, 139
Kurzwell, Ray (futurist), 214

• L •

labeled line coding, 130–131
Langan, Chris (autodidact), 338
language
brain hemisphere specialization for, 237–238
consciousness, relation to, 218–219, 239–240
left brain capacity, 18
processing areas, 114
translation, 241
language skills, 39, 218
lateral geniculate nucleus (LGN), 294
lateral inhibition, 87–89

lateral prefrontal cortex, 325. *See also* working memory
"lazy eye," 100
L-dopa, 177
learning. *See also* memory; plasticity
  appropriate behavior, brain areas for, 213
  classical, 261
  computer-controlled, 349
  development, 259
  disorder, 280
  environmental adaptation, role in, 260–261
  food avoidance, 125
  hippocampus, role of, 271–277, 334
  improvement, 280–281
  influences, 191, 280, 281
  neuron activity during, 49–50
  olfaction and taste, role in, 132–133
  reward/punishment feedback, 281
  synaptic modification, 264–271, 333–334
left hemisphere of the brain, 27
left side interpreter, 240, 241
left-handedness, language processing areas in, 114
lemniscal pathway, 74
LGN (lateral geniculate nucleus), 294
Libet, Benjamin (neuroscientist), 170–172
life functions, control of, 36
ligand-gated receptor, 51, 329
light
  photons, 85
  scattering, 63
limb, loss of feeling in, 79–80
limb position reflex, 151, 153
limbic system. *See also* thalamus
  anatomy, 31, 210
  emotional processing, 209–213
  function, 32–35
linguistic intelligence, 200
Lisman, John (researcher), 251
local recovery from brain damage, 204
local reflex, 74
locomotion. *See also* gait
  description and examples, 142–144
  succession of reflexes, 155–156
locus coeruleus, 184
Loftus, Elizabeth (psychologist), 279
logic gate, computer, 265–266
logical-mathematical intelligence, 200
long-term memory
  as knowing, 275–276
  working memory, transfer from, 272, 274
long-term potentiation, 271, 275
lung fish evolution, 11

• M •

MacLean, Paul (neuroscientist), 211
macular degeneration, 100
magnesium in neural membranes, 53–54
magnetic field
  in brain imaging, 45–46
  earth's, sensing, 16
magnetoencephalography (MEG), 45–46
magnocellular ganglion cell, 90
malleus, 106–107
mammal
  brain development, 224, 244
  neocortex size, 245–246
maternal stress, 308
McCulloch, Warren (neurophysiologist), 267–268
McCulloch-Pitts neuron, 267–268
measles, German, 118
mechanoreceptor of touch
  anatomy, 69, 71
  function and morphology, 70–72
  types, 69–70
medial geniculate nucleus, 112
medial superior temporal (MST) area of the cerebral cortex, 97
mediodorsal thalamus, 123
meditation as metacognition, 207
medulla, 37, 184
MEG (magnetoencephalography), 45–46
meiosis, 284
Meissner's corpuscle, 69, 70
Melzack and Wall gate theory, 79
membrane potential, 54–55
memory. *See also* learning; working memory
  consciousness, impact of, 219–222
  environmental adaptation, role in, 260–261
  episodic compared to semantic, 18–19, 276–277
  hippocampus, role of, 191, 210, 235–236, 271–277
  loss, 278–279, 302, 311
  matrix, 273
  neuron loss, impact of, 339
  olfaction and taste, role in, 132–133
  rehearsal, 275
  repressed and false, 279
  strong emotions, 209
  structures, 33–35
  synaptic modification, 333–334
  traces, 264–265

men
cognitive processing, 38–40
pain tolerance, 81
stress response, 186
Meniere's disease, 118
mental illness. *See* neurological and mental dysfunction
merkel disk receptor, 69
mesocortex, 32
metabolic disorder as cause of blindness, 100
metabotropic receptor, 51, 52, 329
metacognition, 207
microelectrode recording, 44, 61–62
microglial cell, 61
midbrain, 35–36
middle ear
anatomy, 104
function, 106–107
middle temporal (MT) area of the cerebral cortex, 97
migraine headaches, 337
migratory precursor cell, 287
Miller, George (psychologist), 250
minicolumn
definition, 13–14, 288
neuron connectivity, 228–229
mirror neuron, 173–174
mitochondria, 285
mixed left-hander, 114
module of the nervous system, 14, 15
monoamine hypothesis of depression, 310
monofont text, purpose of, 3
mood, 77–78. *See also* behavior; emotion
"moral judgment" processing, 212
morphine, 78
motion parallax, 97
motor control. *See* movement
motor homunculus map, 28, 162–163
motor learning, 157
motor neuron. *See also* alpha motor neuron; movement
definition, 137
function, 13, 16–17, 51
viral diseases, 147–148
movement. *See also* motor neuron
action or thought first, question of, 170–172
complex motor behavior, 144
controlling, 140–144
coordinated, 37, 74
disorder, 176–177
dorsal stream, role of, 97
goal planning, 163–164
inability to judge, 97
locomotion, 142–144, 155–156
mechanical stimulation of, 148
multicellular organisms, 10–11
non-voluntary, 140–141
planned and coordinated, 139–140
plants and animals compared, 15
predicting limb location during, 158–159
process of, 59, 169–170
sensing, 73
sequencing, cortical areas for, 168
types, 17, 138–140
moving target phenomenon, 241
MPTP, discovery of, 303
MRI, diffusion, 46
MST (medial superior temporal) area of the cerebral cortex, 97
MT (middle temporal) area of the cerebral cortex, 97
multicellular organism
cell specialization and communication, 10
movement, coordinated, 10–11
multimodal, definition of, 77
multiple sclerosis, 57, 303
multitasking, 167–168
muscle. *See also* movement
brain, direct control by, 162
contraction of, 146
disorder, 147, 195
extensor-flexor pairs, 151–152
function, 42
innervation, 41
overshoot control, 153
primary motor cortex, message sent from, 162–163
types, 17
muscle fiber, action potential of, 144–146
muscle spindle receptor, 139
music recognition, 115, 203
musical intelligence, 200
myasthenia gravis, 147
myelinated axons, 57
myosin stimulation, 42

naloxone, 78–79
nanobot, treatment with, 350
narcolepsy, 195
nasal retina, 91
naturalistic intelligence, 200

Necker cube, 101
negative symptom of schizophrenia, 312
neglect (vision impairment), 221–222
neocerebellum, 158, 169
neocortex
  anatomy, 24, 27–30, 92, 225, 247
  connectivity, universal compared to small-world, 227–228
  description, 13
  development, 11, 244–247, 284–291
  etymology, 25
  frontal operculum cortex, 132
  function, 17, 25, 319–320
  gray and white matter, 14, 226, 227
  importance, 232
  insula, 132
  minicolumns, 228–229
  size differences, 17, 244–246
  structure, six-layered, 229–231, 244–246, 289, 290
  sulcus and gyrus, 26–27, 226
nerve, definition of, 91
nervous system. *See also* central nervous system; glial cell; neuron; peripheral nervous system
  adaptation compared to facilitation, 262–263
  circuits, segments and modules, 13–14
  compensation, 261
  disorder. *See* neurological and mental dysfunction
  embryonic, 284–286
  evolution, 9–11
  function, 12–19
  key points for understanding, 1
  organization, 12, 15
net negative charge, 54
neural activity recording, 44, 61–63
neural membrane, 53–54, 56
neural prosthesis, 21–22
neural transplant, 21
neurological and mental dysfunction. *See also* *specific dysfunction*
  causes and types, 19–20, 304–314
  developmental, 296–300
  treatment, 20–22, 306, 314–316
neuron. *See also* action potential, neuron; axon; dendrite; *specific neuron type*
  branches, 14
  classes, 88
  computation, 265–266
  definition and description, 13, 47–48
  differentiation from other cells, 48
  electricity, role of, 14–15, 53–58
  electrotonic spread, 331
  excitement, 50
  function, 12–13, 23
  homeostasis, 332–333
  ion channels, computing with, 330
  ligand-gated receptors, 51, 329
  loss, 338–339
  regeneration, 261, 342
  size limitations, overcoming, 327–328
  specialization of, 50–51, 327
  structure, 48
  synaptic modification, 333–334
  voltage-dependent sodium channels, 56
neuronal-modulation therapy, 311–312
neuroprosthesis, 21–22, 347–349
neuroscience, systems, 309
neurotransmitter
  functional classes, 52–53
  hormones, evolution from, 53
  ionotropic and metabotropic receptors, comparison of, 51
  release, path to, 57–58
neurotransmitter-gated receptor, 51
nicotine, self-medication with, 313
nicotinic acetylcholine receptor, 145
night blindness, 100
NMDA receptor
  coincidence detection, 271
  description, 268–271
  hippocampus, role in, 33
  schizophrenia, role in, 315
NMR (nuclear magnetic resonance), 45
node of Ranvier, 57, 332
non-REM (NREM) sleep, 189–190
non-voluntary movement, 140–141
norepinephrine, 43, 53
NREM (non-REM) sleep, 189–190
nuclear magnetic resonance (NMR), 45
nucleus of the solitary tract (NST), 129, 131–132, 183, 184
nutrition for cognitive function, 301–302

objective tinnitus, 118
obsessive compulsive disorder (OCD), 313–314
obstructive sleep apnea, 196
occipital lobe
  description, 28
  function, 30, 94, 226
  thalamic input, 95–98

OCD (obsessive compulsive disorder), 313–314
odorant, 50, 122, 126
off-center ganglion cell, 90
olfaction, 119, 132–133
olfactory bulb projections, 121–124
olfactory glomeruli, 121
olfactory mucosa, 120
olfactory neuron, 50, 134
olfactory receptor, 120
olfactory system, 122
oligodendrocyte, 60
on-center ganglion cell, 90
"ontogeny recapitulates phylogeny," inaccuracy of, 285–286
open-angle glaucoma, 100
open-loop reflex. *See* withdrawal reflex
operant conditioning, 263
opioid, endogenous, 78
optic chiasm, 91
optic flow, 97
optic nerve, 91
optic tract, 93
optical imaging, 46, 62–63
OR gates, 265–266
orbitofrontal cortex
  emotions, control of, 209
  flavor and food avoidance, 125
  function, 33, 122, 184, 253–255
  mediodorsal thalamus and, loop between, 123
  memory, role in, 212–213
Organ of Corti, 107, 108
organization and conventions used in this book, 2–5
otosclerosis, 117–118
outer ear
  anatomy, 104
  function, 105–106
overshoot control in muscles, 153
owl, auditory localization in barn, 117
oxytocin, 40, 185

• P •

Pacinian corpuscle, 69, 70
pain
  brain, lack of receptors in the, 337
  chronic, 80–81
  cognitive distraction to alleviate, 79
  importance of, 79–80
  management, 78–79
  mood, impact on, 77–78, 80
  perception, 72, 77, 80–81
  reflex loop, 143
papilla, 126–127
paralysis
  comas compared to, 216
  neuroprostheses for, 347–348
  spinal cord injury, 148, 176
paraplegia, 176
parasympathetic nervous system, 43, 181, 182
paresis, 148, 176
"parietal eye," 188
parietal lobe
  anatomy, 75
  description, 27
  function, 30, 226
  "touch map" of the body, 69
Parkinson's disease
  aging, effect of, 303
  overshoot control, defective, 153
  restless leg syndrome (RLS), association with, 196
  substantia nigra degeneration, 35, 176–177
  treatments, 21, 310
"Part of Tens" chapters, purpose of, 3
parvocellular ganglion cell, 90
passive sense of touch, 68
patch-clamp electrode, 62
pedicles, photoreceptor, 86
Peek, Kim (savant), 341
Penfield, Wilder (neurosurgeon), 28, 342
perception, subliminal, 221
peripheral nerve, axon regeneration in, 60
peripheral nervous system. *See also* autonomic nervous system
  components, 23, 40
  function, 40
  reflex testing, 42
peripheral neuropathy, 79–80
peripheral vision, loss of, 100
perseveration, 252–253
PET (positron emission tomography), 45
phantom limb syndrome, 77
pharmacological therapy. *See also specific drug*
  advancements, 20–21
  overview, 314–316
  schizophrenia, 313
  seizures, 312
phase lock, 115
phenotype, change in, 297
pheromone, human, 126
phi function, 267
photon of light, 85

photoreceptor
bipolar and horizontal cells, connection to, 88
function, 50, 330
intrinsic, 95
location, 87
types, 85
phototransduction, 85–86
phrenology
discredited, 201
map, 202
overview, 12
phylogeny, 285–286
pictorial depth cue, 98
pinna, 104, 105, 116–117
Pitts, Walter (logician), 267–268
place cells, 276
placebo effect, 78–79
planning behavior
motor neurons, relation to, 51
prefrontal cortex, impact of, 17, 28
plasticity
cortical maps, development of, 291–295
description, 283, 287, 291
synapses, 264
pluripotent stem cell, 287–288
pointer, 251
polio, 147–148
pons, 36
Ponzo illusion, 101
position, sensing, 73
positive symptom of schizophrenia, 312
positron emission tomography (PET), 45
posterior, definition of, 25
posterior cingulate cortex, 256
postsynaptic receptor, 49
post-traumatic stress disorder (PTSD), 309
potassium in neural membrane, 53–54
potentiation, long-term, 271, 275
potentiometric dye, 63
practice, mental, 281
prayer as metacognition, 207
prediction, intelligence as power for, 200
prefrontal cortex
description, 28
function, 17–18, 19
motor control areas, 163
planning behavior, impact on, 17, 28, 243
size and goal-making, 246–247
premotor cortex, 168
pressure on skin, response to, 70
presynaptic cell, 49
primary auditory complex, 112, 113
primary cortex, 233
primary insomnia, 195
primary motor cortex, 28, 162–163
primate neocortex, 245
priming, 221
prokaryote life, 9
proprioception, 139
proprioceptor, 73, 149
prosody in language, processing, 114
prosthesis, neural, 21–22, 347–349
protanopia, 99
pseudounipolar classification, 70
psychiatry, field of, 306
psychoactive substance, naturally occurring, 316
psychological source of pain, 80
psychological stress, response to chronic, 186
PTSD (post-traumatic stress disorder), 309
pulvinar, 231, 320
pupil dilation, control of, 95
putamen, 35
pyramidal cell, 37, 229–231
pyriform cortex, 122, 123–124

quadriplegia, 176

rabies, 148
radial glia, 288, 289
Rain Man (movie), 206, 341
rapid eye movement (REM) sleep, 215–216
receptive field, 70
reciprocal processing, 234
red-green color blindness, 99
reelin, defect in, 298
reflex
closed-loop, 150–153
description, 138–139
hierarchy of, 153
limb position, 151, 153
local, output circuits for, 74
pain reflex loop, 143
spinal, 41–42, 74
stretch, 149
vestibulospinal, 154–155
withdrawal, 141–142, 150
"refrigerator mother," 307

regeneration of axon, 60
relay cell, thalamic, 93–94
REM (rapid eye movement) sleep, 190–192, 215–216
"remember" icon, purpose of, 5
"the remembered present," consciousness as, 214
"research" icon, purpose of, 6
restless leg syndrome (RLS), 195–196
reticular formation, 36
retina. *See also* photoreceptor
  adaptation of the, 87
  definition and function, 84, 85
  Edinger-Westphal nucleus, output to the, 95
  output, 91–95
  vision prostheses in, 347
retinal detachment, 100
retinal ganglion cell, 89–91, 188
retinal slip, 94
retinitis pigmentosa, 100
retinotopic map, 292
retrovirus, 344
Rett syndrome, 307, 343–344
rhodopsin, 86
right hemisphere of the brain, 27, 114, 115
right-handedness, 114
ringing in the ear, 118
risk-taking behavior, 254–255
RLS (restless leg syndrome), 195–196
rod, eye, 85
rostral, definition of, 25
rubella, maternal, 118
Ruffini corpuscle, 69, 70

• S •

saccade, 35–36, 94
Sagan, Carl (astrophysicist), 211
sarcasm, inability to discern, 114, 238
sarcoplasmic reticulum, 146
savant, 206, 340, 341–342
schizophrenia
  cause, theories on, 307, 313
  genetics of, 297
  symptoms, 312–313
  treatment, 313
schwann cell, 60
SCN. *See* suprachiasmatic nucleus (SCN)
"search for the engram," 264, 333
second messenger effect, 269
secondary tumor of the brain, 304
seizure, 52, 312, 337–338
selective ion channel, 48
selective serotonin reuptake inhibitor (SSRI), 310, 314, 315–316
self-medication, 313
semantic memory, 19
semi-circular canal, 155
senile dementia, 302
sensation, loss of, 79–80
sensitization, 264
sensory information processing, 30
sensory neuron, 13, 16, 329–330
sensory processing hierarchy, 232–236
sensory-specific satiety, 134
serotonin level, 310, 314
Serrell, Orlando (savant), 341
sexual behavior, 126
sharp intracellular electrode, 62
Sherrington, Charles Scott, Sir (scientist), 40, 143
shock treatment, 311
short-term memory. *See* working memory
sidebars, explanation of, 3
sign language, 218
silver stain, 43–44
simple cell, 231
single extracellular microelectrode, 61
single photon emission computed tomography (SPECT), 45
"singularity" point, 214
skin. *See also* homunculus map; somatosensory perception
  properties, 68–69
  receptor. *See* somatosensory receptor
sleep. *See also* circadian rhythm
  deprivation, 191
  disorder, 195–196
  learning, impact on, 281
  purpose of, 187
sleep apnea, 196
sleep cycle, 189–192, 193, 215–216
sleep-on cell, 194
sleepwalking, 196
slow neuromodulator, 53
slow-wave sleep, 194
small-world interconnection scheme, 228
smell
  aversion, 124
  response to, 50
  sense of. *See* olfaction

smooth muscle, 17, 141
"sneak up on it" approach to studying consciousness, 215
social consequence, 255
social interaction, mediation of, 208
sodium in neural membrane, 53–54
sodium-potassium ATPase pump, 54
soma, definition of, 48
somatic marker hypothesis, 209
somatosensory cortex, skin map on the, 76–77
somatosensory perception
  passive and active, 68
  "touch map" of the body, 69
somatosensory receptor. *See also* cortical map
  densities, 77
  function, 50, 68–69
  mechanoreceptors, 69–72
  output to the brain, 73–77
  position and movement receptors, 73
  temperature and pain receptors, 72, 73
somatostatin, 53, 185
somnambulism, 196
sound. *See also* auditory processing
  amplitude, 108–109
  auditory cortex, path to the, 111
  azimuth, computing, 115–116
  capturing and decoding, 103, 104–109
  central auditory projections, 110–115
  complex patterns of, 113–115
  elevation, detecting, 116–117
  frequency, 106, 108–109
  localization, 115–117
sour taste, 128
spatial intelligence, 200
spatial skill, gender differences in, 39
spatiotopic organization, 30
SPECT (single photon emission computed tomography), 45
speed in motor control, 143
Sperry, Roger (scientist), 240, 338
spike, neuron, 50, 55–57
spinal cord
  brain and, transition between, 35–37
  circuits, 73–77
  comparator neural circuit, 152–153
  description, 14
  function, 40–41
  injuries, 148, 176, 347–348
  message pathways within, 42
  neuroprostheses, 347–348
  segments, 41
spinal reflex, 41–42, 74
spindle cell, 174–175
spinocerebellum, 158, 169
spinothalmic pathway, 74
split-brain experiment, 337–338
SQUID (super-conducting quantum interference device), 46
SSRI. *See* selective serotonin reuptake inhibitor (SSRI)
stem cell, 287–288, 342, 345
stereoisomer, definition of, 86
stirrup bone, 106–107
stress, 43, 186, 308
stretch receptor, 139
stretch reflex, 149
stretching of skin, response to, 70
striate cortex, 95
striated muscle, 17, 145
striatum, 34, 177
strokes, 204, 303–304
strong left-hander, 114
Stroop task, 212
structure from motion experiment, 98
subjective tinnitus, 118
subliminal perception, 221
substance P, 53
substantia nigra
  degeneration, 35, 176–177
  function, 35, 303, 325
  location, 36
subthalamic nucleus, 35, 310
sulci and gyri of the neocortex, 26–27, 226
Sullivan, Ann (teacher), 218
super-conducting quantum interference device (SQUID), 46
superior colliculus, 35, 94
superior olivary nucleus, 110–111
superior temporal lobe, 112
supplementary motor cortex, 168
suprachiasmatic nucleus (SCN), 94–95, 187, 195
surgery for seizure control, 312
sympathetic nervous system
  fight or flight response, 42–43, 182–183
  function, 43, 181, 182
synapse
  definition, 50
  electrical and chemical compared, 328
  learning, changes during, 264–271
  modification of, 333–334
  plasticity, research on, 264

synaptic transmission, 57–58, 120
synaptic weight, 267–268
systems neuroscience, 309

• T •

taste, sense of
  impaired, 133
  learning and memory in, 132–133
  olfaction, relation to, 119
  receptor cells, 126–127
  satiety, 134
taste buds, 50, 126–127
taste coding, 129–131
tastes, four basic, 127–129
"technical stuff" icon, purpose of, 5
tectum, 35–36
telencephalon, definition of, 27
temperature, 31, 72
temporal lobe, 28, 30, 226
temporal retina, 91
teratogen, 299
terminal bouton, 50
TGA (transient global amnesia), 278
thalamus. *See also* limbic system
  anatomy, 31, 92
  auditory process, relation to, 110–111
  etymology, 31
  function, 32, 235, 320
  input, 91–94, 95–98
  integration and gating, 235
  lateral geniculate nucleus (LGN), 294
  mediodorsal, 123
  output, 233–235
  relay cells, 93–94
theta wave, 193, 194
"third eye," 188
thought. *See also* consciousness
  acting without, 257–258
  or actions first, question of, 170–172
  sensory processing hierarchies, 232–236
tinnitus, 118
"tip" icon, purpose of, 5
tongue, anatomy of, 127
tonotopic frequency map, 112, 293
touch
  receptor. *See* somatosensory receptor
  sense of. *See* somatosensory perception
"touch map" of the body, 69
tract tracing, 43–44
transcranial magnetic (TMS) and transcranial direct current stimulation (tDCS), 278, 311–312, 346
transient global amnesia (TGA), 278
transient receptor potential (TRP) channel, 72
transplant, neural, 21
treatment for neurological and mental dysfunction, 20–21, 306, 314–316. *See also* deep brain stimulation (DBS)
tremor, cause of, 153
trisomy 21, 307
tritanopia, 99
triune brain, 211
TRP (transient receptor potential) channel, 72
tumor, brain, 304
tunnel vision, 100
two-point discrimination, 77
typical antipsychotic medication, 315

• U •

"ultramale" brain configuration, 39
umami, 127
unconsciousness, 217
universal interconnection scheme, 227–228
upbringing as factor in intelligence, 205
utilization behavior, 246

• V •

V1 (visual area 1), 30, 95–96, 220
V1 - 3 complex, 96–97
V4 (visual area 4), damage to, 99
vasopressin, 185
vegetative state, 216
ventral root, 41
ventral stream, 97–98
ventroposteromedial thalamus (VPM), 132
VEP (visual evoked potential), 44
vertebrate
  brain development, 224, 244
  definition, 86
  intelligence, awareness as, 207
  neocortex size, 245
vestibular function, 118, 139
vestibular system, 155
vestibulocerebellum, 158, 169
vestibulospinal reflex, 154–155
veto process, 172, 213–214

vision, 86–87. *See also* eye
  impaired, 98–100, 345
  lateral inhibition, 87–89
  neglect, 221–222
  peripheral, loss of, 100
  processing, 30, 83, 239
  prostheses, 347
  retinotopic map, 292
  stem cell research to restore, 345
visual area 1 (V1), 30, 95–96, 220
visual evoked potentials (VEPs), 44
visual illusion, 101–102
visual map, 30
visual prosthesis, 21–22
visual tracking, control of, 94
voltage across neuron membrane, 55
voltage-dependent sodium channel, 56
voluntary movement, 17, 138
vomeronasal organ, 126
von Economo neuron, 174–175
VPM (ventroposteromedial thalamus), 132

waking state, 215–216
walking, reflexes in, 162
Watson, James (molecular biologist), 296
Weisel, Torsten (scientist), 96, 231
Weiskrantz, Lawrence (psychologist), 220
Wernicke's area, 114, 218, 238, 322
"what" pathway, 98
"where" pathway, 97–98
white matter, 14, 226, 227
Williams syndrome, 308
Wisconsin Card Sorting task, 252–253
withdrawal reflex
  activation, 141–142
  ballistic nature, 150
  research, 264
  "violence" of, 150
women
  cognitive processing, 38–40
  pain tolerance, 81
  stress response, 186
"word salad," 114, 238
working memory
  brain processes managing, 248–250
  description, 165–166, 248
  limits of, 250–252
  long-term memory, transfer to, 272, 274
  loss of, 217
  neuron, 249, 250
  perseveration, 252–253